Russia's 'Age of Silver'

Russia's 'Age of Silver'
Precious-metal production and economic growth in the eighteenth century

Ian Blanchard

Routledge

London and New York

First published 1989
by Routledge
11 New Fetter Lane, London EC4P 4EE
29 West 35th Street, New York, NY 10001

Set by Hope Services, Abingdon
Printed in Great Britain by TJ Press (Padstow) Ltd
Padstow, Cornwall

British Library Cataloguing in Publication Data
Blanchard, Ian
   Russia's 'Age of Silver'
   1 Russia. Silver-mining industries, 1700–1800
   I. Title
   338.2'7421'0947
ISBN 0 415 00831 X

Library of Congress Cataloging in Publication Data
Blanchard, Ian.
   Russia's 'Age of Silver' / Ian Blanchard.
      p. cm.
   Bibliography: p.
   Includes index.
   ISBN 0–415–00831–X
   1. Precious metals—Soviet Union—History—18th century.
2. Silver industry—Soviet Union—History—18th century. 3. Gold
industry—Soviet Union—History—18th century. 4. Export controls—
Soviet Union—History—18th century. 5. Soviet Union—Economic
policy. 6. Soviet Union—Economic conditions—To 1861. I. Title.
HG285.S65B57 1989
338.2'7421'0947—dc19   88–31495

Contents

Contents

List of figures

List of maps and plans

List of tables

Appendices

Acknowledgements

During the spring of 1984, after many years of research into the history of the mining and metallurgical industries during the medieval and early modern periods, and after having settled down to write the initial studies on the subject concerning the early medieval industry, I was transported into a new and remote world by a letter from a close friend in Germany, Ekkehard Westermann. In it he outlined his proposals for a session he was organizing at the forthcoming Ninth International Congress of Economic History which was to be held at Berne in the summer of 1986. His idea, which was to link two previously distinct and discrete historiographical fields concerning European mining activity before and after the Thirty Years War, was an exciting one and his invitation to participate in the session, providing an overview of European mining developments in the period 1650–1800, afforded an irresistible challenge for two reasons. First, as we were both aware, although investigations into European mining and metallurgy during the period 1450–1650 were well developed such was not the case with regard to the subsequent two centuries, and although there was a strong contemporary literature on mining matters during the years 1650–1850 which laid the foundations for the brilliant discussions evoked by the bi-metallic crisis of the late nineteenth century, thereafter the subject was almost totally neglected. In compiling a study of the European industry in this period, accordingly, one was entering almost totally unexplored territory. Second, the position of the European precious-metal industries at this time was an anomalous one. For nearly a century prior to 1650 European production had been subordinated to that of South and Central America, and its renaissance during the years 1670–1770 posed fascinating questions about the relationship between the two sectors of the international industry and about the nature of the so-called Latin American mining 'crisis'. Questions which had exercised my mind, on and off, for almost a decade were thus brought to the fore, and in accepting Ekkehard's invitation I began to draw together numerous strands of research undertaken over a not inconsiderable period of time.

My interest in the subject had first been evoked in the mid-1970s when, having completed my own researches on the industry for the period before 1650, I was afforded the opportunity of discussing with Stephen Fisher his then current research on Latin American production of precious metals. The thoughts provoked by these discussions were moreover consolidated at this time by the appearance of numerous articles by Michel Morineau (subsequently collected together in his *Incroyables gazettes et fabuleux métaux: les retours des trésors americains d'après les gazettes hollandaises (XVIe–XVIIIe siècles)*, Cambridge, 1985) concerning the shipments of specie from the Americas during the years 1650–1800. Together, these researchers had shattered the then fashionable view of a late-seventeenth- and early-eighteenth-century 'crisis' in the trans-Atlantic trade in precious metals and in the process had removed an explanation for the revival of European production during that period – a supply-induced specie shortage at European mints. The time for a major reappraisal of the whole problem had clearly arrived. At this point, moreover, simple curiosity was turned into the necessity for a real research effort by Bill Parker who placed this problem at the very centre of a project concerning the development of resource-saving technology in the mining and metallurgical industries of the pre-modern world. Thus began, in preparation for a memorable conference on Natural Resources and Economic Development which was held at Bellagio in the spring of 1977 as a prelude to the session on that subject convened at the Seventh International Congress of Economic History held in Edinburgh in 1978, my researches into mining and metallurgical production in the period 1650–1850. Few projects could have had a more pleasant beginning. Spring sunshine and early strawberries consumed on the banks of Lake Como coupled with lively academic debate made it a most memorable occasion and the encouragement and kindness shown me by the organizers, Bill Parker and Antoni Maczak, and all present at the conference are gratefully recognized. Nor has there been any lack of similar support as the researches continued over the following eight years. Those who have answered my queries or suggested possible sources of information are legion, but latterly, after the main outlines of the subject had become clear to me, the expert knowledge of Hans-Joachim Kraschewski on the Harz mines and of Bjorn Ivar Berg and Knut Sprauten on Norwegian mining, communicated to me either personally or by receipt of their unpublished papers, has proven invaluable to me in confirming tenatative hypotheses and in establishing the importance of Russian production in eighteenth-century European output.

The significant role played by the Russian producer within the European industry during the late eighteenth century had already become apparent from researches undertaken in the late 1970s and provided yet another reason for my continuing interest in the subject as it brought together the two basic themes of my own researches into medieval and early modern history, particularly with regard to mining and metallurgy, and Russian history which had previously evolved autonomously of each other. It also brought my researches into the sphere of interest of another group of historians with whom I had long enjoyed close relations, the chroniclers of economic relations in Russia and the northern Seas. Already at the Bellagio conference both Karl-Gustav Hildebrand and Sven-Eric Åstrom showed great interest in this aspect of my work and encouraged my efforts whilst over a much longer period stretching back to the early 1970s I have been privileged to benefit from the very particular attention of the late Artur Attmann, who graciously shared the fruits of his own researches into the international metal trades with me. Without his kindly interest in my researches as they slowly evolved and his constant willingness to provide information and criticism this book would certainly never have been written. His perspicacity in perceiving the significance of my researches to the history of northern trade, long before I had, was certainly revealed in discussions at the Third International Conference of the Association Internationale d'Histoire des Mers Nordiques de l'Europe, held at Utrecht in the summer of 1982, and it is partly in response to a promise made to Christian Ahlstrom, Richard Unger, and others present at the conference, to write up my findings on Russian precious-metal production, that this volume now appears. If the Utrecht conference revealed to me the importance of my researches to the history of northern trade in the eighteenth century, however, it also posed major questions about the place of Russian silver production in the evolution of the contemporary Russian economy, and it was only through the generosity of Jennifer Newman, who unstintingly placed the fruits of her researches at my disposal, that I was able to perceive the present study within the context of eighteenth-century Russian economic development. I can only hope that this independent confirmation, based upon the investigation of Russian precious metal production, minting, and prices, of her revolutionary thesis on Russian economic development, will in some small part repay the time she has spent answering my incessant questions.

Thus, many have been the persons, in the fields of mining, trade, and Russian history, whose generosity in sharing their researches, and interest in my own, have smoothed the path of my research since its inception, yet it would be churlish to close without reference to two others: George Hammerley, whose interest and encouragement of my work on mining history over more years than either of us would care to remember, is gratefully acknowledged; and Olga Crisp, who has always sustained my interest in Russian economic history and has encouraged my efforts when my own faith in them was flagging.

As I have previously suggested, quoting one of my colleagues, 'history is a co-operative and kindly profession' and in an age of changing academic values I have always been privileged to work with colleagues who epitomized this ideal. It is to all of them that this book is dedicated.

Newlees Farm
Spring 1988

Part One

International Precious-Metal Production and Distribution, 1670–1770

Chapter One

The South and Central American mining 'crisis'

For a century prior to *c.* 1670 South and Central American silver production had dominated international specie markets. The delicate price equilibrium between product (silver and copper) and input (lead) markets, upon which the fortunes of the previously dominant central European *Saigerhändler* had depended, was seriously disturbed. From about 1570–1670 central European production had declined and in its place there had emerged a series of independent industries – Swedish copper, British lead, and Latin American silver – which operated within autonomous market structures.[1]

The principal cause of these changes, which established a Latin American hegemony over international gold and silver markets, lay in the introduction of a new technology – the amalgamation process – into the silver industry, which not only resulted in input substitution, shattering the symbiotic relationship between the lead and silver–copper producers by substituting a new input supply system distributing mercury to the main mining centres, but also gave birth to an entirely new industrial resource base exploiting silver haloids (argentite and cerargyrite), almost unknown in Europe, but abundantly available in the New World.[2] Initially introduced from the Old World, where it had long been employed in gold production and from the fourteenth century had been utilized in the reduction of Austrian native silver ores to resolve input supply problems caused by the lead crisis of the early 1550s, the new technology's early history had been a highly chequered one.[3] Born of market conditions characterized by high lead prices, stabilization in this metal's price, and an enhancement in the cost of mercury from the main workings at Almaden had doomed Latin American production during the years 1557–72 to a severe crisis.[4] From about the 1570s, however, the advantages of the new technological nexus became apparent. Supply bottlenecks in the Old World mercury mine were resolved as the New World was called in to redress the balance, and the cinnabar deposits of Huancavelica, discovered by the Spaniards in 1564, were opened up. For more than two decades the shallow, open-cast workings of *la Descubridora* provided a cheap and plentiful supply of the liquid metal (figures 1.1 and 1.5), allowing the Ibero-American silver producer to sweep all before him.[5] A new production pattern had been established which would imprint its stamp upon the industry for the next quarter of a millennium. Freed from its subordination to the Old World sources of mercury by the abundant supplies of the liquid metal which issued forth from the mines of Huancavelica, the Latin American industry now realized the full potential of the new technology. The range of exploitable ores was extended, as the new silver haloids containing between 15 and 225 ounces per ton of ore, which had previously been unresponsive to smelting techniques because of their 'dryness', were worked and extraction rates were enhanced beyond the 40 per cent levels of the Old World cupelation processes to about 70 per cent of the ores' metallic content (figure 1.6).[6] Rich ores and high extraction rates thus allowed Latin American producers to create abundant quantities of cheap silver which, from about 1570, became not only the principal source of European supply but, passing on, also served to provision Asiatic markets.[7]

Adoption of the new technology had thus established the Latin American producers' dominance of world specie markets, but it had also rendered them dependent upon the availability of cheap mercury. Having freed itself from its tutelage to one input supplier, the new industry of the

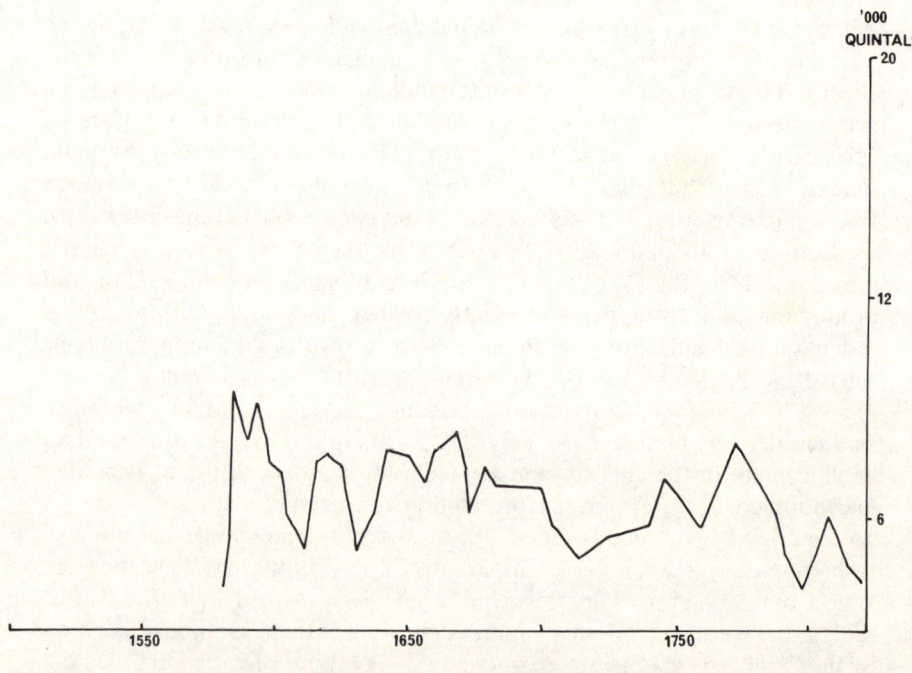

Figure 1.1 Mercury production: Huancavelica

1570s had fallen captive to another. Yet, save for brief periods, at no time during the history of the amalgamation process was the position of South and Central American precious-metal producers in aggregate threatened by protracted shortages in mercury supply (figures 1.1.–1.4).[8] As resource-based production cycles of lengthening periodicity succeeded each other, short-term inter-cyclical crises always gave way to production booms superimposed upon a series of long cycles which maintained output at remarkably high levels. As has already been noted, the first of these production cycles, from about 1565–1615, was associated with the opening up of the Huancavelica mine which ensured a phase, lasting more than two decades, when cheap and plentiful supplies of the liquid metal were produced from the open-cast workings, causing production at Almaden to stagnate as price leadership was assumed by the Peruvian mine. The boom did not last, however, and from 1582 the characteristic pattern of resource depletion and increasing costs may be observed at Huancavelica. Price increases in 1583–5 temporarily retarded the process of production decline, limiting the rate of decline in Peruvian output and enhancing output at Almaden, but production from the giant of the Andes was beginning to display marked instability as workmen plunged deeper into the sandstone country rock. From the 1590s the deeper galleries had to be

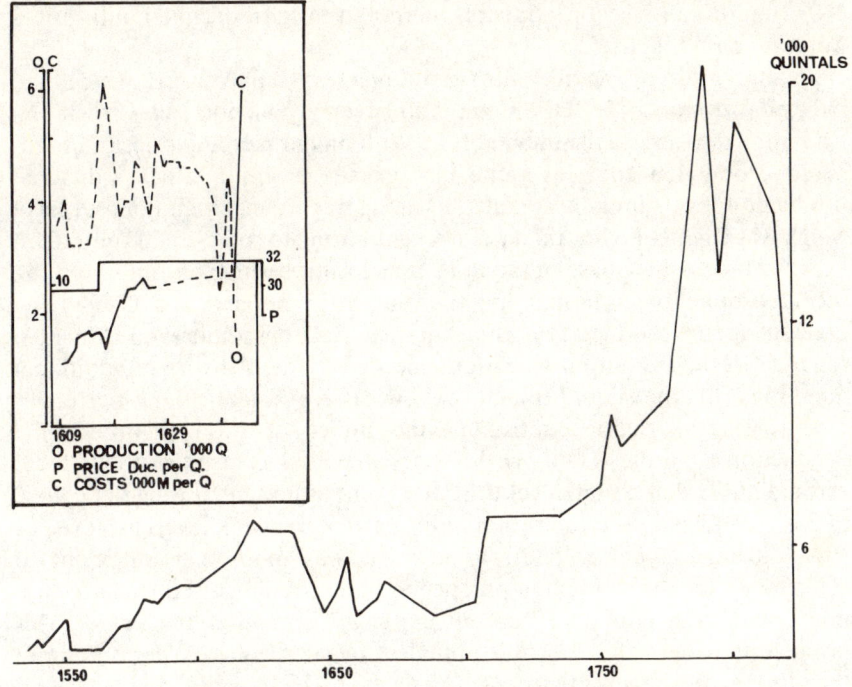

Figure 1.2 Mercury production: Almaden

closed down due to the presence of poisonous gas, and in 1607 production, already declining, collapsed with the subsidence of the major workings. The first upswing in mercury production, having attained a maximum output of *c.* 10,700 quintals a year, thus came to an end in *c.* 1615 in crisis conditions.[9]

The Spanish Crown, unwilling to increase prices further and thereby retard silver and gold production, thus cast around for other sources of supply. Not for the last time Peruvian officials looked to China to make up the deficiencies of existing production centres – but without success.[10] Seemingly only change from within could lead to a resurrection of the old Ibero-American amalgamated silver and gold mining and metallurgical complex. For the producers at Huancavelica this meant moving into a new high-cost phase of production as the renovation of the workings and the sinking of new ventilation shafts begun in 1609 enhanced production costs. Accordingly, the Crown was faced with either raising prices or being satisfied with diminished levels of mercury production. Neither course of action was acceptable and when the Fugger, anticipating the cost reductions arising from the introduction of the reverbatory furnace at Almaden, offered delivery at seven ducats a quintal below the prevailing price, the Crown leapt at the opportunity. The hopes proffered by the Germans were illusory, however, and after a brief revival at Almaden in 1609 (figure 1.2, inset) production there and in Peru declined until prices were increased in 1615.[11]

Finally, after an acute, short-term mercury-supply crisis which had induced a dramatic decline in gold and silver production, the Crown had been forced to accept the inevitable.[12] With higher mercury prices a broad-based production complex came into existence within which individual production levels initially steadied at an aggregate maximum output level which was slightly higher than before, amounting to some 13,700 quintals a year.[13] During this phase in the industry's history, which encompassed the second production cycle spanning the years 1615–50, the onus of supplying precious-metal producers in the Americas fell once more on Almaden (figure 1.2) whose output was supplemented by the delivery of additional quantities of the liquid metal by successive asientists (Albertinelli, Oberholz, and Balbi) controlling the mines of Idria (figure 1.3).[14] Production from these Old World mines, having increased to about 1623, was maintained at a high level to *c.* 1645, but at no time during the course of the second cycle did their joint output attain the peak levels achieved in Peru during the early 1580s. Even so it was more than sufficient to compensate for the shortfall in production at the reconstructed Huancavelica mine, and at maximum levels of aggregate output in the early 1620s production rose to about 13,700 quintals a year, or nearly 30 per cent more than the previous peak. Once again, however, the boom in the Old World production centres, which had sustained the upswing of the years 1615–23,

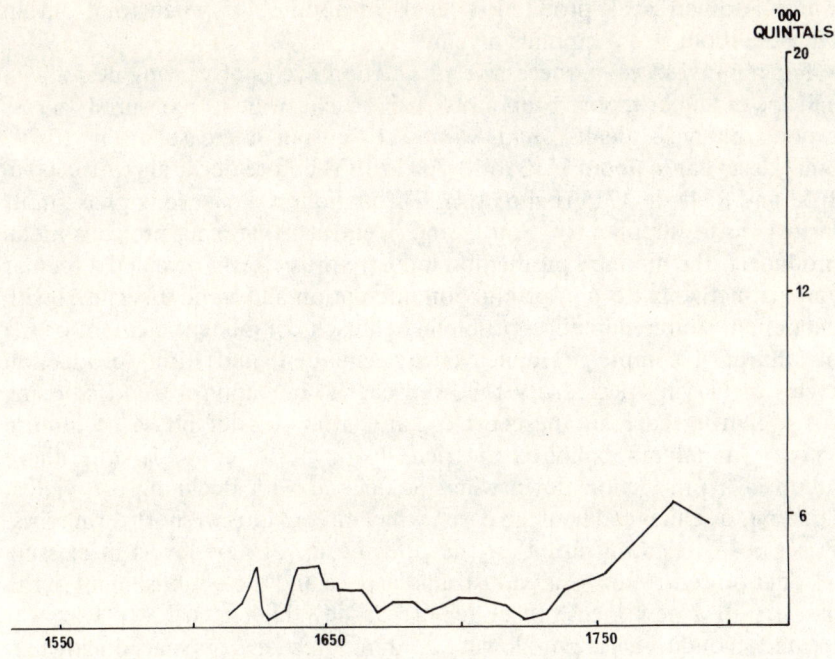

Figure 1.3   Mercury production: Idria

soon thereafter gave way to decline, production falling at firstly slowly to about 1645 before collapsing in the late 1640s.

In this instance, however, as the second production cycle drew to its close, there was no abrupt inter-cyclical crisis. Improved market mechanisms allowed a smooth transition to be made between the second and third production cycles. With the restructuring of production at the Peruvian mine after 1642 output from that centre steadily increased to compensate for the decline in the Old World mines. At the peak of this cycle, which ran its course during the years from about 1630–1715, total output amounted to no more than 10,500 quintals a year. In terms of the long cycle, which had witnessed increasing output between the production cycle peaks of 1583 and 1623, attaining a maximum total production level of 13,700 quintals a year at the latter date, the third production-cycle peak in about 1653 merely marked a stage in the long-term production downswing.

The fourth resource-based production cycle, which ran its course in a truncated form between 1690 and 1730, witnessed the culmination of this long-term downward trend in output. Once again, as production at Huancavelica, which had sustained the upswing of the third cycle, finally collapsed after a period of slow and protracted decline, a revival at Almaden paved the way for the transition to the fourth production cycle which attained peak production levels in about 1715 when total output reached about 8,900 quintals a year.

For about 140 years successive production cycles of varying periodicity had been superimposed upon a long cycle which, measured across production-cycle peaks, had witnessed output increase from 10,700 quintals a year in about 1583 to 13,700 in 1623 before declining to 10,500 in 1653 and 8,900 in 1715 (figure 1.4). Throughout this first long phase in its history as a supplier of South and Central American precious-metal producers, the mercury production industry thus clearly revealed a regular and distinctive pattern of production fluctuation and trend associated with endemic resource depletion problems. In each component element of the metallurgical complex (Huancavelica, Almaden, and Idria) production cycles of varying periodicity followed each other upon a trend of rising costs. During each of these cycles, and after a brief phase of intense activity, as miners exploited the richest and most accessible ores, there followed a production downswing, associated with declining ore yields, flooding, or enhanced haulage costs, which precipitated a short-term crisis. Prices rose, resulting in either the opening up of new levels in existing production centres or in an industrial diaspora and the establishment of the industry in a new locus. In either case a new higher cost structure was created, conditional upon lower yielding ores and/or spacially distant deposits. The pattern of fluctuation then repeated itself. Fluctuations thus succeeded each other, at a rate dependent upon the periodicity of the production cycle, upon a trend of rising costs, falling yields and/or

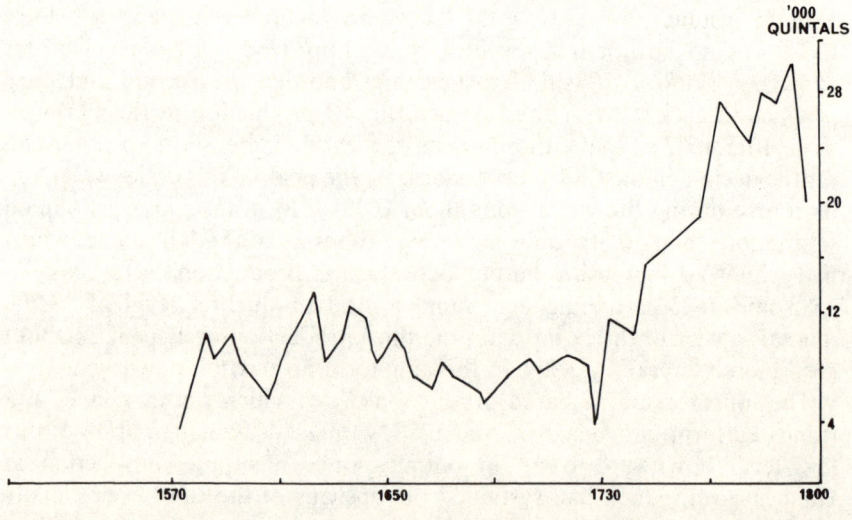

Figure 1.4   Total mercury production

geographically more distant workings until the limits of prevailing mining and metallurgical technology were reached when the crisis precipitated by the production-cycle downswing merged into a general crisis.

The production-cycle pattern, described above, is very clearly revealed during the first phase in the industry's history, spanning the years 1575–1615. After a rapid expansion in output which carried production at Almaden and Huancavelica to a level of some 10,700 quintals a year in about 1583, and sustained it at that amount during the next decade, both workings displayed signs of acute resource depletion. At the ancient mine of Pozo at Almaden the workings were gradually abandoned between 1590 and 1615 having attained a depth of 209 metres and in some places 250 metres, at which point they ceased to be exploitable because of the increase in costs and the difficulties of shoring and draining the galleries.[15] At Huancavelica during the same years similar problems were apparent. Here by the late 1590s the principal layer of ore was dipping too deep in the sandstone country rock for the previously dominant open-cast methods of mining to be continued. Miners accordingly began to excavate subterranean workings which, without proper supervision, became labyrinthine in character and subject to penetration by poisonous gases. So bad indeed was the situation in the workings that in 1604 they were temporarily closed. As surface working was resumed, however, the old underground workings posed a major hazard, and when in 1607 one of the galleries was penetrated a major disaster occurred. Heavy rains washed a deluge into the galleries and blocked the main entrance of all the underground workings, bringing production to a halt.[16] Thus came to an end the first production cycle, both of the major production centres – at Almaden and Huancavelica – after an initial phase of rapid expansion, being subject to rapidly rising costs, as mines became deeper and ores poorer, which, at prevailing prices, caused production to decline.

Now began the task of renovation. In 1606 work was begun on a major adit to drain the deeper galleries at Huancavelica, and in 1609 a team of experts arrived there from Almaden, under whose supervision ventilation shafts were cut into the deeper levels, making previously gas-filled galleries available for exploitation.[17] Yet this was little more than a patching job, leaving the major problem of ore exhaustion unresolved, so that whilst production recovered to some 7,000 quintals a year in 1615–16 it rarely, if ever, thereafter exceeded 5,000 quintals a year before the completion of the adit in 1642, some thirty-six years after work had begun on it, when a new mineral resource base was opened up at Huancavelica. As far as the history of the second production cycle is concerned events at Almaden were of much greater significance. Here workmen abandoned the old Pozo mine, and began to exploit the western veins of the mining complex and more particularly the rich deposits of San Pedro y San Domingo.[18] Thanks to a heavy capital investment programme a new high-cost production complex had been brought into existence, which in the higher-price conditions prevailing after 1615 produced increasing quantities of mercury, total output within the industry rising to some 11,000 quintals a year on average in the late 1610s. From about 1620, however, at both Huancavelica and Almaden, diminishing returns once more set in and it was only through an industrial diaspora, incorporating the Idrian mines into the American supply network, that the boom continued into the early 1620s and output was maintained at high levels until 1645.[19] When the Balkan producer was subject to the same forces which had wrought havoc in Spain and Peru a quarter of a century earlier, however, the second production cycle drew to its close and the Crown was forced once more to raise prices to producers.[20]

In Spain the effects of this price rise, which ushered in the third production cycle, were ephemeral. In spite of a massive effort, involving some 45,000 peasants retimbering the workings which had fallen into decay during the final years of the Fugger lease, the new administrator of the *Real Hacienda*, who from 1646 controlled the Almaden mine, was only capable of increasing production for some ten years, and having attained an average annual output of 4,050 quintals during the years 1653–5 production collapsed. Henceforth until the mine was gutted by fire in 1693 Almaden played an insignificant role in world supply of the liquid metal; individual workmen, making their own provision for the excavation of ore and evacuation of water, produced on average some 1,815 quintals annually during this period, or an amount roughly comparable with contemporary production in the Idrian mine.[21] The Old World production centres were now totally eclipsed by Huancavelica which from 1642 became the primary source of mercury supply to the New World silver industry. Yet again by a heavy investment programme bringing to a successful conclusion work on the adit which had begun in 1606, the problems which had beset the industry in the 1640s were resolved, but only at the cost of enhanced capital costs which necessitated the payment of higher prices for the liquid metal. The shaft, some 520 m long, improved ventilation of the deeper levels enormously; but its most dramatic effect was to open up hitherto unexploited deposits, allowing producers not only to overcome the major difficulty which had been encountered in 1648 – the loss of the principal vein, on account of a geological fault – but even to increase output to some 6,500 quintals a year from 1650–4 and sustain it close to that level for nearly half a century. Nor did the deterioration in the quality of the ores raised during this period seriously threaten the prosperity of the Peruvian mine. In the late 1630s an improved system of extracting mercury from the ore was invented by one Lope de Saavedra Barba and generally adopted at Huancavelica. This system, employing a series of water-cooled condensors, made the use of these low-grade ores possible as well as reducing the emission of toxic fumes and effecting great savings in labour and fuel.[22] A combination of capital investment and technological innovation thus allowed a high level of output to be maintained at Huancavelica from about 1650 to 1690 and it was only as the century drew to its close that resource depletion problems began to appear at the Peruvian mine, bringing the third production cycle to an end.

In the resultant mercury-supply crisis, which assumed catastrophic proportions during the years after 1690, men cast about frantically in their search for new deposits of the metal, the prize ultimately falling to prospectors from Almaden whose efforts had been given an additional impetus by the disastrous fire of 1693 which had gutted the existing workings on the San Pedro y San Domingo vein. Before the decade was out a new mine – the Concepcion Vieja – had been opened up at Almadenejos some 12 km east-south-east of Almaden. Of far greater significance, however, was the discovery in September 1697 of rich deposits of cinnabar, far to the east of the old collapsed workings, in a tenement located above the township of Almaden near the Retamar castle. A new mine had been discovered which would dominate the fortunes of the industry during the fourth and final phase of the first long-cycle which would run its course during the years 1690–1730. Already before 1697 drew to a close the first pit had been sunk in the new location, taking the name of San Antonio, and with the discovery of the main deposit in the next year pits proliferated, creating the great open-cast Mine de Castillo, above and in the midst of the township, where production increased steadily to nearly 4,000 quintals a year after 1700. Problems of flooding appeared as the new century dawned, but did not bring the boom to a halt. In 1703 work was begun on an adit – the Socavon del Castillo – which was driven some 207 m and entered the pits of San Antonio in 1707, allowing production to continue and indeed to increase to about 4,300 quintals a year between 1709 and 1725 at a time when the whole industry could support an output of no more than 8,900 quintals a year.[23] The Spanish boom of the early eighteenth century had thus been a not unspectacular one but it was based on very limited mineral reserves, and production, which had begun so promisingly only slightly more than a quarter of a century before, collapsed in 1725 bringing production to a halt and the first long cycle to an end.

For more than a century and a half, from about 1570 to 1730, successive production cycles of varying periodicity had been superimposed upon a long cycle of mercury production. During each of these fluctuations, after a brief upswing, output inevitably declined as resource-depletion problems emerged. On each successive occasion these problems were resolved as producers either opened up new levels in existing mines or created new workings on virgin deposits. Such work, however, required heavy investment and each successive production cycle saw costs rise as mining operations became more capital intensive until in 1715 production was a third less and prices 50 per cent more than a century before. To those contemporaries familiar with the recent history of the industry a serious crisis was emerging, but in reality, as far as mercury supplies were concerned, conditions were not so very different from those prevailing in the 1580s when Spanish American silver had first established its dominance over world specie markets; and, although such contemporaries were unaware of it, a new long cycle was about to begin which would increase production and reduce prices to levels which would previously have been regarded as inconceivable.

Whilst from about 1730 to 1850 the old resource-based production cycles thus once more ran their course (*c.* 1725–55, 1755–85, 1785–1815, and 1815–45) they did so now in an environment which was totally transformed. The duration of each fluctuation was shorter, amounting to about thirty years, and each was superimposed upon a new long cycle which pushed production up to unprecedented heights. Measured between fluctuation peaks, output within the industry increased from about 8,900 quintals a year in *c.* 1715 to 15,600 in *c.* 1745, 27,200 in *c.* 1775, and 29,800 in *c.* 1798 before declining to *c.* 23,000 quintals a year in about 1837. During the course of this long cycle, moreover, the industry underwent a fundamental structural metamorphosis as both the Idrian and Peruvian mines were displaced as major production centres. Their contribution to total industrial output, which amounted to 40–60 per cent (average 50 per cent) during the first production cycle of 1725–55, thereafter declined to 33 per cent during the second and third, before finally collapsing during the fourth, when, producing on average about 1,670 and 2,000 quintals a year respectively, their share in total output sank to below 20 per cent.[24] The story of the second great long cycle in mercury production of the years 1725–1850 was thus essentially the story of the Spanish mines.

Few at Almaden, during the second quarter of the eighteenth century, could have anticipated, however, the dominance they would soon achieve within the industry. During this period workmen excavating at a depth of 80–113 yards, after a phase of working either sterile or highly irregular strata, finally broke through to the virgin San Francisco and San Nicholas veins and to the south penetrated to deposits which were named San Julian at the time but which subsequently proved to be the upper reaches of the rich San Diego vein. On the basis of these new sources of ore, production expanded rapidly but at maximum output levels of about 9,000 quintals of mercury a year in the early 1750s producers at Almaden were still barely keeping ahead of those competitors who, for the past quarter of a century, had stood poised to take advantage of any weaknesses in their production organization.[25] As the second long cycle thus began, market conditions, characterized by an uneasy equilibrium between the principal mining centres, continued in much the same form as had existed during the final phases of the first. Even though output levels now exceeded those of the

1620s, commercial – like industrial – organization in the major mining centres remained cast in a seventeenth-century mould. Excavation work at Almaden continued to take place on an individualistic basis, ore being passed from gallery to gallery before being raised to the surface. Problems of water evacuation were resolved in a similar manner. Passed from hand to hand in leather buckets it was ultimately disposed of through small shafts, known as *resolladeros*, located at the major work-points which also served to ventilate the galleries. Unregulated and unplanned, the workings in the mine assumed a labyrinthine character which was quite unsuitable for the efficient exploitation of the rich mineral reserves. Production potential and organization were no longer compatible with each other, and when from *c.* 1745 production at Almaden expanded to compensate for decline elsewhere existing mining technology there was stretched to the limit – and beyond, for an outbreak of fire in 1755, fanned by the uncontrolled breeze which blew through the galleries, turned into a conflagration which burnt continuously for some two and a half years, bringing production at Almaden to a halt and the first production cycle of the new long cycle to a close.[26]

Only in 1758 could the work of reconstruction begin, but on this occasion there was no return to past mining practices. The German mining engineers brought in to undertake the task during the years 1758–60 introduced a new order and systemization into the exploitation of the deposits and laid the foundations of a mining enterprise which would dominate world mercury markets for almost a century. From the newly constructed San Theodoro shaft, tunnels were driven north and north-east to tap the deposits first discovered during the initial production cycles: the northerly San Nicholas and San Francisco veins and the easterly San Diego which revealed its full splendour at a depth of 154 yards.[27] Henceforth, the story of production at Almaden was essentially conditioned by the quality of the deposits discovered as men probed deeper and deeper into what, at least until the 1790s, were highly irregular veins. Phases of rapid expansion in output thus alternated with ones of diminished production. During the second production cycle, therefore, the opening of the fifth level at a depth of *c.* 186 yards revealed rich deposits, in some cases almost 8.5 yards thick, allowing production to expand to some 18,200 quintals a year and the mine to be worked until banks of quartzite impeded further lateral operations. Attempts to drive tunnels into the impediment proved futile and were abandoned after penetrating between 11 and 33 yards. At this point the sixth level was opened up: a poor thing of little worth which could sustain an annual output of only 11,800 quintals during the early 1780s. Below this level were rich, almost vertical veins of great regularity but before they could be worked effectively a major problem had to be resolved, for in attempting to establish seventh-level working miners found that their tunnels were flooded. The solution to the problem was provided by the installation in 1791 of a Watt pump engine which evacuated the water, and allowed the third production cycle to realize its full potential, output rising to almost 20,000 quintals a year on average during the years 1798–9. By the exploitation of richer and more extensive veins at greater depths miners at Almaden had established that centre's dominance in an industry which, at the close of the eighteenth century, was producing unprecedented quantities of mercury.

Even as the third production cycle reached its peak during the closing years of the eighteenth century, however, changes in the mine's resource base prevented further expansion to compensate for decline elsewhere, and ushered in the long-cyclical downswing in production. During operations at the seventh level it was discovered that the principal veins displayed a slight tendency to dip to the south, a feature which became more pronounced as over the next half-century the workings were driven to even greater depths. The effect of this was to accentuate the importance of the more southerly workings at the expense of the more northerly ones. Both the great regularity of the veins at this depth and the great length of workable deposit allowed a high level of ore extraction, but as the minerals raised at the San Pedro y San Diego workings became progressively richer those from the San Nicholas and San Francisco became poorer, a feature with which the prevailing smelting technology could not cope.[28] Production, in 'normal' circumstances, accordingly, stabilized at the high production levels of the late eighteenth century. During the fourth production cycle, however, conditions were rarely 'normal'. The collapse of Central and South American precious-metal production during the Wars of Independence precipitated a short demand-induced crisis within the industry, output at Almaden falling to below 7,000 quintals a year during the period 1810–15 before recovery in the 1820s carried it back towards the 20,000 quintal level. Before the full potential of the mine could once more be realized, however, market conditions in Europe were transformed when in 1831 the London branch of the House of Rothschild obtained from the Spanish Crown, as security against the repayment of a loan, control over the distribution of mercury from the Almaden mine. The Vienna branch at the same time having obtained a lease of the Idrian mine, the Rothschilds exercised a virtual world monopoly of quicksilver which they exploited vigorously. Prices which in Mexico had stabilized at their eighteenth-century level of c. 50 pesos a quintal during the 1820s rose rapidly to more than 120 during the 1840s.[29] The second great long cycle thus drew to a close in a mercury-supply crisis induced by the monopolistic practices of the Rothschilds, but as before the crisis was an ephemeral one. With the opening up of the Californian deposits at New Almaden in 1848 a new long cycle was about to begin which would push production up to ever higher levels and prices to ever lower ones. . . .[30]

From about 1570, therefore, international precious-metal markets had been transformed as a new group of Central and South American producers utilizing a new technology – the amalgamation process – completely altered production relationships within the industry. A new resource base, encompassing high-grade silver haloids, was opened up. Metal extraction rates were enhanced and abundant quantities of gold and silver were produced. In creating this new Central and South American industrial complex a supply network to channel mercury to the major mining centres was established, and during the years 1570–1870 the secular trend measured between long-cycle peaks in 1623, 1798, and 1865 in the supply of the liquid metal was ever upwards (figure 1.4), and its price, both in nominal and real terms, ever down (figure 1.5). During the inter-cyclical trough which extended from about 1623 to 1775 there was some contraction in supplies of mercury to the mines and enhancement in its price but neither were significant. Throughout most of the history of the amalgamation process precious-metal producers in the Americas enjoyed an increasing supply of mercury at falling long-term prices.

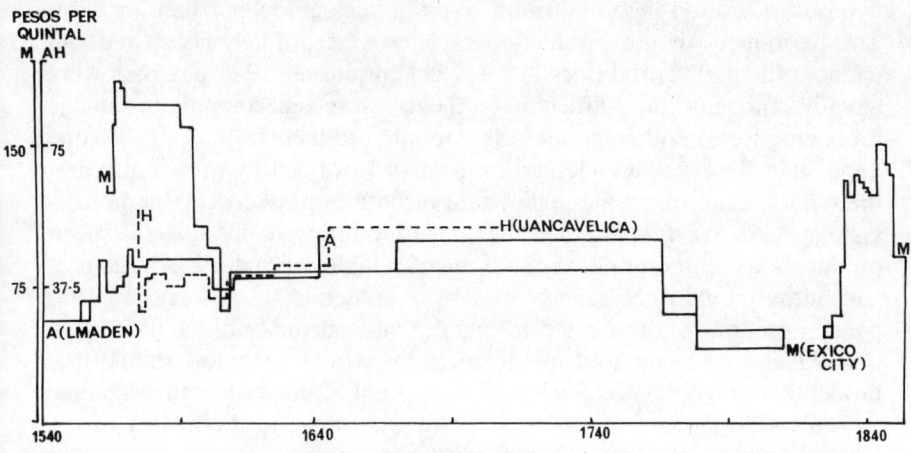

Figure 1.5 Mercury prices

The post-1623 crisis in production and export of silver from the Americas was thus not a result of a deficiency in mercury supplies. Its causation lay in indigenous resource-base problems and inter-sectoral redistribution of mercury supplies in response to changes in resource productivity.

Even during the first production cycle of the industry, operating on the basis of the amalgamation process, which ran its course during the years 1565–1615, such problems had been apparent. With the opening up of the easily worked *cerro rico* at Potosi, high-grade argentite or native silver was discovered which could be converted into the precious metal at textbook rates of mercury consumption (1.25 lb of mercury per 1 lb of silver). In such circumstances the entire output of mercury from Huancavelica was diverted thither, allowing silver production once more to resume the growth path, commenced in the period 1548–57, which was to carry output from the mine to its peak in 1589 of 7.9 million pesos.[31] The Mexican producer was not so fortunate, plagued by a higher proportion of 'rebellious' ores he could attract much lesser quantities of the reduction agent, predominantly from Almaden, and in conditions of shortage and enhanced prices the system of mercury allocation within New Spain required rationalization. The northern and central *reales*, due to the incidence of transport costs falling on the supplier, were starved of mercury. Accordingly, production at centres like Zacatecas stagnated whilst output at Real del Monte or Pachuca, where the mines were more easily supplied from the capital, boomed. Even here, however, the effects of mercury shortage made themselves felt. By restricting producers to the exploitation of higher-grade ores (yielding 70 oz per ton in comparison with the exploitative minimum at Potosi of 15 oz) the search for such deposits was intensified and mines rapidly became deeper than in Peru. The periodicity of the production cycle was accordingly shortened. As early as the 1590s producers at Real del Monte and Pachuca thus were already encountering difficulties. Yields were uneconomic, drainage problems were endemic, and as labour productivity fell it became impossible to secure an adequate supply of workmen by means of either the *repartimento* or by the inadequate incentives provided by the *partido* system.[32] As the first production cycle thus drew to its close Mexican producers were operating on the basis of a high-cost industrial structure, and output could only be sustained by a reduction of mercury prices, a policy initiated by the Crown in *c.* 1590 and pursued until 1605/8 when quicksilver was being sold in Mexico City at a price below the cost of production at Almaden. Such a situation could not last, and when the Crown raised mercury prices in 1615 the first phase in the history of the Central and South American mines came to an end.

As conditions of input-supply price equilibrium were established between 1615 and 1625 a reorientation of existing networks for channelling mercury to producers took place, available supplies of the liquid metal being directed to the contemporaneously highest-yield mines. Thus the produce of Huancavelica continued to be sent to Potosi where, as the resource base became more restricted and mercury consumption rates (*correspondencia*) increased from textbook levels to 1.5 lb of mercury per 1 lb of silver, production of the white metal stabilized at about 5 million pesos a year or about a third less than previous peak output levels.[33] In Mexico, however, the time had come for the frontier to move on. The opening years of the seventeenth century thus witnessed the migration of the industry to the central and northern mining districts – Zacatecas,

Guanajuato, and Parral – where mercury consumption rates were comparable with Potosi and production had not yet entered upon the high-cost phase of deep mining found further south. Here production enjoyed a bonanza during the years 1600–25, rising to a peak of some 4.4 million pesos a year.[34] Long before the inter-cyclical recession in mercury production caused a contraction in supplies of the liquid metal, and whilst the redistribution of quicksilver between mines took place within the base of the silver industry, output of the white precious metal declined. As the amount of mercury available to producers increased from about 10,700 quintals a year in the early 1580s to 13,700 quintals forty years later, or by 28 per cent, the output of silver in the main Central and South American production centres fell from c. 11 million pesos a year to 9.8 million pesos (or by 10 per cent). The cause lay in a deterioration of input productivity within the industry as mercury consumption rates increased from the classic text book levels prevailing before c. 1610 to c. 1.65lb of mercury per 1lb of silver during the years 1615–35 (figure 1.6).[35]

The remaining years of the first long cycle from about 1630 to 1730, encompassing the downswing of the second sub-cycle and the whole of the

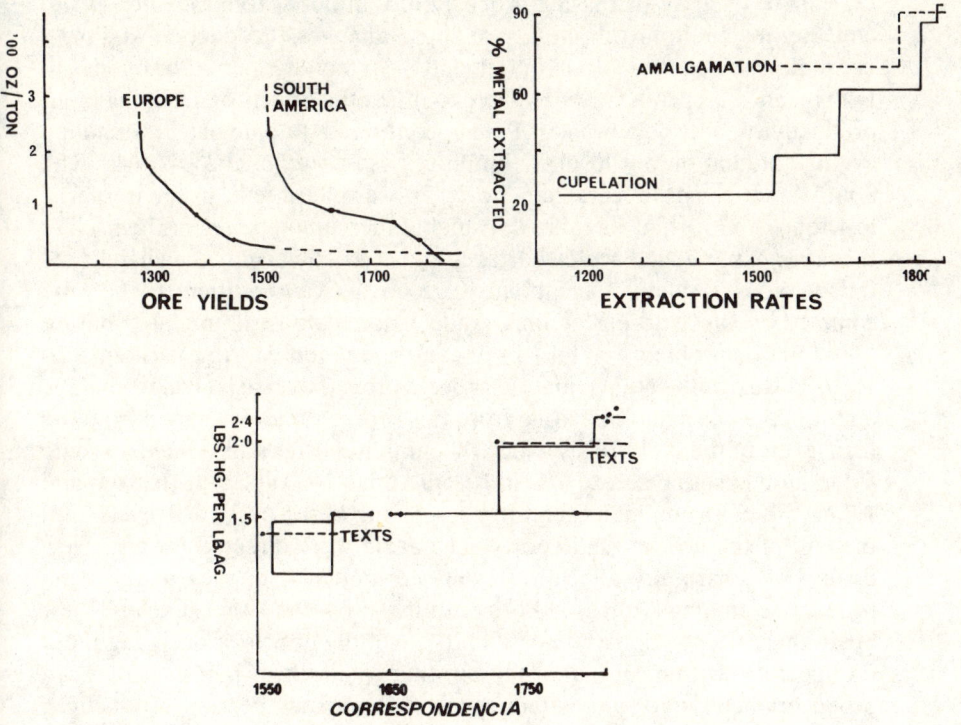

Figure 1.6 Silver: ore yields, extraction rates, *correspondencia*

third and fourth, saw the ultimate culmination of this trend. Mercury consumption rates continued to deteriorate. At Potosi, these years witnessed a transformation of the industrial resource base as an increasing proportion of resistant ores caused mercury consumption rates, which had stabilized at *c*. 1.65 lb of mercury to 1 lb of silver in 1635, to increase rapidly thereafter. In New Spain the situation was broadly similar. Able to maintain pre-existing levels at Zacatecas and Guanajuato until about 1640, during the next decade the amount of mercury used per unit increased rapidly to 2.1 lb of mercury to 1 lb of silver. Viewed in long-term perspective encompassing the years 1585–1715, therefore, declining input productivity threatened a major reduction in silver production. In 1585 and 1715 roughly comparable amounts of quicksilver were available to producers, but at contemporary *correspondencia* they were likely to yield in 1715 only 55 per cent of the 1585 silver output.

In response to this emergent crisis the system of mercury distribution was, accordingly, rationalized. By diverting supplies of the reducing agent to the most productive mining centres the impact of these changes could to some extent be alleviated. Thus until *c*. 1640 Potosi became the main recipient of supplies, monopolizing the output of Huancavelica and drawing to itself an increasing proportion of imports from the Old World, and thereby limiting the impact of these changes. Production at Potosi declined, but the ability to maintain and even increase mercury supplies to 1640 meant that producers were able to offset the effects of declining input productivity.[36] Production at Potosi continued to fluctuate around an average annual output level of 5 million pesos from *c*. 1625 to 1640. The South American producers' gain, however, was their Mexican counterparts' loss for as the former engrossed available mercury supplies to themselves the latter experienced acute shortages and restrictions in the availability of the precious reduction agent and high prices caused them to become confined to the working of high-grade haloids whose intractable nature posed major problems. Whilst Peruvian production was thus sustained to 1640, in Mexico the amalgamation process proved increasingly uneconomic and supplies of silver emanating from this source rapidly declined from the high levels of the early 1620s, when the mining centres had yielded up some 4.4 million pesos a year, to 4.1. in 1625/9, 3.6 in 1630/4, 2.3 in 1635/44, and 1.7 in 1645/9, production thereafter stabilizing at the post-1640 levels until the end of the century as imports once again were directed towards New Spain.[37] By carefully allocating available mercury supplies to the most productive mining centres – Potosi until *c*. 1640 and the mines of New Spain thereafter – it proved possible to offset the effects of declining input productivity within the industry so that, had the Spanish American producers been able to maintain their share of total mercury production, silver output would have stabilized in 1715 at a slightly higher level than would have prevailed in the absence of such reallocation, amounting to 58 rather than 55 per cent of 1585 levels.[38] Even so the crisis would have been a severe one arising out of the incapacity of the amalgamation process to cope with the increasing proportion of intractable ores being raised.

As has already been observed, these problems first made themselves felt in New Spain where after *c.* 1625 the effects of a deterioration in mercury consumption rates were coupled with shortages and high prices of quicksilver.[39] Under such circumstances the amalgamation process was uneconomic and was displaced by the lead cupelation techniques of an earlier age which enjoyed two distinct advantages in the working of high-grade ores. First, it was cheaper utilizing what had become a low-cost input. Second, it allowed the exploitation of previously unattractive 'rebellious' lead ores. Operating, accordingly, on a new resource base the Mexican producer assumed a new role in a restructured Central and South American industrial complex. Production of silver in the central and northern 'frontier' towns continued but now the proportion of smelted ores to those processed with mercury increased. At Zacatecas, Guanajuato, and Parral the production cycles thus ran their course on the basis of the 'new' technology, causing the creation of a 'new' supply network for the all-important smelting reagent – lead. Initially this involved tapping a pre-existing supply system which had been used intermittently (1580–90, 1610–14) to channel low-cost English lead via Cadiz to the New World whenever mercury supplies had been interrupted[40] Prior to 1625, however, the necessity of drawing on such supplies had not been great and it was only after this date that it came into its own, permitting the supplementation of dwindling supplies of amalgamation silver (which fell from 4.25 million pesos a year in 1620/9, to 3.6 in 1630/4, and 1.0 million in 1635/9) with about 1 million pesos annually of smelted silver.[41] With the emergence of crisis conditions in the English lead industry during the years 1640–70, however, this trade came to a halt, just as declining input productivity affecting producers employing the amalgamation process fundamentally undermined that sector of the industry.[42] The mid-seventeenth century thus witnessed an acute crisis in the Mexican silver industry, only marginally alleviated by the rediversion of Old World mercury supplies there, which left its imprint on the export trade.[43] It also saw a fundamental structural change within that industry as a New World lead industry was born.[44] New centres of production, only marginally exploited before 1640, in Nuevo Leon (San Pedro de Boca, San Gregorio, and Las Salinas), Chihuahua (Mapimi, Todos Santos, and Santa Barbara), and Durango (San Luis Potosi and Mazapil) were created. Old centres of silver–lead production discovered a valuable trade in lead by-products, *greta* (lithage) and *cendrada* (hearth lead). The years 1640–70 were difficult ones for the Mexican producer but by about 1655/60 a new industrial complex had emerged wherein the diminutive supplies of 'unrebellious' ores continued to be reduced by the amalgamation process but the predominant group of argentiferous lead sulphides was smelted, producing some 4.8 million pesos of silver a year on average during the years 1660–1680 to 1680–1710.[45]

A process of structural change within the Central and South American silver industry thus went a long way towards resolving endemic indigenous resource depletion problems. By concentrating available mercury supplies in the most productive silver mines – Potosi to about 1640 and those of New Spain thereafter – it proved possible to sustain output at the Peruvian workings at the high levels of the early 1620s until 1640, production fluctuating at 5–6 million pesos a year, and then to effect an expansion in New Spain to compensate for the decline in production at Potosi caused by the redistribution of mercury supplies. Aggregate production of amalgamation silver, having fallen from 9.8 million pesos a year in 1620/4 to 6.9 in 1635/9, thus stabilized thereafter at 6–7.5 million pesos to 1660. The restructuring of the Mexican industry, to accommodate the introduction of cupelation techniques in order to utilize the increasing quantities of 'rebellious' lead sulphides being raised, moreover, enhanced the output of smelted silver therein from about 1.25 million pesos a year on average between 1625/9 and 1635/9 to 3.4 million 1655/9. Viewed in a long-term perspective, therefore, production of silver measured between cyclical peaks in 1623 and 1660 was thus sustained at a level of slightly less than 10 million pesos a year. Nor was there any apparent reason why this situation should change during the remaining years of the seventeenth century or the opening decades of the eighteenth. Production at Potosi might be reduced as operations were affected by the culmination of the inter-cyclical downswing in mercury production and the diversion of Old World supplies to New Spain, but, had the latter sector of the industry been able to maintain its share of Old World quicksilver production, the Mexican industry would have had the capacity to compensate for any shortfall and total silver production would have been maintained at a level of 9–10 million pesos a year to c. 1715.[46]

In the event, however, between about 1660 and 1730 Central and South American silver production contracted significantly, falling from almost 10 million pesos a year to 6.8–7.8 million between 1680/9 and 1701/10 and not recovering to its former level until the 1730s (figure 1.7).[47] This crisis, which left its imprint on world specie markets, was largely a result of the decline at Potosi, where silver production fell from a level of 5–6 million pesos a year before 1640 to 3 million from 1660 to 1690, before collapsing to 1–2 million in the early eighteenth century; and also the absence of compensatory growth in Mexico, where output increased from 2 million pesos to 4.6 and 5.5 over the same period. On this occasion, however, the causation lay not in any indigenous resource-base problems at the mines but was due to a redistribution of mercury supplies for the first time beyond the base of the Central and South American silver industry. Until the 1640s

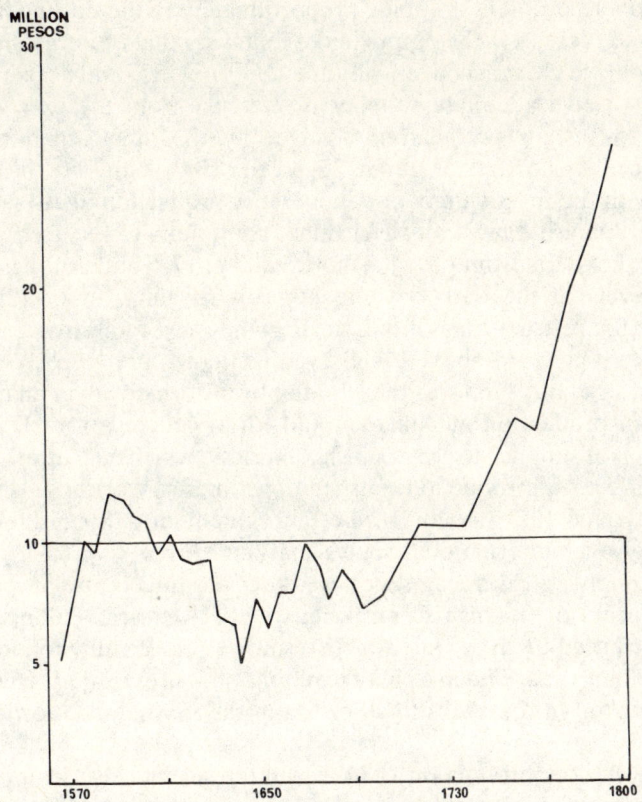

Figure 1.7 Central and South American silver production

by far the largest part (72–85 per cent) of the output of the Old World mercury mines at Almaden and Idria had been exported to the Spanish Americas, the residual production of about 1,000–2,000 quintals a year passing to Venice from whence it was distributed to African and Asian markets.[48] Henceforth, however, from *c.* 1640 to 1690 it was the former trade which bore the full brunt of the decline in aggregate Old World production. Exports to markets other than the Spanish American silver mines were sustained, fluctuating still between 1,000 and 2,000 quintals a year, but now increasing their share of the trans-Atlantic traffic from 15–28 per cent (the average during the period 1605/9–1635/9 was 20 per cent) to 35–50 per cent. Mercury supplies to the Americas thus contracted sharply and whilst Mexico benefited from the diversion of supplies previously destined for Peru, imports stabilized at only 1,750–2,500 quintals a year or some 500–1,000 quintals a year less than it would have received had it maintained its share of the trade (figure 1.8).[49] Whilst the Andean mine thus became totally dependent upon Huancavelica for mercury, silver production accordingly dropping proportionately to the decline in imports to 1660 and thereafter the output curve following that of Peruvian mercury production which reached an all-time low in 1720, Mexican producers failed to receive adequate supplies of mercury from the Old World, as imports to sustain production of silver by the amalgamation process experienced a shortfall of about 25–30 per cent. Nor did the situation change with the massive increase in mercury production at Almaden after 1700, the Spanish American silver mines continuing to receive only 50–65 per cent of exports from the Old World until c. 1775 and not regaining the import levels of the early seventeenth century until about 1750 when, thanks to the restructuring of the Mexican industry, total silver production was some 50 per cent above early-seventeenth-century levels.

Only after about 1765 was the situation transformed when an enormous increase in production at Almaden and Idria, coupled from 1775 with a rediversion of supplies to the Spanish Americas, which henceforth received 80–90 per cent of production, pushed down mercury prices to an all-time low and paved the way for a re-establishment of a Spanish American hegemony over international specie markets. From c. 1765 to 1810 the onus of supplying European specie markets was thus borne once more by the producers of Spanish Central and South America. Falling mercury prices permitted them to widen their resource base steadily to incorporate lower-yielding ores. The margin of exploitation was extended, accordingly, from 70 oz/ton in the 1740s to 40 oz/ton in the 1790s and 20 oz/ton in the 1810s.

Moreover, low production costs permitted an increase in unit mining costs, the product of a closed 'frontier'. Expansion took place not on the basis of a geographical diaspora (the only new production centres to emerge were Bolanos and Catorce which played a minor role in the boom) but through a revitalization of old workings. Zacatecas, Real del Monte, and above all Guanajuato continued to number amongst the leading producers of the colony until the end of the boom.[50] Intensification rather than extension provided the key to growth. Shafts were sunk deeper. Shafts of 270–370 yards (equivalent to the depth of the deepest Saxon mines) were not exceptional in the 1790s and the Valenciana, which by 1810 had attained a depth of 635 yards, was declared by von Humboldt to be the deepest point then known in the entire world.[51] Capital costs were thus enhanced as they were also by the driving of adits to drain the deep workings. From the 1740s old adits were extended and new ones begun in almost every mining centre. During the years 1739–68 one was driven at the Veta Vizcaina mine of the Count of Regla in Real del Monte which ultimately attained a length of 2,881 yards. Less ambitious perhaps than the Reglas Moran was the drainage tunnel constructed for Jose Vincente de Anza at Tehultepec (477 yards) or that driven at the Crown's expense at Potosi during the years 1789–1810 but, whether large or small, in centre after centre such drainage schemes had to be undertaken, the cost only being limited by increases in productivity amongst those engaged in 'dead-work', associated with the use of blasting powder.[52] Nor was it only capital costs which increased with depth. Haulage capacity of horse whims (*malcacates*) also increased and with it capital and variable costs. More whims and more horses to operate them were required. In 1801 the Veta Vizcaina required 28 whims operated by 1,200 horses and 400 men to extract water and ore from workings below the Moran adit. Annual expenditure amounted to 125,000 pesos including outlays on 5,000 quarters of maize to feed the horses.[53] Thus as mines became deeper haulage capacity increased (at Real del Monte by some four times during the eighteenth century) and operating costs rose. This problem was insoluble even within the framework of the latest technology. Attempts to introduce steam engines to drain the lowest levels of the Vizcaina vein during the 1820s and 1830s by English engineers, arrogant in their blind faith in 'progress', proved a complete débâcle – the units had neither the capacity nor power to undertake the task.[54] Unit costs per load of ore raised thus increased whilst the metallic content of the ores decreased, but in total production costs this rise in mining expenses was more than offset by the decline in outlays on refining. The key to prevailing boom conditions during the years 1760–1810 was cheap mercury and as prices of the reduction agent fell the effects of both lower refining costs and an extension of the ore base were to cause a reduction in the practice of smelting. Already by the 1770s the days when more than half the industry's output had been smelted were long past. At both Zacatecas and Guanajuato in the early 1770s only 30 per cent of output was obtained by smelting and thereafter the proportion steadily declined until in the 1800s it amounted to no more than 10–15 per cent.[55] As more and more ore was subjected to the amalgamation process, however, so the amount of intractable ore treated increased and mercury consumption rates rose, rapidly reattaining the levels prevailing in the early 1640s (2.1 lb Hg/1 lb Ag) before the culling of lead sulphides and their reservation for smelting had effected a reduction. Yet during the late eighteenth century there was no simple return to the *status quo ante* 1645, for in the interval the resource base had continued to deteriorate, so that by 1800 consumption rates of 2.4–2.9 lb Hg/1 lb Ag were common at Zacatecas, Bolanos, San Luis Potosi, Real del Monte, and Potosi (figure 1.6). Relentlessly, therefore, during the years 1760–1810 productivity in both mining and refining sectors of the Mexican silver industry declined. Yet rapidly falling mercury prices until *c.* 1820 more than offset rising costs and paved the way for a massive expansion of silver production providing such production centres as Guanajuato, where mercury consumption rates had risen by no more than half since the sixteenth century, with the lowest unit input costs ever and stabilizing costs at centres like Zacatecas and Real del Monte at levels unknown since the 1680s.

With respect to long-term trends in both input prices and unit production costs the mines of Spanish America thus continued to enjoy that superiority over their European counterparts which had first been established in the 1570s and which would endure until at least the 1820s. Measured between long-cyclical peaks in 1623 and 1801, the secular trend in silver production was ever upwards. Nor did the inter-cyclical recession in mercury output, whose effects were exacerbated by a decline in input productivity, pose a major threat to this boom. Structural changes within the Mexican industry, as 'rebellious' ores were processed by older cupelation techniques, ensured that the reduced aggregate levels of mercury output were more than sufficient to sustain Central and South American silver production levels during these years. The Spanish American silver crises of 1625–55 and, particularly, of 1670–1760 (which first became apparent during the years 1670–90 and assumed catastrophic proportions between 1690 and 1720 before a slow recovery began) were thus not caused by either indigenous resource-base problems within the industry or external aggregate input supply ones but were a result of mercury-supply shortages caused by a diversion of quantities of the liquid metal beyond the base of the Central and South American silver industry.[56]

From about 1640 the produce of the Old World mercury mines at Almaden and Idria passed increasingly to markets other than those provisioning the silver mines of Spanish America. Initially these markets were traditional ones in Africa and Asia which were increasingly supplied through Amsterdam rather than Venice, but even within this existing market framework a new force was at work leading to a redistribution in mercury supplies – a search for gold which at this time attained its highest price in terms of silver.[57] Small production centres in such areas as West Africa were revitalized but perhaps the most significant contemporary development was in New Granada where the auriferous quartz gravels could be made to yield up about 0.6 lb of gold for each 1 lb of mercury employed.[58] These workings, which had enjoyed a brief period of prosperity during the 1580s before a disastrous reduction in their Indian workforce caused their decline, had lain dormant since the beginning of the seventeenth century. From *c.* 1615, however, production began once more, slow growth turning into boom conditions from *c.* 1623, and exports of mercury to New Granada rapidly expanding to *c.* 1640. In a trade to markets outside the base of the Spanish American silver mines, amounting to about 1,200 quintals of mercury a year, the gold workings of New Granada now absorbed almost 20 per cent of supplies. Nor does this situation seem to have changed significantly during the subsequent years of acute crisis between 1640 and 1655. Firm data on mercury exports to New Granada during these years is unfortunately lacking but even the most conservative estimate would suggest that the gold workings, like other markets outside the Spanish American silver industry, sustained supplies during these difficult years. Only after 1655 did exports to New Granada decline, falling from perhaps 250 quintals a year to *c.* 150 quintals in 1660/4 and 60 in 1665/9 (figure 1.9).[59] For the first time, therefore, during the crisis years 1625–55, there had been a major diversion of mercury supplies beyond the base of the Spanish American silver industry. As total Old World mercury production, emanating from the mines of Almaden and Idria, declined, traditional markets maintained supplies ceding ground only to the gold workings of New Granada where imports increased, allowing production of the yellow metal to grow from about 1 million pesos a year in 1615/19 to perhaps as much as 4 million a year in 1650/4 before decline set in.[60] The effects of these changes on Central and South American silver producers and particularly those of Mexico, who were struggling to effect an internal transformation of their industry, were disastrous. The recovery of Mexican amalgamated silver production, envisaged in the redistribution of available imports from Potosi to New Spain, aborted as producers experienced a chronic shortfall in mercury supplies of about 500–1,000 quintals a year which now passed elsewhere.

Nor did this situation change during the second great crisis of 1670–1760, although during the interregnum between the decline of New Granada and the rise of Brazil a 'new' star rose as the gold producers of sub-Saharan Africa, long eclipsed by their Central and South American rivals, once more briefly held the centre of the stage.[61] During these years from *c.* 1660–1710 exports of mercury to markets other than those supplying the Spanish American silver mines continued at the slightly higher level of about 1,500 quintals a year, or slightly less than half of Old World production, but that portion passing to the gold fields was now dispersed predominantly to African producers working beyond the Niger bend where production of gold increased from about 1,500 kg (1.5 million pesos) a year in *c.* 1660 to 4,875 kg (2.9 million pesos) a year on average in 1690/1710.[62] Small quantities of mercury were distributed elsewhere: to New Granada where residual gold production amounted to *c.* 0.67 million pesos a year in the 1690s and to the new deposits of Brazil which produced an annual average output of 1.65 million pesos between 1695 and 1710.[63] In total, by the beginning of the eighteenth century gold production amounted to *c.* 5.2 million pesos a year, or only slightly less than in the 1650s. During the years 1670–1710 gold production and trade thus rose to levels comparable with those attained during the previous crisis of 1625–55, Africa becoming the major supplier. By far the largest part of this African production (3,245 kg or 1.9 million pesos a year) passed to Europe by way of the trans-Saharan routes controlled from Morocco, the terminal points of these routes – Tangier and Ceuta – attracting the covetous attentions of the European nations. Only a diminutive flow passed to the west African coast

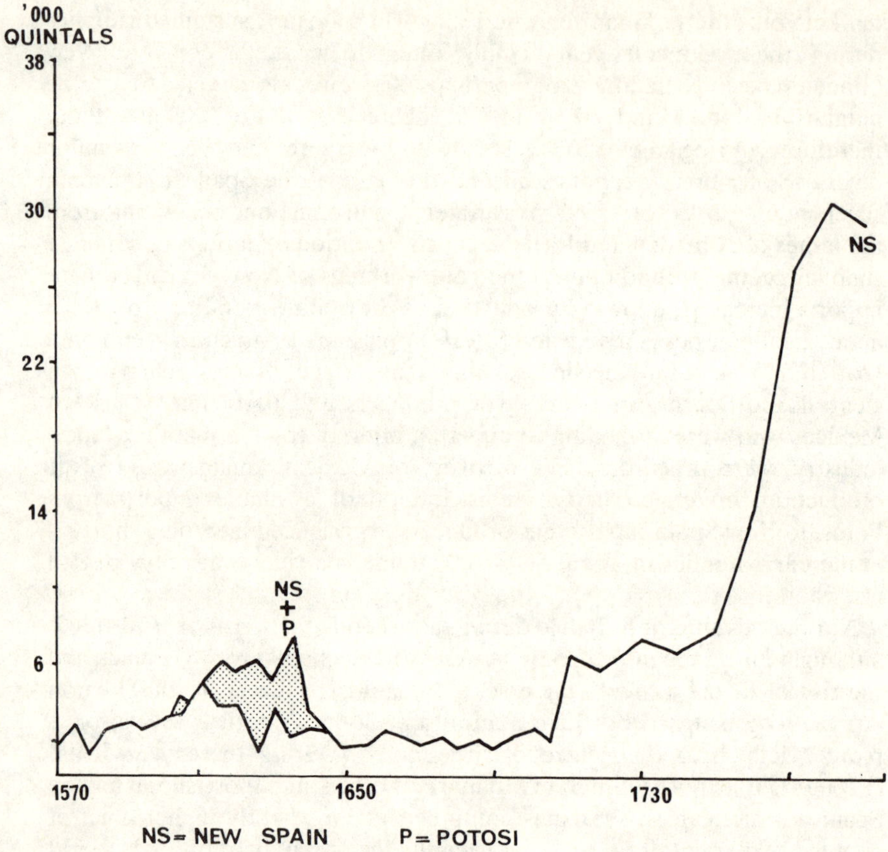

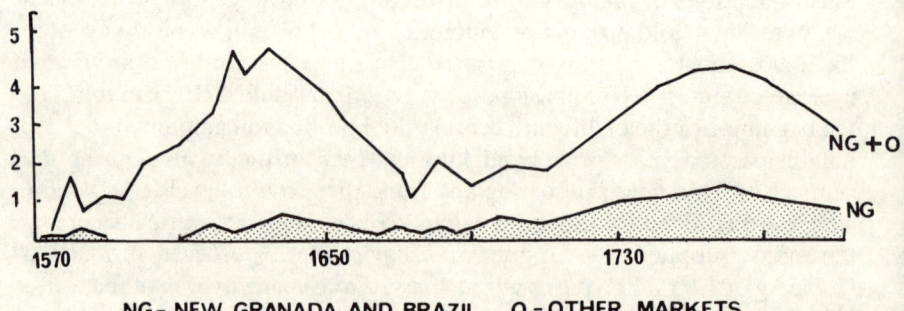

'000
QUINTALS

NS= NEW SPAIN   P= POTOSI

Figure 1.8   Exports of mercury to Spanish American silver producers

NG= NEW GRANADA AND BRAZIL   O = OTHER MARKETS

Figure 1.9   Exports of mercury to New Granada and Brazil

where the trade grew from about 1,300 kg a year in *c*. 1660 to 1628 kg or 1 million pesos a year in 1690/1710, attracting a host of Dutch, English, Brandenburgers, Danish, Portuguese, and French traders to pick up the crumbs from the Moroccan emperor's table.[64] Thus, whilst silver production at Potosi declined from 4–5.8 million pesos a year before 1660 to 3 million on average between 1660 and 1690 and 1–2 million in the early eighteenth century, and Mexican production failed to rise much above the 4 million pesos attained in 1660 before the end of the century when it rose to 5.5 million pesos a year, causing aggregate production to fall from almost 10 million pesos a year to between 6.8 and 7.8 million, gold production increased to compensate, rising from about 2.5 million pesos a year in the 1660s to 5.2 million during the years 1690–1710.

Nor did this situation change after 1695 as the discovery of the rich Brazilian gold fields once more established the South American producers' hegemony in the production of the yellow metal and paved the way for the revival of an active trans-Atlantic gold trade. That Latin American–European pattern of commercial and industrial development, which had been disrupted during the years 1670–95, was once more in accord and was being played out with the framework of a mercury distribution network whose salient features had been established long before. Markets for quicksilver, other than those servicing the requirements of Spanish American silver producers, continued to engross 45–50 per cent of the output of the Old World mercury mines, although now that portion of this trade destined for the gold industry passed increasingly to the workings of Portuguese America. Unlike during the first 'crisis' (*c*. 1625–55) or the first phase of the second (*c*. 1670–95), however, on this occasion (*c*. 1695–1760) the distributional pattern was superimposed not upon a declining output trend but, particularly from *c*. 1730, upon a rising one. As the second long cycle in mercury production began its dramatic upswing and the balance of output within the industry moved in favour of Almaden, exports of mercury from the Old World to the gold fields of the New thus steadily increased, providing the necessary inputs for a rapidly expanding industry.

Old decayed mining centres in the Spanish colonies were revived, gold production in New Granada increasing from *c*. 0.67 million pesos a year in 1690/9 to 2.7 million pesos a year in the period 1748/53, but most spectacular was the emergence of new production centres in the Portuguese colony of Brazil.[65] Production here dated back to the opening decade of the seventeenth century when the São Paulo deposits had first been opened up, but the workings of this south-eastern province were never of any great importance and it was only with the discovery of the placer gold deposits in the Rio das Velhas region of the Brazilian highlands north of Rio de Janeiro, within the future captaincy of Minas Gerais, during the 1690s that the Brazilian gold industry came into its own.[66] During the subsequent years from 1695 to 1735 the colony was gripped with gold fever. New mining towns were established at Ouro Preto and Villa Rica and the labour force of the industry expanded rapidly, pushing up slave prices to unprecedented heights.[67] On the rich hillside, known as the Morro de Pashchoal da Silva, outside the latter town, there were some 3,500 slaves (or 46 per cent of the overall provincial slave population) at work in 1719 excavating the auriferous gravels for the numerous *lavras* established there.[68] Activity was frenzied and production expanded rapidly from about 72 kg a year (0.04 million pesos) in 1710/13 to *c.* 2,296 kg a year (1.3 million pesos) in 1714/25, and *c.* 3,675 kg a year (2.2 million pesos) in 1726/35.[69] In the speculative atmosphere of these years when the whole colony was gripped with gold fever, however, the workings of Minas Gerais were only one of many centres feeding the gold trade. Prospecting activity was frenzied and in the 1700s new deposits were discovered in the Rio das Contas region of Bahia and in the Jacobinas of Goias which dominated the industry with an annual output of *c.* 2,700 kg (1.2 million pesos) during the opening years of the eighteenth century. The life of these workings was short, however, and their decline during the 1720s and final extinction in *c.* 1730 marked the end of the dominance of mining activity on the frontier. Prospecting in the Matto Grosso in the 1710s might lead to the discovery of the rich Cuiba deposits, which from 1718 became the mainstay of production outside of Minas Gerais, just as the opening up of the Fanado and Aracuahi workings in the late 1720s and those of the river Guapore in the early 1730s might contribute to the flagging fortunes of the frontier industry, but activity there from 1715 contributed less and less to total production, output falling to 1,400 kg a year (0.83 million pesos) in 1715/25 and 772 kg a year (0.45 million pesos) in 1726/35. As the industry thus experienced an initial phase of rapid expansion, total output rising from 1.24 million pesos a year in 1700/14 to 2.13 million pesos a year in 1715/25 and 2.65 million a year in 1725/35, the mines of Minas Gerais gradually assumed a dominant position within the Brazilian gold industry until during the years 1726–35 they contributed 83 per cent of total output. Yet this was merely a prelude to even greater things and during the subsequent boom years 1735–1751 they reigned supreme. A working population of perhaps 200,000 within the province, increasingly maintained by natural increase rather than slave imports (which fell from a peak of *c.* 7,360 annually in 1739/41 to *c.* 5,900 in the early 1750s), produced each year about 10,000 kg of gold (5.9 million pesos). This was, however, the high point of production in Minas Gerais and as production there declined to 7,650 kg a year (4.5 million pesos) in 1752/62, 6,618 kg a year in 1763/73, and *c.* 5,500 kg a year 1770/80, the onus of maintaining output once more fell upon the prospectors. Nor were they found wanting. New sites from as far afield as Ceara, Sergipe, and Goias were reported in the 1750s which on being brought into production contributed some 2,900 kg a year

(1.75 million pesos) on average during the years 1752/77. The boom thus continued, industrial output rising to 6.25 million pesos a year in 1752/62 and with an output of 5.6 million pesos a year, still remaining at or near the level of the 1740s into the period 1763/73. Only after about 1770 did the frontier captaincies of Goias and Matto Grosso face an irreversible collapse in their mining economies, leading to a catastrophic decline in Brazilian gold production and a loss by that country of its ascendency in world specie markets. The pendulum was from 1770 once more swinging in favour of the Central and South American silver producer who, able to engross an increasing share of available mercury supplies, henceforth went from strength to strength.

During successive 'crises' in 1625–1655 and 1670–1760 Spanish American silver producers had thus suffered from the effects of mercury-supply shortages caused by a diversion of supplies of the liquid metal beyond the base of their industry. Consumers in markets outside the network supplying the Central and South American silver industry increased their share of Old World mercury production and amongst their numbers gold producers enjoyed a very special position. Each downswing in Spanish American silver production was matched, accordingly, by a countervailing increase in gold production, each successive 'crisis' witnessing an increase in the yellow metal's share in total specie production.

Viewed in their entirety the precious metal industries of Central and South America thus experienced only one major crisis during the years from about 1570, when they first established their ascendency in world specie markets to 1820, when they began to cede supremacy to others. Production and exports, in spite of a brief interruption between *c.* 1670 and 1720 when the onus of maintaining output shifted ephemerally to Africa, continued on a path of secular expansion. The main mining centres, having attained an output of almost 12 million pesos a year during the first decade of the seventeenth century, maintained or even on occasion surpassed that level until about 1660 when production, declining over the next decade to 11 million pesos a year, entered upon a phase of acute difficulties as the onus of maintaining European specie supplies passed from America to Africa. By 1720, however, normality had returned, and, as the first long cycle drew to its close, Central and South American precious metal mining enterprises, producing about 12 million pesos' worth of output a year, once more established their dominance over world specie markets. With the beginning of the second great long cycle in *c.* 1730/4 this dominance was assured as output rose to a new high level equilibrium in excess of 43 million pesos a year during the first decade of the nineteenth century (figure 1.10).[70]

If total precious metal production in Central and South America continued to expand during the years 1570–1810 with only a brief interruption between 1670 and 1720, when the onus of maintaining output

29

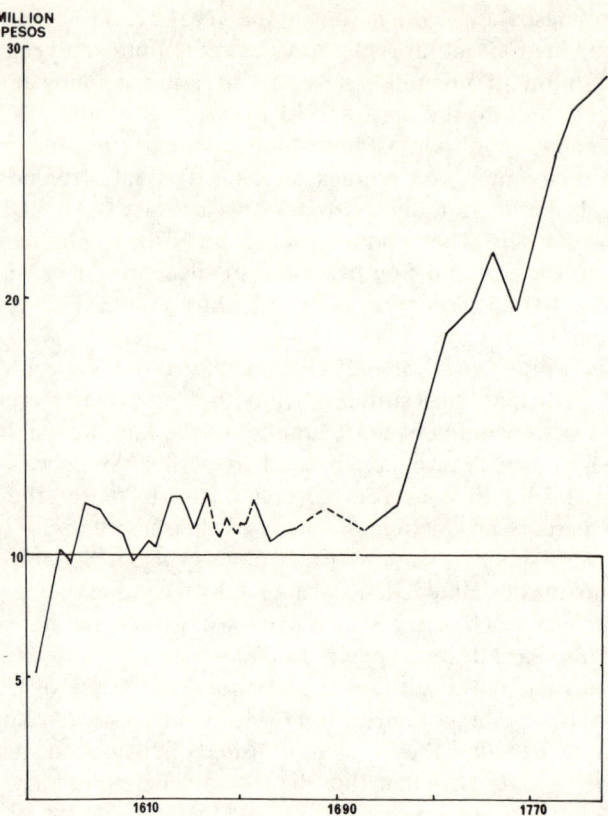

Figure 1.10 Total precious-metal production: Central and South America and west Africa

levels shifted to Africa, however, the industries producing these metals did undergo a process of major structural change over the same period. Falling productivity in the silver industry, associated with increasingly deep mining, declining ore yields, and the excavation of an increasing proportion of minerals which were intransigent to working by the amalgamation process, coupled with rising prices for the products of the low-productivity gold industry, created by about 1625 an uneasy equilibrium between the two sectors of the precious metal industry.[71] By that date the overwhelming importance of silver – that drudge of the sixteenth-century world monetary system – had become a thing of the past. During subsequent 'crises' in c. 1625–55 and 1670–1760, as mercury supplies were redeployed beyond the base of the silver industry in accord with relative productivity levels, gold, for the first time, posed a significant challenge to the dominance of silver, increasing its share of total specie production from 27 to 42 per cent and establishing a trend which ultimately would be realized in the aftermath of the Central and South American mining crisis of 1810–30 when, borne upon an industrial diaspora which carried the gold industry first to Russia and then to the lands of the Pacific basin (California, Alaska, and Australia), it established its absolute supremacy on world specie markets contributing almost 80 per cent of total supplies in the 1850s.

The 'crises' of c. 1625–55 and 1670–1760 thus marked the beginning of a process of major structural change within the industries supplying gold and silver to world markets which could not but affect European markets for specie and the monetary systems dependent upon them. During these years European monetary stocks were transformed. Diminishing silver flows from the Americas, creating shortages and enhancing free market prices, caused a diminishing flow of silver for coining as mint prices lagged. As supplies of gold increased on the other hand and prices fell, more and more of that metal was minted. A *de facto* silver mono-metallism was giving way to a true bi-metallic system embracing both metals. This tendency, already apparent during the first 'crisis' of 1625–55, became particularly pronounced during the second, although initially during the years 1670–1720, as European traders scrabbled for available supplies of African gold, mint-masters were forced to enhance both metals' mint price in an attempt to alleviate overall short-term supply deficiencies. By 1720, however, the crisis was past and, as monetary stocks once more began to grow, the value of silver relative to gold was increased, and the bi-metallic ratio, which had risen over the previous half century from 14.25 to 15.5:1, fell from that level to 15.1:1.[72] In no case, however, amongst the nations receiving Spanish and Portuguese specie was the adjustment sufficient, and silver remained grotesquely undervalued, causing a collapse in the minting of that metal. The impact of these changes was dramatic and nowhere more so than in the United Provinces.[73] Down to 1720 a *de facto* silver mono-metallism had reigned therein, the gulden providing the basic circulating medium. Then steadily as free market silver prices rose and gold prices fell the country was denuded of its coinage. Where once there had been only silver, gold now stood in its place, causing merchants to demand the gold ryder as their standard or at least requiring a statement as to the tenderability of the ducat. This situation was not exceptional. All over Europe, as an acute shortage of silver coins prevailed and the market for that metal was characterized by an acute instability, merchants adopted the relatively stable gold as the basis for their transactions. Only in the north, amongst the lands bordering the Baltic, was the situation different. Here, although a few gold issues were minted, silver maintained its dominance. The reason lay in the availability of abundant local supplies of the white metal deriving from a buoyant, broadly based, European industry. For the first time in more than a century, during the years 1670–1760, significant quantities of European silver were entering the market.

Chapter Two

The European producers' response

As has already been suggested, the massive inflow of Central and South American silver onto European specie markets after 1570 utterly destroyed the pre-existing European industry based upon the *Saigerprozess*.[1] The delicate price equilibrium between produce (silver and copper) and input (lead) markets, upon which the fortunes of European silver producers depended, was seriously disturbed. As silver prices fell, relative to both input costs and by-product prices, the industry underwent an internal metamorphosis which totally altered its market position. Enhanced copper prices in relation to silver increasingly made that metal the primary objective of exploitation whilst the declining importance of silver in total returns ultimately made the refining of the copper to obtain that metal unviable and rendered the use of lead in that stage of production unnecessary. The close symbiotic relationship which for a century had bound together lead and copper/silver producers was split asunder. The focus of silver production, as has been indicated, shifted elsewhere – to Central and South America. The erstwhile European *Saigerhändler* was transformed into a copper producer who now had to compete with other producers without the benefit of cross-subsidization of his wares from silver sales. The lead producer lost his major market and now had to vend his wares to new customers. For a century from 1570 to 1670 copper, lead, and silver producers in Europe developed independently of each other, each operating within autonomous market structures.[2]

Copper

In the case of copper production the events of the 1570s merely exacerbated a trend which had been apparent since at least 1527. Until that time rapidly rising production in Europe had ushered in an epoch of abundant and cheap copper on European markets. Only when output temporarily declined, due to stoppages in the all-important English lead trade (in 1509–12, 1522–3, and 1525–7) had producers, other than those of central Europe, been able to vend their wares.[3] Thus in 1522–3 and 1526 small shipments of Swedish copper to the Netherlands may be discerned but, apart from the royal consignment of the summer of 1525, which evoked Jacob Welser's astonishment, they were of little significance.[4] Central European output dominated international markets.

Such was not the case, however, after 1527. As silver production shifted to Saxony, copper and silver production in Hungary, the Alpenlands, Mansfeld, and the non-Saxon Erzgebirge commenced on the path of long-term decline, opening up opportunities for copper producers elsewhere.[5] Swedish production and exports accordingly grew. The rate of increase was greatest when the central European industry was beset with difficulties in 1534–8, 1542–4, 1550, 1555–64, and 1569–74, but not in 1551–4 when Swedish production was impeded by the flooding of the main workings known as the *Blankstöten*.[6] When the central European production cycle reasserted itself – in 1539–42, 1544–9, and 1565–6/8 – Swedish production and exports retreated. Even in 1572, however, its contribution to European supply was slight. In spite of the drainage of the *Blankstöten* and the opening up of new workings (such as *Johannisgruva* and *Drottninggruva*) on the southern slopes of the 'copper mountain' during the 1550s, yields only occasionally exceeded 1–1.5 per cent and production growth was slight, output never rising above 200 tons a year, or half the level of the 1490s when Swedish production had first been eclipsed by that of central Europe. In terms of present reality such quantities were insignificant. Nor were future prospects other than bleak, particularly as in the boom conditions of 1569–74 Swedish production began to display signs of marked instability.[7] Even the newly established English industry, working Cumberland ores (chalcopyrite $CuFeS_2$, chalcocite Cu_2S) of *c.* 44 per cent metal content seemed to offer more enticing prospects. Growth was rapid, particularly in the early 1570s, but output never exceeded, even in favourable conditions, 60 or 62 tons. When the central European industry was not beset by difficulties induced by lead supply shortages (1565–8), English production could make no headway, for in spite of the existence of ores whose copper content was comparable with that of central Europe (yielding 8.3 per cent black copper in comparison with Tyrolean 6 per cent, Erzgebirge and Hungarian 7.1–8.6 per cent) their silver yield was slight (0.04 grams/kg in comparison with 12 and 2 respectively).[8] To the 1570s, whether in conditions of expansion or decline, the central European producer retained price leadership. Thus as production in the lands between the Rhine and Elbe declined, falling to 2,370 tons during the 1570s or about a third of the peak levels of the 1520s, prices rose (figure 2.1). An epoch of dear copper was dawning.

Nor did the decline of central European production and the rise of Sweden to a position of world dominance between 1574 and 1650 do anything to alleviate the situation. Production might expand steadily up to the 1650s until, with an average annual output during that decade of

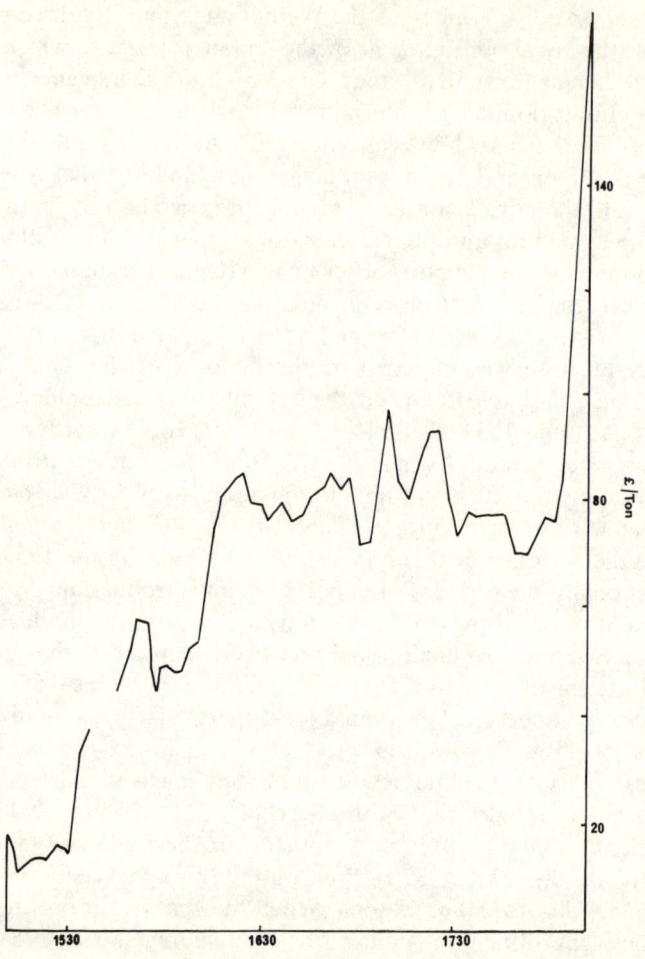

Figure 2.1   Copper prices

c. 2,100 tons, Sweden contributed 80 per cent of indigenously produced European supply, but at that date total European production was no greater than in the 1570s and in the market conditions of the mid-seventeenth century fell far short of demand (figure 2.2). Its entry into the market, on any significant scale, in 1574, had been dependent on the opening up of the *Bondestöten* with its rich ores of red copper (Cu_2O), initially yielding about 6.5 per cent copper, excavated from shallow open-cast workings in conditions of high prices. Its continuing expansion therein in conditions of declining yields, which fell to 3.5–4 per cent in the 1610s and 2–2.5 per cent in the 1650s and 1680s, was dependent upon increasing demand, rising prices, and the absence of alternative lower cost sources of

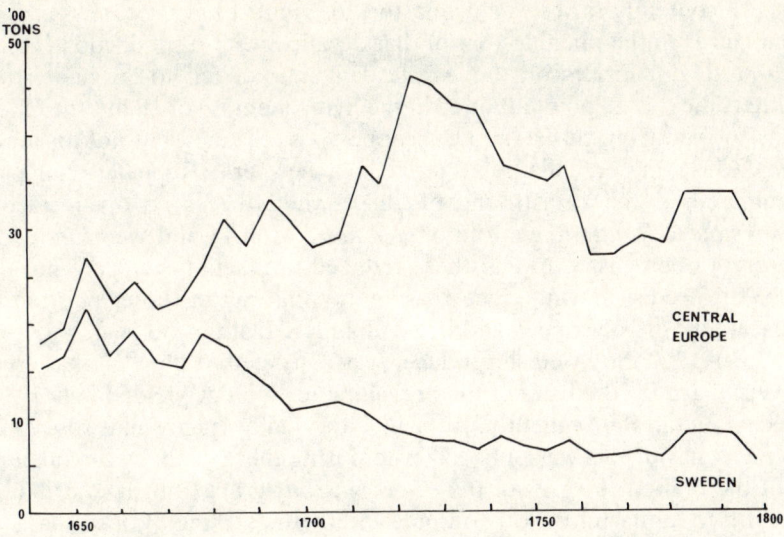

'00 TONS

Figure 2.2 Central European and Swedish copper production

supply. In the mid-seventeenth century such alternative sources were certainly lacking. On the basis of 2.5 per cent extraction rates from easily excavated mixed copper and iron sulphides ($3Cu_2S + Fe_2S_3$ of 46 per cent metal content) the Swedes could sell on the Amsterdam market at 40–50 gulden per zentner or about 10 per cent below the price of Norwegian copper (yielding 2 per cent from gravels $CuFe_2$ of 35 per cent metal content) or Mansfeld (returning a similar amount of copper from gravels with 8–9 per cent metal content), and could undersell on their domestic markets English copper (which since 1615 had been reduced to the exploitation of Cumberland gravels of 25 per cent metal content yielding 6 per cent copper), Tyrolean (exploiting similar gravels in deep deposits with no more than 22–24 per cent copper content), or Russian (produced from very low grade – 6 per cent – Urals gravel). The days when relatively shallow Hungarian or Alpine mines would yield white or red ores of 48–88 per cent metal content were long past. Gravels mined at much greater depth in 1650 contained only 22–5 per cent metal whilst most shallow workings yielded ores of 6–9 per cent metal content. In such circumstances the great shallow beds of 'high' grade gravels and sulphides of Scandia, even when exploited on the basis of the inefficient *Sulu-ofen*, came into their own, but the consumer paid the price, for the product of these mines cost seven and a half times as much as that manufactured, 130 years earlier, by the *Saigerhändler* from argentiferous copper ores which would yield 6–9 per cent metal.[9]

If long-term trends are of interest to historians, however, rarely do they concern contemporaries, for whom the 1650s must have seemed years of abundance. In the opening year of that decade Swedish production rose to 3,067 tons and prices on the Amsterdam market fell to 36 gulden per zentner, the lowest point attained in the living memory of all but the oldest men. Viewed from this standpoint copper was 'cheap' – but not for long.

In 1655 a major collapse ushered in a new era. Repeated collapses culminated in the catastrophe of 1689, when the three open cavities (*Blankstöten*, *Bondestöten*, and *Skeppsstöten*) fell in and were united in one great open cast working; these reduced production but were not the cause of Swedish decline. The reasons for that lay in the exhaustion of surface deposits of copper and iron sulphides. During the years 1655–80 and 1690–1700 the Swedish producer was restricted to the exploitation of gravels which on the basis of the prevailing technology yielded only 1–1.5 per cent metal. Subsequently, from 1699 to 1724 average yields rose to 2–2.8 per cent but this was only achieved during these years by the opening up of the limited reserves of the *Måns Nilssons-gruva*, which contributed one-fifth to a quarter of total output, and by the sinking of deep shafts at the opening of the century (Wredeshaft in 1697 and the Karl XII shaft in 1699) and during the subsequent two decades (the Karl XI shaft in 1707 and the Fleming shaft in 1722). In the short run yields were increased but production, and particularly capital, costs enhanced. In the long run yields declined and Swedish production dwindled into insignificance.

From 1655 the European market thus entered into a new period of rising prices, as the margin of exploitation was extended to incorporate surface ores which would yield 2 per cent copper, under the prevailing technologies; or the products of deeper workings which would return 5–6 per cent. The exploitation of these 'low-grade' European ore beds was delayed, however, by the intrusion into the market of an extra-European supplier. As prices on the Amsterdam market rose, in the course of 1655, from 36 to 56 gulden per zentner, the *Herren XIII*, governing body of the Dutch East India Company, ordered as much Japanese copper as could be despatched by its agents without endangering its 'country trade'.[10] At the new price level it had become viable to ship these wares, via Formosa, to the European market and, as long as Japanese prices did not rise above 35 gulden per zentner or European fall below 56 gulden, providing the margin which Pieter Van Dam regarded as necessary to cover the cost of freight, interest, and risk in shipment, a new supplier would service the needs of European consumers. Thus from 1655 to 1672 shipments from Formosa fluctuated in accord with the adequacies of Swedish supply, maintaining an aggregate level of supply available to consumers acquiring copper on the Amsterdam market of 2,000–2,400 tons a year to 1665 before it fell to 1,600–2,000 tons annually between 1665 and 1672. With the recovery of Swedish production in the late 1670s and 1680s such shipments ceased, but the fate of the Japanese producer was already sealed by that date. The large shipments to the Coromandel coast in the early 1670s signalled the beginning of the end of the trade in Japanese copper to Europe. Required to produce an extra 900 tons to meet the demands of both Asiatic and European consumers, the weaknesses of the Ashio mines (on Hondo island, north of Tokyo) were revealed. Costs rose as diminishing returns set in. Thus whenever the mines were henceforth required to make up for the deficiencies of European supply, in the years 1688–98, 1709–18, 1719–22, or after 1735, their contribution became smaller and smaller.

Thrown back, during the crisis of 1688–98, on indigenous sources of high-cost supply the European market was subjected to great stress and fragmented. New production centres sprang up on the basis of low-yield deposits which were seriously threatened when a semblance of normality returned to the Swedish production centre after 1698, and where governments attempted to protect these 'hot-house' growths of the last decade, as in England and Russia, by using prohibitive import tariffs imposed in 1697 the crisis pending since 1655 was finally precipitated.[12] Capacity in Russia grew, but because of markedly enhanced prices it was not until the early 1750s that rising demand allowed consumption to reattain the levels of the 1690s, when Swedish imports had amounted to 275–300 tons a year (figure 2.3).[13] Chronic underutilization of capacity, occasioned by an inability to enter the world market, thus characterized the high-cost industry, but at least until the late 1720s it enjoyed a domestic market free from competition. When, however, the rich white copper (Cu_2S) ores of the Altai were opened up by the Demidov's between 1727 and 1745, and particularly after the disruption in the workings there occasioned by the fire which swept through the Kolyvan-Voskressensk plant in 1732, a crisis was precipitated for the Uralian producer. In the short term the state intervened to protect the domainal plants to the west, by seizing the Demidov works and suppressing production there; but a

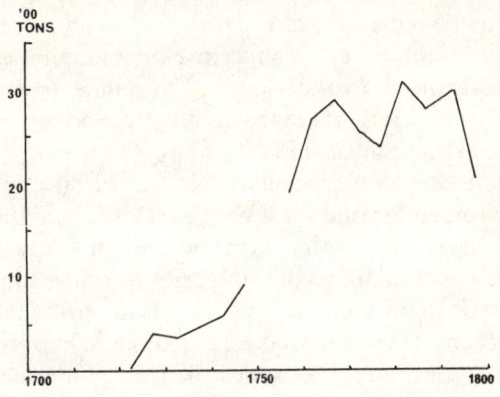

Figure 2.3   Russian copper production

more permanent solution to the problems of the industry was required.[14] If the high-cost products of the Uralian gravels could not be sold abroad and enhanced prices on the domestic market retarded consumption there, only one solution remained: for the state to buy the wares of its own creation. From the 1730s to the 1800s anything up to 80 per cent of total output passed to the imperial mints at Ekaterinburg (founded in 1735) and Kolyvan (established in 1765).[15] Russian consumers thus paid prematurely, from the 1690s, the price of dependence on copper derived from low-yield indigenous gravels.

Nor was the impact of the English tariff of 1697 markedly different. Imports, which had been running at about 2,500 tons a year in the 1660s, dwindled to insignificance. During the two years 1699 and 1700 an average of barely 50 tons of copper entered English ports.[16] Prices rose by some 40 per cent above the 'free-trade' level of the early 1690s and the consumer became dependent on copper manufactured from the deep mined ores of Cornwall which would yield only 5.4 per cent of refined metal. Production expanded, but down to 1730, even when unimpeded by the existence of competitors, Cornish producers could only sell 200–300 tons of the high-priced metal to domestic consumers.[17] When such a competitor emerged, through exemption from the tariff, and prices fell, the industry faced disaster. Thus when, between 1703 and 1713, Bengali copper shipped by the English East India Company entered the English market, manufacturers experienced acute difficulties.[18] Their response, as in Russia, was to lobby the state to buy their wares. During the 1700s and 1710s there was continual agitation that the coinage should be manufactured from 'high-quality' English copper, ultimately resulting in the contract of 1717 by which the government undertook to purchase the equivalent of three years' output at prices approximating to those of the late 1690s.[19] Before the contract had run its course, however, East Indian imports ceased. In 1720 the English industry once more dominated domestic markets and consumers once more had to be satisfied with 300 tons of high-priced, indigenous copper. Once more, however, high prices and loopholes in the tariff engendered imports from exotic climes – this time from Barbary and Aleppo. Barbary copper imports rose steadily, from 50 tons in 1720, to 132 tons in 1725–6, and 145 tons in 1730.[20]

Cornish sales on the domestic market declined correspondingly and smelters once more cast around for a market. Denied anything more than the most peripheral contacts with Europe because of the high-cost of their wares, they turned once more to the state, not this time to purchase their wares but to ensure them a captive market elsewhere within which they could sell their expensive wares. Thus emerged the re-export controversies of the years 1704–30, as a result of which the foreign producer was denied the drawback, thereby affording his English counterpart an advantage in supplying the colonial trades. Accordingly, during the years down to 1730 not only Russian and English consumers 'enjoyed' the dubious benefits of buying copper of domestic provenance but that 'privilege' was even extended by the English to consumers in the West Indies and west Africa.[21] Nor did this situation change much over the next thirty years, although for reasons largely exogenous to the British economy the competitive balance between local and alien sources of supply gradually moved in favour of the domestic product. In England changes internal to the industry effected a marginal improvement in the producers' situation. As a result of a reduction in mining costs, concomitant upon the introduction of the Newcomen engine into four of the larger Cornish mines (Wheal Fortune, Wheal Rose, Chacewater, and Polgooth), prices fell heavily (by 20 per cent) in the late 1730s but ultimately stabilized at 15–16 per cent below the peak levels of the early 1720s.[22] Domestic consumption increased and, now operating on a more price elastic supply curve than the North African producer, the English producer gradually increased his share of the market. However, his displacement of competition after 1740 owed more to rising costs in Barbary and the Middle East than greater efficiency in England. By 1750 the situation of dependence on 'low grade' ores, yielding 2 per cent copper from surface deposits and 5–5.5 per cent from deeper deposits, which had seemed imminent a century earlier, had become a very present reality. The only difference from the circumstances in the 1650s was that now there were no Asiatic riches to be drawn upon to alleviate the situation. Asiatic copper mining was in the mid-eighteenth century subject to the same fate as that of north Africa and the Middle East. Rising costs, which had led to the exclusion of Japanese copper from European markets in the 1670s, continued unrelentingly to the 1730s when the balance of intercontinental exchange swung in favour of the European producer. Even before that date the Japanese hegemony in Asiatic markets had been briefly challenged by the, as yet unexplained, Bengali boom of 1703–1714 but that phenomenon had been an ephemeral one, and from 1714 to 1728 the *status quo ante* had been restored. The crisis of the years 1728–33, when prices in almost every eastern port rose by 25–30 per cent, found the producer of Japanese copper incapable of response, costs at the Ashio mines rocketed, and even 'high-cost' English copper became relatively cheap. Shipments from England to the east increased. The English East India Company made its first significant shipments in 1731 but it was not until the 1750s, when the English product enjoyed a small competitive advantage over its rivals, that English shipments increased markedly to about 500 tons a year.[23] In the 1750s therefore, England had become the purveyor of high-cost copper not only to its own consumers but also to those of Africa, the Near East, and Asia.

From 1760 to 1781 the situation was temporarily relieved as English output rose from *c.* 950 tons to 4,000 tons, but this was a passing phenomenon, owing nothing to any technological resolution of the basic problem but merely resulting from the opening up of the Anglesey mines whose undepleted copper gravels yielded 6 per cent refined metals (figure 2.4).[24] Even so in the 1770s English consumption was restored for the first time in eighty years to the levels of the early 1690s, whilst the East Indies received shipments from England equivalent to the largest shipments ever despatched from the Ashio mines. Spectacular but, as already noted, shortlived, the boom in the Welsh 'copper mountain' soon gave way to decline as from about 1775 the workings began to display symptoms of diminishing returns. In the 1780s the *status quo ante* was restored and Williams, who had been responsible for the opening up of the cornucopia of the preceding decade, began to fear imports from abroad. Nor was his assessment of the situation unrealistic. During 1780–98, as Cornish ore

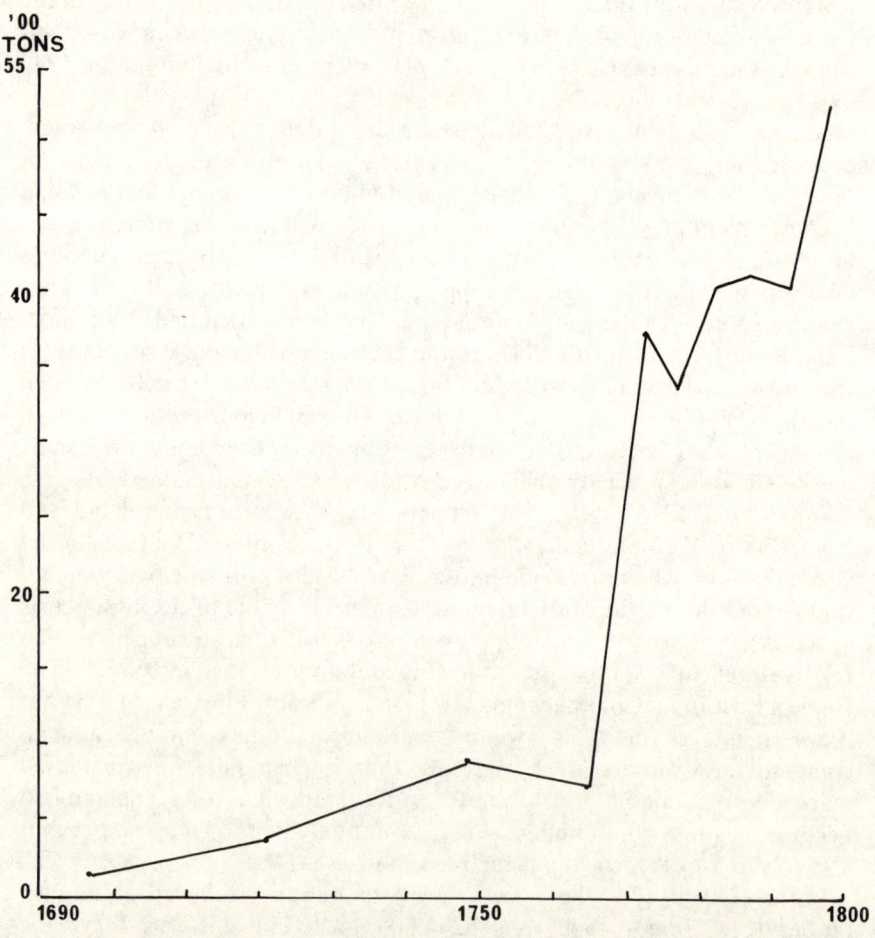

Figure 2.4 English copper production, 1696–1800

yields fell to 5 per cent, refined metal and Welsh to 2 per cent, output fell to 3,800 tons a year and prices rose, opening up opportunities for those alien producers whose competition Williams and his fellow smelters so feared. The period, accordingly, witnessed the beginnings of an import boom as merchants drew on a constantly changing supply network to satisfy the demand of English consumers for copper.[25] Perhaps the most significant event, however, occurred in 1787 when, for the first time, Spanish copper began to undersell English on the domestic market, raising profound fears amongst the members of the English copper syndicates. Only in one respect was their analysis of the prevailing market situation incorrect as they mistook the point of shipment for the point of origin of the copper now entering English ports: most of the copper shipped from Cadiz was Chilean in origin. The story of the protected English industry was drawing to its close just as that of the Chilean industry, which would dominate markets in the nineteenth century, was beginning.[26]

By protecting the 'hot-house' enterprises created during the crisis of 1688–98 Russian and English governments precipitated the catastrophe which had been hanging over the European industry since the mid-1650s. Forced back onto the consumption of high-priced wares, the product of low-quality ore deposits yielding up to 2 per cent refined copper from surface deposits and 5–5.5 per cent from deeper ones, the customer experienced a dramatic loss of welfare. Sales of copper to private customers in both countries fell catastrophically in the opening years of the eighteenth century, to about one-tenth the levels of the early 1690s, and only increased slowly thereafter with enhanced incomes, the rate of increase in Russia far outstripping that in England until in the 1760s almost two and a half times more copper (1,100 tons) was sold there than in England (450 tons), before the latter nation began to outstrip its eastern counterpart, domestic consumption rising to some 3,500 tons a year or roughly double the level of sales to private customers in Russia.[27]

Those who remained dependent on the Amsterdam market enjoyed a completely different experience. In spite of the tribulations of the opening years of the 1690s, when increased exports from Norway failed to compensate for Swedish decline, their fortitude and loyalty did not go unrewarded. A fortunate conjuncture in copper and silver markets, where prices rose markedly, coupled, as will be shown, with a dramatic fall in lead prices between 1670 and 1730 brought a new producer – the central European *Saigerhändler* – once more into the market and, particularly with their integration into Amsterdam's commercial network in *c.* 1700, caused a transformation of conditions therein. From *c.* 1701 to 1730, as prices in England and Russia equilibrated at the high levels first established in 1698 and consumers experienced a significant loss of welfare, those who had remained true to the Amsterdam market witnessed a dramatic fall in prices. During these years prices on the Amsterdam market fell relentlessly, interrupted only by the dislocations occasioned by Swedish hostilities in the 1710s, to ultimately stabilize at a level which, if higher than that prevailing in the 1650s, was still about half that reigning on Russian and English markets. In such circumstances the products of Mansfeld, the Harz, and Erzgebirge, where they were not excluded by tariffs, assumed a position of complete market dominance and, as a result, continental European customers enjoyed a completely different market experience from their English and Russian counterparts. The latter, as has already been shown, perforce had to make do with diminutive quantities of high-priced copper. The former, on the other hand, as incomes rose from the abysmally low levels of the mid-seventeenth century and copper prices fell, bought more and more of the metal, consuming by *c.* 1725 by far the largest part of a continental European industrial output of about 5,000 tons a year.[28] Nor were they significantly worse off than their Russian and English counterparts, who since the 1690s had become dependent on products manufactured from high-priced indigenous copper, when that conjuncture came to an end in the 1720s and declining Harz production ushered in a period of rising prices during *c.* 1725–60. In spite of an enhancement of continental European prices the periphery made little or no headway in the market, exports thence never rising above 150 tons a year or about 3–4 per cent of continental supplies.[29] As markets moved into an uneasy equilibrium, therefore, the central European *Saigerhändler* continued to hold the centre of the stage, maintaining the position of chief provisioner to continental European customers which they had first assumed in *c.* 1700.

Only after about 1760 did this situation begin to change as rising demand, through enhancing prices between about 1700 and 1830, constantly shifted the margin of exploitation into lower and lower grade ores, and falling silver and rising lead prices once more undermined the central European producers' position.[30] The onus of satisfying increasing European demand shifted to England where, as has been shown, the opening up of the Anglesey deposits paved the way for a production boom which pushed output up to about 4,000 tons a year in the early 1770s. The Welsh bonanza was, however, an ephemeral phenomenon and as production declined, not reattaining its earlier levels until about 1806–10, a host of minor producers operating on progressively slighter resource bases rose to challenge the English position. From about 1775 to 1798, in the aftermath of the English onslaught on continental European preserves, Norwegian and Mansfeld producers successively came to the fore.[31] With the depletion of the Thuringian deposits, whose metallic content fell to 2.6 per cent yielding 0.6 per cent refined metal in the 1830s, it was Russia's turn. From 1800 Russian production and exports grew until the 1820s, when not only in absolute but also in growth terms England once more assumed the centre of the stage, but now on the basis of a much depleted resource base. By 1850 'free-market' prices were some 50 per cent higher than in the 1770s and the story of the European contribution to the history of the international copper industry was drawing to its close.

Even as this final phase in its development was just beginning the position of the European industry had been challenged. From their first entry, via the doors of Cadiz, in the 1780s Chilean products had begun to increase their share of the market which rose to some 8–11 per cent in the 1820s, 18–19 per cent during the reassertion of English ascendency, and, as the European margin of exploitation shifted from 3.6 and 1.5 per cent levels for deep and shallow workings respectively in the 1850s, to 38 per cent at the close of that decade.[32] By the 1850s the Chilean challenge had become a reality, its industry now leading an American onslaught on European markets.

The years after *c*. 1670 had thus witnessed a renaissance in the fortunes of the central European *Saigerhändler* who rose to a position of dominance within the European industry by 1700 and continued to be the major supplier of those markets from which they were not excluded by tariffs until at least the 1760s. Even then they were not completely ousted from their important position within the European industry. On occasion, as during the 1780s, they could still offer a significant challenge whenever the meteoric rise of the English industry faltered. Behind this transformation in the position of the central European producers during the years 1670–1760 were the high silver prices, occasioned by the South and Central American mining 'crisis', the equally high copper prices caused by the collapse of Swedish production after 1655 and the cessation of Japanese imports after *c*. 1670, and the abundant availability of cheap English lead between *c*. 1670 and 1750.

Lead

If the events of the 1570s had merely exacerbated existing trends within the central European copper industry, hastening its ultimate demise, in the lead industry they operated in a similar way but with the effect of raising the English industry to a position of absolute dominance within Europe. For some forty years prior to 1570 the English industry had experienced acute market difficulties as the availability of alternative supplies, in the form of monastic lead, and periodic crises which disrupted the export trade, in 1550–4, 1555–64, and 1569–74, depressed prices on English lead markets.[33] For those industrialists who were struggling with their own acute resource-depletion problems the impact of these changes was disastrous. Production within many lead fields completely closed down and in Derbyshire and Somerset, where it continued, mining activity was reduced to a low ebb. In such circumstances the loss of the central European industrial market for lead after 1570, as the *Saigerhändler* of the region experienced a catastrophic reversal of their fortunes, merely exacerbated an already bad situation. Faced with this prolonged and acute market crisis, however, producers had responded by initiating a process of technological innovation.[34] In the 1540s the introduction of furnace smelting on the Mendips, whilst failing to improve on the metallic extraction rates of the prevailing 'bole' technology, had freed the industry from the constraints of its existing resource base. Previously the industry had been confined to the smelting of large pieces ('bing') of virtually pure galena. Now it could operate on the basis of the smaller pieces ('wash ore' and 'riddlings') which had been discarded over the centuries and which now could be smelted. The industry had evolved a way of weathering the storm. During the years of depressed lead prices, however, it ensured little more than survival. Only when the impact of monastic lead sales had been spent and the export trade was unimpeded did the Mendip producer come into his own, output which had fluctuated between 7 and 75 tons a year between 1536 and 1550 thereafter rising during each successive period of free trade (in 1555/6, 1565–8, and 1575–80) to ultimately attain a level of some 272 tons a year in 1578/9.[35] Nor during these years had the Derbyshire industry, still operating on the basis of the old 'bole' technology until about 1570, remained immune from the effects of the intermittent trade revival. Output, which probably never exceeded 290 tons a year between 1536 and 1550, thereafter rose slowly during successive periods of uninterrupted trade until during the late 1560s it may have amounted to about 320 tons a year or a level comparable with that prevailing during the pre-crisis years.[36] With the introduction of the new technology into Derbyshire, however, the situation was transformed. A new industry, employing a new technology which united in a close symbiotic relationship the new ore hearth and the old 'black-work' oven, provided the basis for a rapid expansion of production. By the end of the decade capacity had been brought into existence in Derbyshire which was capable of producing some 3,000 tons of lead a year and although it was only partially utilized at that time, with output amounting to no more than 750 tons a year, the potential for future growth was enormous.[37]

As the first flood of Central and South American silver thus flowed into Europe, undermining the position of the central European *Saigerhändler*, setting their industry on an irreversible path of decline, and destroying the major market for English lead, the English industry, with an output of about 1,000 tons of lead a year, stood supreme in Europe. Those centres of continental European production, such as the Harz or Poland, which had enjoyed brief periods of prosperity during the years when the English trade had been interrupted, now faced an indomitable competitor and a collapse of their major market.[38] In the face of these dual threats to their position they declined or underwent an internal metamorphosis. English production, on the other hand, went from strength to strength. Output, though lagging significantly behind capacity, increased dramatically, rising from the 1,000 tons a year attained in the late 1570s to about 4,000 tons annually in the 1600s and 12,600 tons a year in the 1630s.[39] Whilst industrial demand, deriving from the central European *Saigerhändler*, declined and the Central and South American silver industry provided only an intermittent (1580–90, 1610–14, and 1625–40) and initially diminutive demand for the metal, therefore, sales to non-industrial consumers increased significantly until during the 1630s in England alone about 2,500 tons of lead each year was being utilized for building, plumbing, or other domestic purposes.[40] Technological change within the industry, reducing costs, limiting the rise in nominal prices, and causing 'real' prices to fall, had greatly extended the use of lead within European society and had afforded its citizens a new standard of comfort. The ability of builders to provide these new amenities, however, was entirely predicated on the capacity of producers to reduce costs and bring about a fall in 'real' lead prices, and when this situation came to an end in the 1640s Europe experienced an acute lead famine which, as has been shown, also had serious implications for New World silver producers.[41]

As English production declined, and renovated Polish workings or new Scottish ones failed to provide adequate compensatory supplies of the metal, prices rocketed upwards, ultimately attaining a peak level in 1670 which was 75 per cent higher than that prevailing during the years 1575–95 and 1610–15 (figure 2.5).[42] The causes of this crisis remain obscure but it owed something at least to the failure of the 'free miners' in the principal lead fields (Derbyshire, Somerset, and South Yorkshire) to invest in the drainage of their workings and to their deliberate attempts to restrict output in order to obtain monopoly profits. Nor during the years 1615–70 were they entirely unsuccessful in the pursuit of these objectives. By the manipulation of their jurisdictional privileges they managed to keep outsiders at arm's length, preventing their participation in the economic rent they derived from the possession of the workings. By adjusting their work patterns in accord with their own distinctive work psychology, moreover, they managed to steadily push up lead prices, allowing them to maintain their anachronistic life style when their contemporaries were losing theirs. Yet having won this battle it was already becoming clear by 1650 that they had lost the war for, having neglected to undertake the necessary drainage of their mines, the mid-century witnessed the wholesale abandonment of workings due to flooding, thereby accelerating the decline in output which had begun as a result of their restrictive practices. By the 1660s the industry lay in ruins. A grandiose legal organization and elaborate system of law was superimposed upon an industry of delapidated mines, improverished workers, and moribund production.[43]

Now began the task of reconstruction but not before the power of the 'free miners', already weakened by economic circumstances, was finally broken by a series of legal reforms which were encapsulated in a new

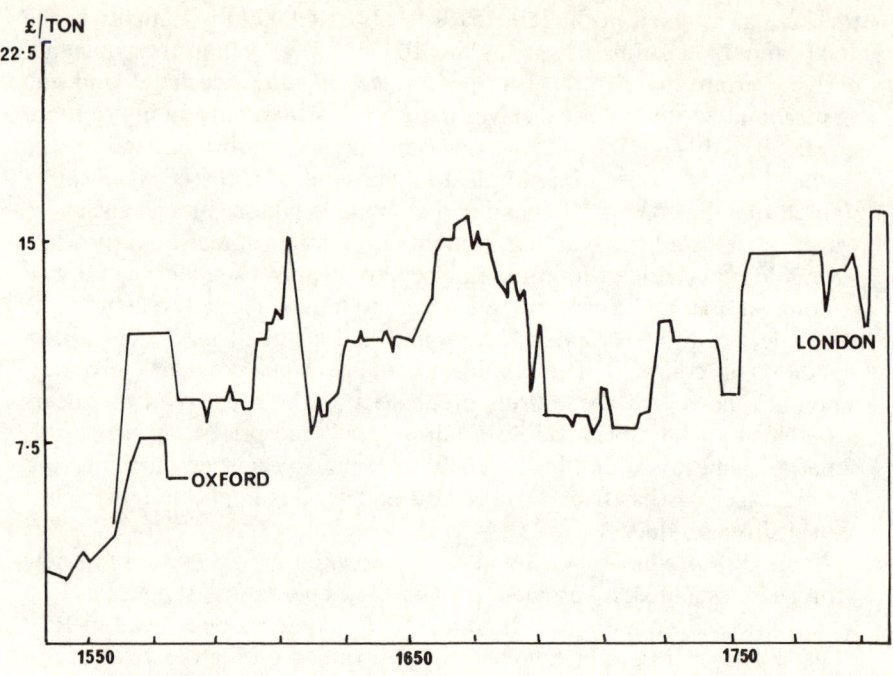

Figure 2.5 Lead prices

corpus of mining laws promulgated in the 1660s.[44] By such measures the way was laid open for outsiders to gain access to the ore fields from which they had for so long been excluded. In the late seventeenth and early eighteenth centuries a host of local investors, aristocrats and gentlemen, traders and farmers, as well as Yorkshire lead merchants and smelters, intent upon the backward integration of their concerns, began to buy into the mines, relegating the independent 'free miners' to the periphery of the industry. The new men, moreover, brought with them a new attitude and displayed a marked willingness to undertake necessary investments in adits or, at the very least, to conclude agreements with 'soughers' to undertake drainage work.[45] Cheap money, occasioned by a lack of alternative investment opportunities, coupled with a suitable legal framework, embodied in the new mining laws, thus paved the way for an investment boom during the years 1670–1760 which resulted in the de-watering of the Derbyshire mines and a massive increase in production which rose to some 10,000 tons of lead per annum during the second quarter of the eighteenth century.[46] Once more the Derbyshire mines reigned supreme within the industry. Yet there was to be no simple return to pre-crisis conditions. Indeed the 'free miners' had only been brought to the conference table and rendered amenable to the demands of would-be investors by the much more fundamental change then taking place within the industry. Even as the crisis had reached its height the smelters were evolving new ways of offsetting the impact of rising ore prices. By major improvements to the basic ore hearth they managed to raise yields to some 0.55 tons of lead per ton of ore, or by about 42 per cent over the rate achieved, in optimal conditions, from the older hearth and over 100 per cent more than was being achieved by that technology in the processing of contemporaneously raised ores (figure 2.6).[47] The impact of these changes was dramatic. Where the new technology was not adopted, as on Mendip, falling lead prices, pushing down ore prices, resulted in a dispersal of the labour force and the closing down of smelting capacity embodying the old hearth/blackwork oven technology.[48] In Derbyshire the situation was somewhat different. As the introduction of the new technology caused lead prices to fall, eliminating the rent element in the 'free miners'' earnings, they experienced major problems as it became increasingly difficult for them to employ auxiliaries and mining operations in the deep stepped workings, which were also plagued by flooding, became increasingly unviable,

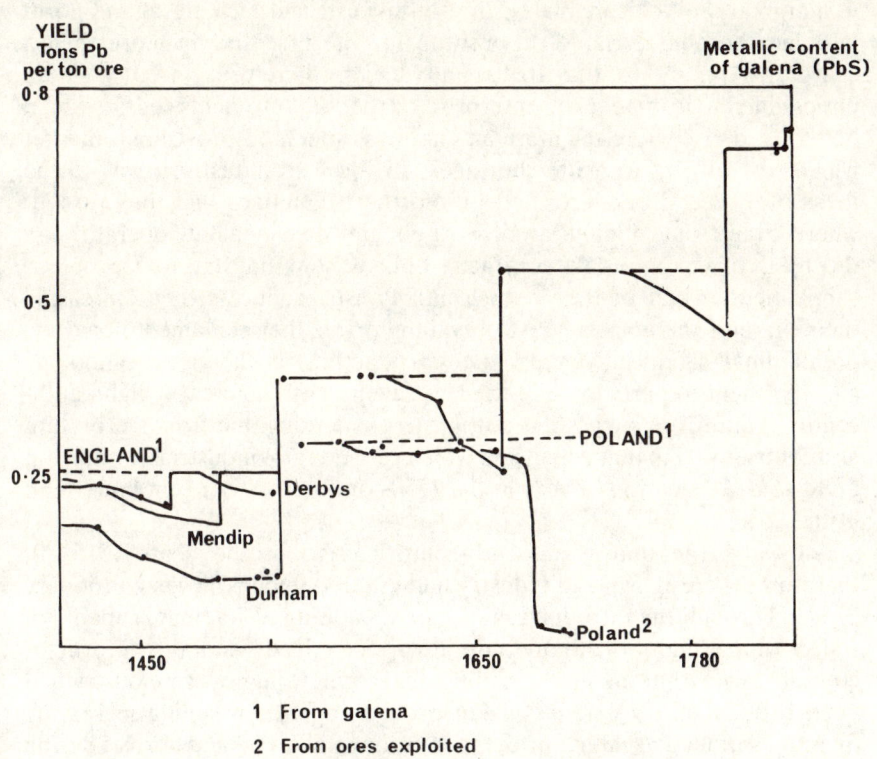

1 From galena

2 From ores exploited

Figure 2.6 Lead extraction rates

leading to a collapse in production. The restructuring of the mines in the 1660s, however, unlike elsewhere in the 'old' industry, did not entirely destroy the old order. Tradesmen who had previously provided working capital to the miners now funded fixed investments and in the restructured mines, which increasingly conformed to the shaft and gallery system, the reorganized teams of face-workers were encompassed in 'coping' arrangements, their higher productivity allowing the employment of additional 'tutworkers' undertaking 'dead work' (sinking shafts, driving levels or gates, and constructing whinzes) without any enhancement of unit mining costs.[49] As lead prices fell and sales of the metal increased, therefore, miners' earnings were maintained at the high levels of the 1660s and 1670s and the industry was able to attract a growing number of workers providing the basis for the boom which, as has been shown, carried Derbyshire production to some 10,000 tons a year in the second quarter of the eighteenth century.[50] Yet the ability of the new technology to cope with the problems posed by the high ore prices of the 1660s affected not only the 'old' production centres. As lead prices had risen, particularly after 1650, and the 'free mining' communities enjoyed supra-normal profits, production in marginal ore fields in Wales, the south-west, and the central and north Pennines became a viable proposition for the first time in more than a century. Already in the 1650s and 1660s, therefore, faced with an impossible environment for enterprise in the 'old' ore fields, businessmen had looked elsewhere for more amenable suppliers to provision a market which was subject to acute shortages. In the lead industry it was to the fields of Wales, Yorkshire, and the northern Pennines that they turned, where, free from 'mining laws', workers and management, operating on the basis of common-law contracts, built a new industry whose output supplemented that of the Scottish and Polish producers then increasing their share of the market.[51] At prevailing prices they remained, however, of marginal significance until the introduction of the new technology allowed them to participate in the late-seventeenth- and early-eighteenth-century mining bonanza, their output supplementing that from Derbyshire and contributing to that expansion which carried total industrial production up to some 21,365 tons a year in the 1750s or about 70 per cent more than in the 1630s.[52]

Following the mid-seventeenth-century crisis of the years 1615–70, therefore, the English lead industry had entered upon a new expansionary phase. Through the introduction of a new smelting technology, capable of transforming high-priced ore into low-priced lead, and the reform of contract law within the major mining fields, a new high-wage industry had been brought into existence which proved capable of selling increasing quantities of lead at falling prices to customers at home and abroad during the years 1670–1750.

Only as the 1750s drew to their close was it becoming apparent that the bonanza was over. Symptomatic of the changes afoot was the decline of Derbyshire production and the displacement of mining activity towards the previously marginal Welsh and north Pennine fields.[53] Acute and ultimately terminal resource depletion problems were beginning to appear which even the improvement in the metallic content of the concentrates delivered to the furnace, from about 50–60 per cent metal to 80 per cent, and the introduction of reverberatory furnaces or 'cupolas', which raised extraction rates to 92 per cent, could do nothing to alleviate. Production, thus, continued to grow but expansion was now entirely predicated on rising prices and the imposition of tariffs to keep out foreign competition.[54] As the final production cycle of the English industry thus ran its course the share of exports in total output steadily dwindled from 75 per cent in the 1750s to 33 per cent in the 1780s and 25 per cent in the 1820s, and industrial sales became increasingly dependent upon the willingness of English consumers to buy ever greater amounts of the over-priced metal.[55] Nor was the position of their continental counterparts very different. As the English product withdrew from continental European markets its place was taken by Spanish lead, a high-priced commodity which in the 1810s was being sold at a price which was almost double that prevailing at the time of the English hegemony in the early eighteenth century.[56]

The 1750s thus witnessed the end of era. For almost two centuries an ongoing process of technological innovation within the English industry, by widening the range of ores which could be smelted and increasing extraction rates, had allowed the exploitation of a constantly deteriorating resource base. Production had expanded from about 600 tons a year to 21,000 tons a year and save during the crisis years 1615–1670 prices had constantly fallen in 'real' terms until during the years 1690–1725 the price (measured in terms of the general price level) was but two-thirds of what it had been in the 1540s. Consumers, particularly during the years 1670–1750, thus enjoyed abundant and cheap supplies of lead and only after that period did their position deteriorate as the process of technological change failed to keep pace with the depletion of ore reserves, causing the rate of production increase to slacken and prices to rise. Nor did the resultant industrial diaspora, as others rushed in to fill the place vacated by English exports to continental European markets, do anything to alleviate the lead 'famine' of the late eighteenth and early nineteenth century. Spanish lead might undersell the English product by about 5–10 per cent but, in comparison with lead prices in the 1720s it was still a very expensive commodity.

Silver

For the first time since the establishment of the Spanish American hegemony over international silver production in c. 1570, the years 1670–

1760 saw the creation of conditions conducive to a revival of European production on the basis of the *Saigerprozess*. High silver prices, occasioned by the South and Central American mining 'crisis'; the equally high copper prices, caused by the collapse of Swedish production in 1655 and the cessation of Japanese imports after *c.* 1670; and the abundant availability of cheap English lead, as a result of that industry's entering upon an expansionary phase between 1670 and 1750: all combined to create a favourable price conjuncture encouraging a renaissance of central European production.[57]

In this respect the environment was quite different from that existing during the first great Central and South American 'mining crisis' of the years 1625–55. On that occasion there had also been an enormous rise in silver prices but increasing lead prices and a 'glutted' copper market had limited the *Saigerhändler* in their ability to respond, and during these years they made no contribution to relieving the prevailing silver shortages. Indeed, the enhancement in lead prices had provided an impetus to the deployment of an even older technology, leading to the exploitation of argentiferous lead by cupelation.[58] High and rising primary (silver) and by-product (lead) prices thus on this occasion had created a speculative atmosphere encouraging men to venture fortunes on the reopening of old workings and establishing new. Old workings were reopened in Bosnia and Serbia, now under Ottoman rule, whilst in western Europe a rash of speculative claims were filed.[59] Few, however, proved as successful as either Sir Hugh Middleton at Darren and Cwmsymlog in the Cardiganshire hills or the Baron de Beausoleil at Brugny.[60] At best, even under conditions of excessively high lead prices, such as existed during the years 1630–70, the production of silver by the cupelation of European argentiferous lead yielded only marginal supplies of the white metal to the market. With the fall in lead prices after 1670 their contribution was insignificant.[61] By that time the mantle of supplying European markets from indigenous sources of silver had passed into other hands.

From 1670 it was to be the *Saigerhändler* of central and east Europe, the heirs to the technological traditions of the period 1470–1570, who were once more to establish their ascendency as indigenous suppliers of European markets, on the basis, as has been shown, of rising copper and silver prices and falling lead prices. The shortfall in Central and South American silver production after 1670, as indicated, enhanced prices of that metal. Prices of copper similarly rose after the collapse of Swedish production in the mid-1650s and rocketed upwards after the cessation of Japanese imports in the 1670s and lead prices plummeted as the English industry entered upon an expansionary phase at the same time, flooding continental European markets with abundant and cheap supplies of the metal. For the first time in more than a century a favourable price conjuncture advantaged central European producers, and they rapidly seized their new opportunities. From 1670 the central European argentiferous copper industries once more resumed the pattern of yield-related production cycles, which had been interrupted during the previous century, and production expanded rapidly. Initially, it was the old production centres of the German states which took the centre of the stage, but structurally the new industry was very different from that of the sixteenth century. Mines whose names had been household words in an earlier age were now all but moribund. The once world-famous Bohemian mines of Joachimstahl and Kuttenberg contributed only a meagre 644 kg of silver a year (c. 24,100 pesos) to European output during the years 1670–1740, whilst the other workings in the kingdom could not muster more than a further 27 kg annually (c. 1,000 pesos).[62] In the Tyrol the legendary Schwaz mines had long before the eighteenth century been eclipsed by those of Salzburg which at their height during the years 1709–27 yielded an output of only 127 kg a year (c. 4,750 pesos) and on average during the years 1670–1740 produced annually about 80 kg (c. 3,000 pesos).[63] The situation was not very different in Hungary where, as the focus of Slovak mining activity shifted from Banska Bystrica (Neusohl) to Smolnik, production fell to 2,500–3,000 kg a year (c. 93,000–112,000 pesos).[64] Change within the industry thus took place during the years 1670–1740 against a blackcloth of diminutive activity in once-famous centres which provided a trickle of silver, amounting to about 3,500 kg a year (c. 130,000 pesos) to the European market. Collectively the whole of this sector of the industry produced little more than the output of the single Norwegian mine at Kongsberg which, during the post-war depression years, amounted to 3,100 kg a year on average.[65] Neither old nor new were of much significance, however, for the main impetus came elsewhere in Oberharz and Saxony which became the main mining districts during the boom years from 1670 to 1740 (figure 2.7). In the former centre the opening up of shallow deposits of argentiferous copper paved the way for a mining bonanza which pushed silver production up from c. 1,725 kg a year (c. 64,500 pesos) during the years 1640–60 to 11,100 kg annually (415,000 pesos) in the early 1720s.[66] In the latter, as the once important Schneeberg, Annaberg, Marienberg, and Weisenthal mines all lapsed into obscurity, growth within the mines of the Freiberger Revier and at Johanngeorgenstadt led to an increase in production which resulted in a rise in total Saxon output from about 300 kg a year (c. 11,450 pesos) in the 1670s to 3,025 kg (c. 113,200 pesos) in the early 1720s.[67] Together the Oberharz and Saxon mines dominated the German mining boom of the years 1670–1740, their combined output rising to about 14,000 kg a year (528,200 pesos) or about 70 per cent of a total central and northern European production of c. 20,500 kg (c. 770,000 pesos).

For more than half a century copper and silver passed by way of the Elbe to Hamburg and Lübeck from whence these wares were either distributed

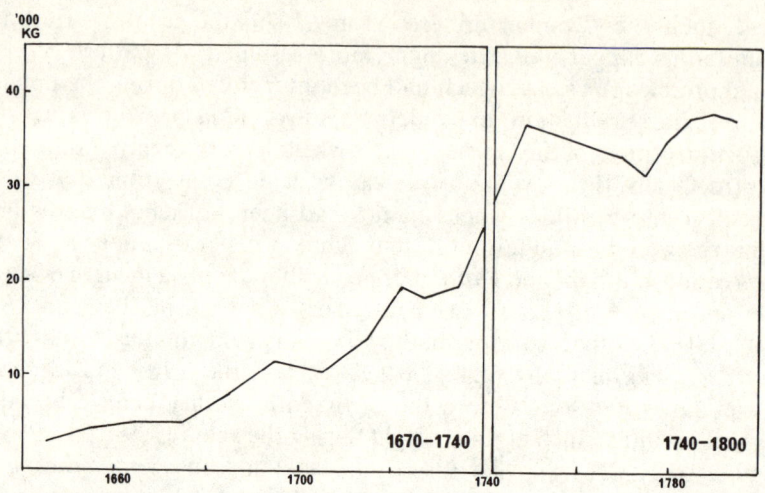

Figure 2.7 Central European silver production, 1670–1740

Figure 2.8 Central and northern European silver production, 1740–1800

south towards the Low Countries or east, the latter flow allowing Lübeck merchants to increase their share of the lucrative Russia trade at the expense of the English and Dutch. Product markets were transformed. The Hanseatic towns ran on adverse balance of trade and exporting silver created a distinct monetary zone in the north. The integration of the German *Saigerhändler* into the commercial network of the western metropolis of Amsterdam from 1701, moreover, bestowed on western European consumers the benefits that their central European counterparts had enjoyed for two decades. From 1701 to 1725 copper prices fell relentlessly, interrupted only by the dislocations occasioned by Swedish hostilities in the 1710s, to ultimately equilibriate at a level which, if higher than that of the early 1650s, was still sufficiently low to ensure market dominance in those areas where the products of the Harz forests and the Erzgebirge were not excluded by tariffs. A fortunate conjuncture in both silver and copper markets during the years 1670–1725 thus once more created the conditions for the establishment of a broadly based central European metallurgical complex which formed the focus of an important European metal-supply network. Over the same period production from this complex expanded rapidly, reaching a level of some 20,500 kg of silver a year (770,000 pesos) before the downswing in output from the Oberharz mines (which began slowly after 1725 before gaining momentum in the 1740s) again precipitated a crisis causing the focus of industrial activity once more to move elsewhere.

As the second production cycle of the European industry thus ran its course during the years 1740–80 and Oberharz–Saxon production declined to about 7,000 kg of silver a year (c. 218,000 pesos), or approximately half the peak levels of the 1720s, the Central European sector of the industry underwent a major transformation. In the aftermath of the Oberharz collapse new production centres in Hungary rose to dominate the central European scene (figure 2.8). The transition was a rapid one. Already in 1744 the mines of Hungary and Siebenbergen were delivering some 20,000 kg of silver annually (747,000 pesos) to the Kremnitz mint and henceforth under 'normal' conditions they provided the primary focus of a new central European metallurgical complex.[68] As the Hungarian star rose, moreover, a host of lesser satellites clustered around it. Old workings, like Joachimsthal, were renovated. New ones, like the unique Norwegian silver mine at Kongsberg, expanded their production. In the Bohemian mine production increased from the pre-1740 level of about 644 kg a year to about 1,000 kg a year between 1740 and 1760.[69] This small incremental gain, however, was dwarfed by changes taking place in Norway. Here production, which had fluctuated between 1,000 and 2,000 kg a year between 1630/9 and 1700/9, first rose to prominence during the Great Northern War when the disruption of the Swedish and German trading systems allowed it to secure a niche in the market supplying almost 3,400 kg of silver a year during the 1710s. Peace, however, caused production to retreat and, as Oberharz–Saxon production attained peak output levels during the years 1720–1740, the amount of silver issuing forth from the Kongsberg mine fell to c. 3,100 kg per annum. Only with the collapse of Oberharz production did output once more increase, rising to some 5,600 kg a year between 1740 and 1760.[70] In 'normal' circumstances during the years 1740–1780, therefore, the Hungarian mines formed the focal point of a central and northern European metallurgical complex which supplied about 32,000–36,000 kg a year (1.2–1.35 million pesos) to European specie markets. When Hungarian production faltered in the 1760s and 1770s, however, the tautness of the existing industrial structure was revealed. At Joachimsthal output increased, but only by about 43 kg annually, and even the greater expansion at Kongsberg, which carried production up to an annual average output of 6410 kg during the years 1760–1780, or some 810 kg a year above the levels of 1740–1760, was insufficient to compensate for the Hungarian shortfall of about 3,100 kg a year, which created acute shortages on European specie markets and sent violent shock waves through European financial and monetary systems.[71] Within the framework of the prevailing technology and resource endowment, the second European production cycle thus saw the north and central European metallurgical complex reach the limits of its capacity. Production, having risen during the first cycle to some 20,500 kg of silver a year (0.77 million pesos), during the second cycle rose once more to 36,700 kg a year

(1.37 million pesos), but only managed to sustain that level for a relatively short time, ultimately stabilizing after 1760 at about 31,000–32,000 kg of silver a year (1.1–1.2 million pesos).

Even as the second great boom of the central and northern European industry peaked and then faltered, however, events were taking place further east which would cast a shadow over the whole European scene. In Russia new mines, which had been opened up at Nerchinsk in 1704 and at Kolyvan in the Altai in 1729, now, during the years 1740–80, rose to a position of dominance within the European industry.[72] Production from this interlocking industrial complex, which had initially rarely exceeded 50 kg of silver a year, from 1740 expanded rapidly, attaining maximum output levels in the 1770s when some 24,500 kg of silver a year (0.92 million pesos) were produced, before decline set in (figure 2.9). When the diminutive supplies of silver derived from the Urals copper mines, which also attained peak output levels in the 1770s with a production of 3,150 tons of copper and perhaps about 7,500 kg of silver a year (0.28 million pesos), are added in, the importance of the Russian industry is clear. In the

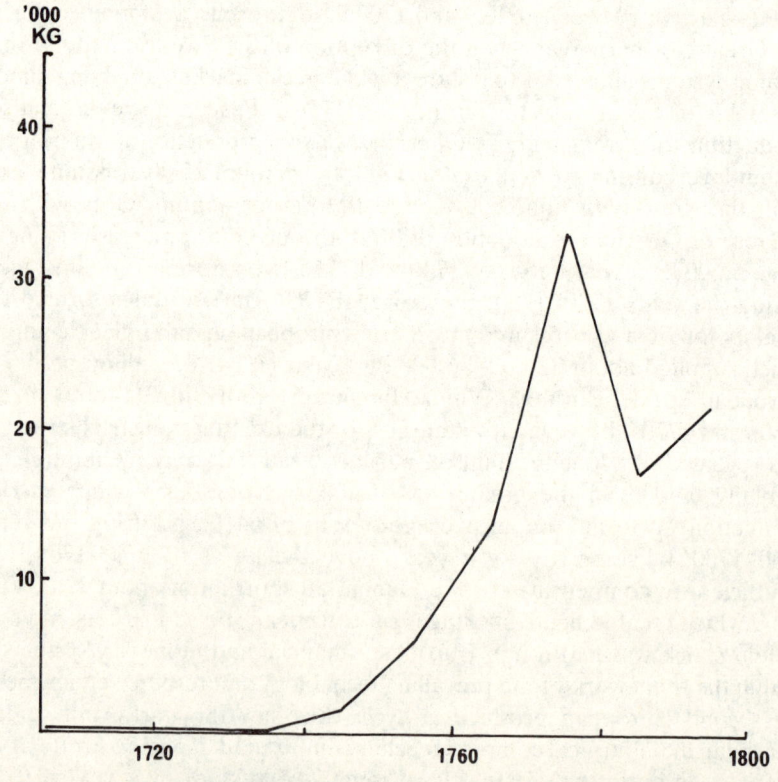

Figure 2.9 Russian silver production, 1700–1800

1770s Russia alone produced as much silver as the whole of the rest of Europe, and contributed about 7 per cent of total European supplies. Yet its impact on western European specie markets was remarkably slight and its position was highly precarious. Thanks to an 'economic revolution' within the Russian countryside that country's exports had been maintained in spite of incipient inflationary pressures and, enjoying a passive balance of trade, Russian silver was not exported to the west, which remained dependent upon Norway and the German lands for indigenously produced supplies of copper and silver.[73] The success of both groups of European producers, moreover, was entirely predicated upon the continuing mercury crisis in the Spanish Americas and the resultant recession in the silver industry there.

With the successful renovation of the Idrian and Spanish mercury mines and the rediversion of supplies previously destined for Brazil after *c.* 1765 the European mining centres were doomed. Born upon a flood of cheap mercury the Spanish American industry between *c.* 1770 and 1810 established its dominance over world silver markets once more. Yet the scenario first played out in 1570 was not to be repeated 200 years later. Certainly between 1770 and 1810 Russian and central/north European production declined but this time the recession in the European industry was only a sectoral one. Innovation, in the form of the barrel amalgamation process, permitted growth in Saxony (Freiberg) and the Bohemian Erzgebirge (Pribram) on the basis of low-grade 'rebellious' ores.[74] The technical problems which had isolated European producers from the benefits of the mercury amalgamation process had been overcome. As long as high-grade ores, exploitable by the patio amalgamation method, existed in the Americas, however, they could not gain the full benefit. This only came in the 1820s when the deterioration of the American resource base created conditions of parity between the American and European industries – and established the necessary preconditions for the gold industry's rise to a position of dominance in world specie markets.

Part Two

Russia. Precious-Metal Production and Economic Development

Chapter Three

Russia. Precious-metal production[1]

Against a background of declining Central and South American silver production the years from about 1670 to 1780 witnessed a revival in the European production of that metal. Central European production, initially concentrated in the mines of Saxony and Oberharz (*c.* 1670–1740) and then in those of Hungary (*c.* 1740–1780), steadily increased, attaining a level of 20,500 kg a year at the height of the first production cycle in *c.* 1725 and 36,700 kg a year at the height of the second in about 1760. Yet even as the central European production cycles ran their course, contributing between 4 and 7 per cent of total European silver supplies, events were taking place to the east which would cast a shadow over the whole European scene as new mines, opened up in the lands of the Russian Empire, rose in the ascendent. Initially discovered during the first production cycle of the European industry, the mines of Nerchinsk in the Amur valley of south-eastern Siberia (opened in 1704), and Kolyvan–Voskressensk in the Altai region of that province (opened in 1729) had at first been of little importance. From the 1740s, however, they went from strength to strength. Formed, at that time, into a massive interlocking mining and metallurgical complex, thereafter, with the Uralian gold mines of Beresov (opened in 1752), they contributed an ever-growing production of precious metals to the second great production cycle of the European industry until, at the height of the production boom in the early 1770s (with a combined output worth 2.2 million roubles, equivalent to 40,000 kg of silver a year), the mines, located in the vast wilderness of Russian East Siberia (map 3.1), produced and delivered to St Petersburg more of the precious metals than were produced in all of the rest of Europe put together.

Thus began the first great period of Russian precious-metal production, Russia's 'Age of Silver', when, for half a century from *c.* 1745 to 1795, a flood of precious metals poured forth from the Siberian mines to the St Petersburg mint for refining and minting into coin, establishing the nation as Europe's primary producer of precious metals. Her position, however, was entirely predicated on the continuing 'crisis' in Latin American silver

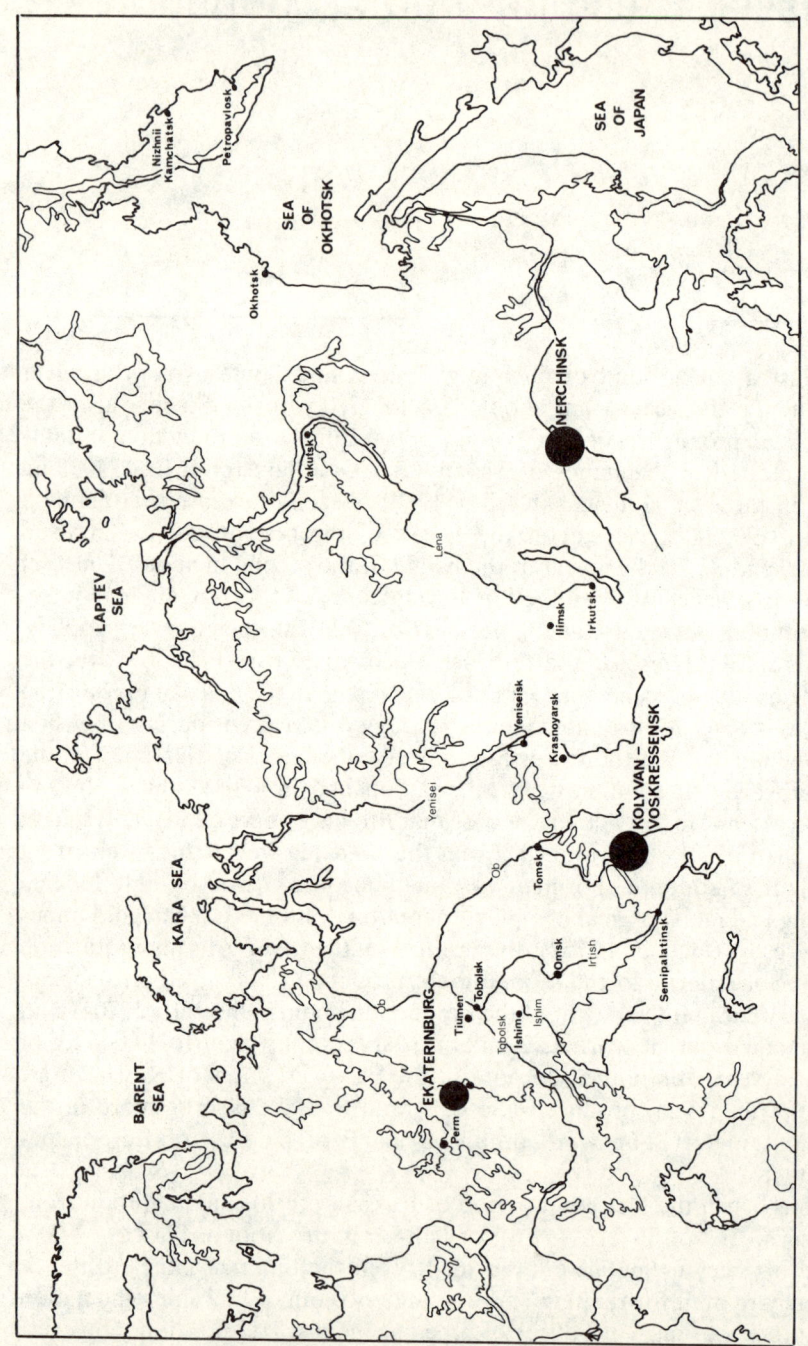

Map 3.1 Siberia (indicating principal precious-metal production centres)

production. With recovery in the Americas after 1760 the European industry was doomed. Production declined and within the increasingly depressed European industry Russia ceded her position of supremacy to Saxony and Bohemia where the introduction of barrel-amalgamation techniques allowed producers to weather the impact of the Central and South American onslaught.[2] From 1760–1830 on the basis of high grade ores, exploitable by the patio-amalgamation methods, the Central and South American silver producers, engrossing in excess of 90 per cent of world output, once more established their hegemony over international specie markets – but for the last time.

During the years c. 1800–30 the deterioration of the American resource base, as the average metal content of ores excavated fell to 20 oz and then 15 oz per ton, undermined the position of indigenous producers, enhancing unit costs of those employing patio-amalgamation techniques and rendering unsmeltable the increasing volume of 'rebellious' ores then being raised.[3] It also opened up new opportunities for European producers of the white metal. Against a background of slowly increasing European (including Russian) silver production by traditional cupelation techniques, the availability of cheap mercury during these years encouraged a vigorous exploitation of European native silver ores, production at Kongsberg (Norway) increasing from 814 kg a year in 1805/15 to 3,216 kg (or half of eighteenth-century peak production) in 1839/43, and new production centres being opened up in the Rhineland.[4] Even more spectacular was new growth at Pribram in Bohemia where the introduction of barrel-amalgamation in the 1780s set this centre on a growth path which pushed up annual production from 250 kg to 31,476 kg in sixty years.[5] From c. 1800 to c. 1830, therefore, as Central and South American silver production declined, European output of that metal increased, the major production centres increasing their share of world output from 8 to 16 per cent. The introduction of barrel-amalgamation techniques from the Old World to the New in the 1830s reversed this trend, but as a new equilibrium was established between the two sectors of the industry there was no return to the conditions of the 1790s.[6] Now operating in conditions of inter-sectoral yield parity the European industry maintained an 11–12 per cent share in world silver production until at least the mid-nineteenth century.

Whilst European producers, in conditions of deteriorating average industrial silver yields, gained ground at the expense of their Latin American counterparts during the years c. 1800–30, however, the silver industry as a whole lost ground to gold. Once again in the 1810s, as during the years 1625–55 and 1670–1760, available mercury supplies were redeployed beyond the base of the silver industry in accord with relative productivity levels; they were directed towards the gold fields of New Granada and Brazil and once more set the gold industry on that path which would ensure its dominance of international specie markets from the

1840s.[7] That pattern of industrial development whereby gold displaced silver in international specie production (which, as has been shown, began in the 1620s) was once again in operation.

Before the new production cycle of the gold industry ran its course, however, events conspired to effect a major structural transformation within the industry. From 1831 there was a major rise in mercury prices occasioned by the Rothschilds' creation of a world monopoly of that metal. The impact of this change on all precious-metal producers was considerable. Within the silver industry it encouraged the diffusion of the barrel-amalgamation technology, which not only allowed the processing of 'rebellious' argentiferous lead sulphides but also reduced mercury consumption rates, and, enhancing input productivity in this sector of the precious-metal industries, reversed that flow of mercury which had previously passed beyond the base of the silver industry.[8] Thus, as Latin American and European silver producers employing the new technology engrossed an increasing share of diminished mercury supplies, production of the white metal recovered to, and stabilized at, the levels of the 1790s, the Latin American producer ceding ground only to European producers deploying the same techniques; whilst the recovery of gold production, discernible during the years from c. 1810–30 in New Granada and Brazil, aborted in conditions of mercury-supply shortages, output stabilizing during the years 1830–50 at about 60 per cent of the level of the 1790s.[9] Amongst those precious-metal producers utilizing quicksilver as an industrial input, relative inter-sectoral changes in productivity in the use of that material (between patio- and barrel-amalgamation producers of silver and between silver producers using the new technology and gold producers) reversed the flow of mercury to the gold fields and undermined the position of producers there. Those same productivity changes were not sufficient, however, to offset completely the impact of rising mercury prices on refining costs, and, as trans-industrial costs of producing gold or silver by one or other of the amalgamation processes increased, producers utilizing these techniques were subject to competition from others employing different methods of metal extraction. Amalgamated gold and silver production which had dominated world output in the 1800s had by the 1840s lost its ascendency, its share of total world production amounting to no more than three-quarters of the whole.

As a result of rising mercury prices in the 1830s and 1840s the development of the precious-metal industries thus took place within a technological milieu which had been totally transformed. Whilst silver production in both Europe and Latin America continued to employ one or other of the variants of the amalgamation process and, engrossing an increasing share of available mercury supplies maintained output at the level of the 1790s, amalgamated gold production, subject to acute input supply shortages, declined. Against a background of aggregate decline in amalgamated precious-metal production acute shortages, especially of gold, began to appear which were only alleviated with the emergence of a new producer of that metal employing 'new' techniques. Against a background of declining extra-European gold supplies, the second great period of Russian precious-metal production, Russia's 'Golden Age' was about to begin. The years 1830–50 witnessed a massive increase in Russian hydraulically-refined gold production which established the nation as the world's primary producer of that metal, and ensured, by the 1840s, gold's supremacy in world precious-metal production. Year after year Russian gold production increased, until in the late 1840s, with an output of 28,750 kg of gold a year, the mines, located in the remote wilderness of north-eastern Siberia, produced more than half of total world gold and a quarter of all precious metals.[10] As in the eighteenth century, however, the industry's position was entirely predicated upon the contemporary crisis in extra-European sources of precious-metal supplies occasioned, in this instance, by high mercury prices arising from the Rothschilds' monopoly of that metal. With the breaking of that monopoly in the late 1840s and early 1850s and the discovery of new gold deposits in the lands of the Pacific basin (California, Alaska, and Australia) the fate of the Russian industry was sealed. After a brief, short-term crisis, output continued to grow but the industry's share of total world gold production fell from $c.$ 52 per cent in the late 1840s to $c.$ 12.5 per cent in the early 1850s.[11]

The eighteenth and nineteenth centuries thus witnessed two great periods of Russian precious-metal production, when, against a background of crisis conditions in extra-European sources of supply, that nation's mines assumed an important position in international specie production (figure 3.1). During the years $c.$ 1745–95, Russia's 'Age of Silver', an ever-growing volume of gold and silver poured from the eastern Siberian mines of Nerchinsk, Kolyvan–Voskressensk, and the Uralian workings of Beresov until, at the height of the boom in the early 1770s with a combined output worth 2.2 million roubles (equivalent to 40,000 kg of silver) a year, Russia established itself as Europe's primary producer of precious metals. Similarly during the years $c.$ 1835–85, Russia's 'Golden Age', ever-increasing quantities of gold poured from the Urals' workings and the remote mines of north-eastern Siberia until, in the late 1840s, with an output of some 28,750 kg of the yellow metal a year, worth about 23.2 million roubles, the nation ranked first amongst world gold producers and second only to Mexico in combined precious metal production. Thanks to the work of Professor Danilevskii, the story of Russian production during the second of these great mining bonanzas of the years from $c.$ 1835–1885 is well known.[12] Such is not the case with regard to the first great mining boom of the years from $c.$ 1745–95 whose story may now be told.

63

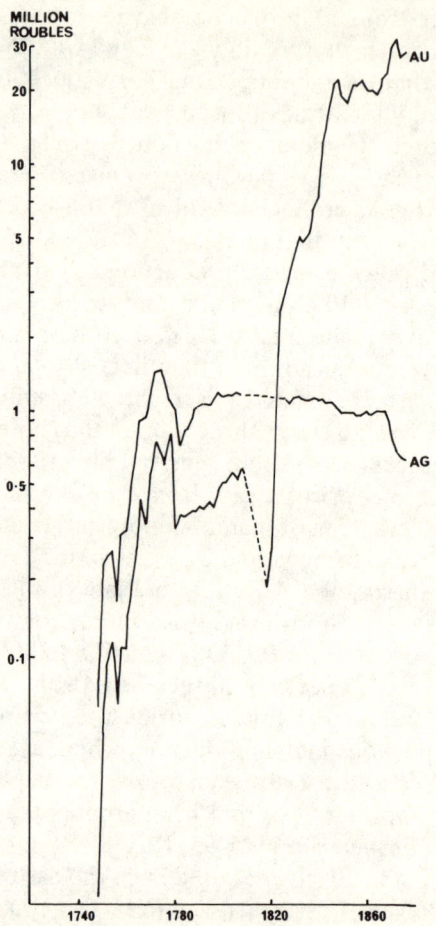

MILLION
ROUBLES

Figure 3.1 Russian gold and silver production, 1740–1876

Silver production in the eighteenth century

Nerchinsk (map 3.2)

The twenty years following the establishment of Ilimsk and Yakutsk in the early 1630s had witnessed an astonishingly rapid penetration by Russians into the inhospitable lands of north-eastern Siberia. Before that decade was out a small party under the leadership of the cossack Moskvitin, proceeding via the Aldan, Maya, and Ulya rivers, had arrived on the shores of the Sea of Okhotsk, and reconnoitring the lands thereabout found forests abounding in sables, rivers teeming with fish, and a land peopled by numerous Tunguses. At the same time a large hunting party under Il'ya Perfir'yev sailed down the Lena to the Arctic Sea, establishing the foundations for the long Russian drive to the north-east. The main body of his party, coasting east, reached the Yana. A detachment, under one Ivan Rebrov, traversed beyond, to the Indigirka, where a settlement, comprising a *zimov'ye* and small *ostrog*, was established as a base of the collection of the fur tribute from the native Yukaquir. Another group, taking the land route to the north-east, settled to the south of Rebrov's centre on the middle reaches of the river, creating the Sredneindigirskoye *zimov'ye*. By the end of the 1630s, the Russians had not only reached the Sea of Okhotsk but were well established in the north-east. In the 1640s the push towards the north-east was even more rapid. In 1642 the cossack Dimitri Zyryan reached the Alazeya and established a *zimov'ye* there. In the next year he and two others reached and sailed up the Kolyma, again establishing winter quarters which became the focus of a major fur trade, supporting the activities of no less than 396 Russians by 1647. Yet even as activity intensified on the Kolima others were moving on. In 1647–8 parties were organized at Nizhne Kolimsk to seek out the land of the river Andyr which, after many trials and tribulations, was reached in 1649. Here a *zimov'ye* was established, and, via exploration of the course of the river, the Pacific attained. Thus, within the brief space of two decades, *c.* 1630–50, as has been established by historians of the Russian fur trade, cossacks and *promyshlenniki*, chasing the elusive sable, had established imperial authority over the lands of east Siberia.[13]

However, even as the intrepid cossack Dezhnev first set eyes on the Pacific and, further south, a small community became established at Okhotsk, events were taking place far to the west which would render the development of the north-east insignificant, ensuring that for more than a century population movement into the area would be minimal.[14] The key to this relative abandonment of the north-east after about 1650 lay in the circulation amongst the Russian settlers in Ilimsk and Yakutsk of Tungus rumours concerning rich lands beyond the Uda river and Lake Baikal which were credited with vast wealth in sable, gold, silver, and grain. Excitement reached fever pitch amongst the denizens of the settlements and interest in expensive and highly speculative ventures to the Pacific coasts waned. Indeed, the north-east was, after 1650, of so little interest to the inhabitants of centres like Yakutsk that even the return of Dezhnev in *c.* 1659 caused no stir and his report was simply left to gather dust, only coming to light when discovered by the historian G.F. Müller in 1736 when it was of little more than antiquarian interest.[15] The whole focus of economic activity had shifted elsewhere. From 1650 it was the lands of Dauria and Amuria, beyond Baikal, which became the primary focus of activity in eastern Siberia, initially, at least, as sources of abundant supplies of sable. Within the decade settlers following the course of the Angara had

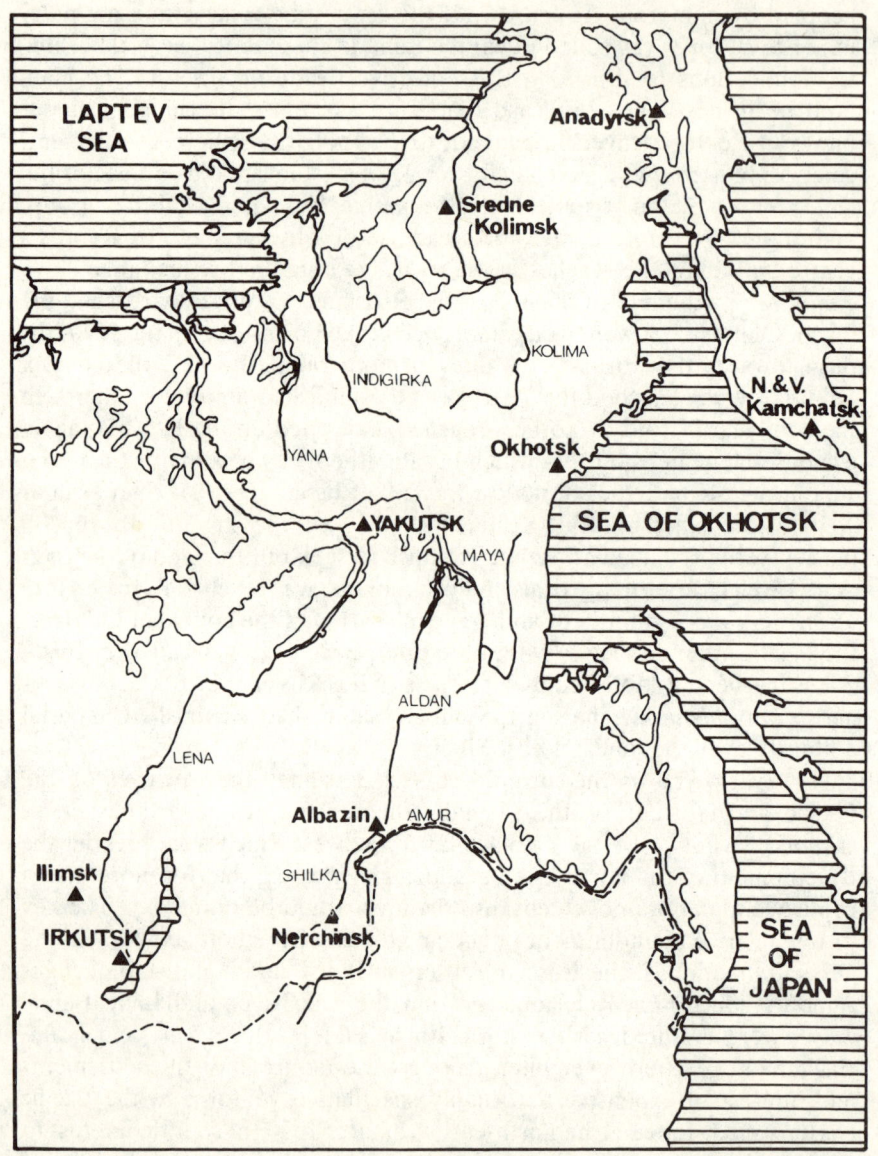

Map 3.2   East Siberia

established themselves at Irkutsk (in 1652) whilst others, passing on and skirting Baikal, carried the line of the river Shilka to establish Nerchinsk (in 1659). In the 1660s members of this group, following the course of the waterway, joined up with others from Yakutsk, who had come to Dauria via the Aldan and Olyokma rivers, and together they built an *ostrog* at

Albazin (in 1665) commanding the confluence of the Shilka and Amur. Within a mere fifteen years of the circulation of the first rumours about the region the Russians had laid the territorial basis of a much more lucrative fur trade than could ever be established in the north-east, and had ushered in Dauria-Amuria's 'First Age of Sable and Grain'. For almost a quarter of a century, from 1665 until 1689, the region was the focus of intense activity for the inhabitants of Ilimsk–Irkutsk and Yakutsk who diverted both their energies and their resources thither and away from the development of the north-east. A steadily growing number of Russian peasants began to settle Dauria, the homeland of the agricultural Daur natives. By the late 1680s some 277 households were established along the Shilka between Nerchinsk and Albazin, who provided grain not only for their own subsistence but also for the growing number of *promyshlenniki* who, at the same time, were penetrating the rich sable reserves of lower Amuria beyond Albazin.[16] On the basis of cheap grain to provision their expeditions and abundant supplies of pelts, Russian trappers exploiting the resources of Dauria–Amuria had, during the years 1650–89, enjoyed a veritable bonanza which caused an almost total diversion of effort and resources from the development of north-eastern Siberia, a region which rapidly faded from men's memories.

As increasing numbers of trappers poured into the lands of the lower Amur beyond Albazin, however, incidents began to occur as the incomers came into conflict with the Chinese, who had recently become subject to the powerful Ching dynasty and who now resolutely resisted any encroachment upon the northern borders of the dynastic Manchu homeland. Every attempt by the Russian trappers and peasants to establish forts and villages on the Amur now evoked a violent reaction from superior Manchu military and naval forces, ultimately culminating in the Chinese sieges of the Albazin *ostrog* in 1685–6 which, carried to a successful conclusion, led to the deportation of the Russian survivors to Mukden and Peking and to the ultimate exclusion of the Russians from Lower Amuria by the terms of the Treaty of Nerchinsk in 1689.[17] As a result of the treaty, the agricultural region of Dauria and the hunting reserves of Lower Amuria were split asunder and the boom which had begun in 1650 came to an end, ushering in twenty years of economic stagnation in Dauria.

With the signing of the treaty in 1689 Nerchinsk and its Daurian hinterland simply became the base for the exploitation of the Russo-Chinese commercial privileges won by Golovin by the treaty of that year, the assembly point for the official caravans which every two years from 1689 to 1722 left for Peking and for the far more numerous traders who illicitly propagated trade to Urga.[18] The impact on the local economy of provisioning the caravans, which might have a complement of as many as 400 people, or the smaller but more numerous private expeditions was far from negligible, but it was markedly subordinated to the impact on the economy of north-eastern Siberia which now experienced a major increase in demand for its furs – the principal export staple of the China trade. In spite of the state monopoly over the export of Siberian furs to China, the treaty of 1689 initiated a phase of intense speculative interest in this commodity which pervaded all ranks of east Siberian society. The merchants of Ilimsk–Irkutsk, Yeniseisk, and Yakutsk, intent on gathering a rich harvest of pelts to support their illegal trade, once more began to stake teams of trappers who, traversing to the north and north-east brought new life to the near-moribund settlements of the frozen taiga. Particularly during the years *c.* 1696 to 1716 the population of the northern zone of east Siberia expanded rapidly. On the lower reaches of the Yenisei numbers tripled. In the Lena basin and the lands extending eastwards they doubled.[19] Everywhere, save for the far north-eastern settlements of the Kolima valley and Chukotka, Russian trappers could be found gathering a rich harvest of sable and other furs to provision the illicit trade in these commodities to China. Nor were they or their backers much impeded in their illegal pursuits by Siberian officials or customs officers for all of these soon discovered that immense personal fortunes could be made from trafficking in illicit furs or from accepting bribes to overlook the contraband carried by both the caravans and private merchants travelling alone. From the lowliest bureaucrat to the governor himself Siberian officialdom connived at the illegal trade which was bringing new life to the economy of the Siberian north-east.[20]

In the boom which, during the years *c.* 1696–1716, saw Dauria totally eclipsed by the rapidly developing north-eastern Siberian lands only the state lost out, being forced to shoulder the costs of private expansion whilst being excluded from the profits reaped by the private merchants and trappers. These costs were not insignificant, for as avaricious trappers began to push hard into the lands of Chukotka, *en route* for Kamchatka, they soon created troubles with the indigenous Koryak and Chuckchi which the state was forced to resolve. Here in the far north-east, as elsewhere, the year after *c.* 1696 had witnessed an inflow of *promyshlenniki*, encouraged by the success of the Atlasov expedition of 1696–9, who took the two- or three-year trip to Kamchatka in search of the plump sable and on arriving there created a series of *ostrogs* – at Verkhne Kamchatsk in 1697, Bolsheretsk in 1700, and Nizhne Kamchatsk in 1701.[21] By 1712 they and the state servitors who accompanied them numbered almost 200 souls, and in passage across the isthmus of the peninsula or within its bounds they soon came into conflict with the indigenes, conflict which in 1703 broke out into open revolt. For more than a decade, from *c.* 1703 to 1714, the state thus became embroiled in a military campaign in the north-east which cost the Russians dear both in terms of casualties and in money maintaining forces on the Kolima to contain the rising. Worse, by being unable to force the passage between Anadyrsk and the Kamchatka outposts in the face of superior Koryak and Chukchi forces, they were also faced during the years 1707–17 with a mutiny amongst the Kamchatka cossacks during which three officers, including Atlasov, were killed. The revolt, moreover, was only brought to an end when the peninsula could once more be provisioned, a task only made possible by not insignificant expenditures during the years 1710–17 developing a seaborne traffic between Okhotsk and the outposts.[22] Private expansion on the last frontier of the Siberian mainland during the years c. 1696–1716 thus involved the state in heavy expenditure. Although it also opened up new rich reserves of sable, once again the state was the very last to benefit. Sable tribute or *yasak* from the Kamchadals never amounted to more than 2,500–3,000 pelts a year at this time.[23] In the fur bonanza of c. 1696–1716 which, yet again, established north-eastern Siberian superiority over Dauria, it was the private entrepreneur who made the greatest gains, engrossing most supplies of furs at source and undercutting the expensively acquired state pelts on the Chinese market.[24]

Yet, in their exclusion of the state, private entrepreneurs and venal officials alike had sown the seed of their own destruction, engendering in government circles an intense hatred of those who profited at the state's expense and an intense resolve to suppress the contraband trade. Individuals caught engaging in the illicit trade were summarily dealt with. The Siberian governor from 1710 to 1719, Prince M.P. Gagarin, who with his relatives had grown rich by illegal participation in the China trade, was simply recalled – charged with theft, nepotism, and ineptitude – tried, and hanged.[25] Having thus made an example of the supreme authority in the province, Peter, moreover, set about reforming the system within which such venality could take place, initiating a series of measures which ultimately culminated in the wholesale reformation of the system in 1727/8. As a result of these measures the state achieved its objective – participation in the boom by monopolization of fur exports to China.

By its suppression of the contraband trade, however, it also ensured that the remaining traffic would be but a pale shadow of what had gone before. A combined official and contraband trade which, according to Lange and Vladislavich, amounted to between 1.5 and 2.25 million roubles' worth of exports (predominantly furs) a year in the early 1720s was now reduced to an official trade worth on average about 100,000 roubles a year in the latter part of that decade and c. 30,000 roubles annually in the 1730s.[26] The state might now monopolize the China trade but in the process it had reduced that commerce to miserable dimensions, precipitating a crisis from which the trade, even in current prices, would not recover until the 1780s.[27] As a result of state intervention to protect its monopoly of fur exports to China the market for that commodity had collapsed, bringing the north-eastern Siberian boom to an end and relegating that region once more, during the years 1725–85, to the position of an economic backwater.

The balance of economic activity between north-eastern Siberia and Dauria was swinging in favour of the latter region during the years 1725–85 as the former underwent a severe crisis associated with the collapse of the Chinese fur trade. The forces which caused this displacement in economic activity were reinforced, moreover, by changes taking place within the Daurian economy, where the discovery of silver in the Nerchinsk (Yabolonoi) mountains in 1704 paved the way for a mining boom of significant dimensions and ushered in Dauria–Amuria's second great 'age', this time of 'Gold and Silver'. For half a century, as abundant supplies of auriferous silver poured forth from the Nerchinsk mines, both manpower and resources were concentrated in eastern Dauria in the lands of the mining *nachal'stva* organized there in 1720, which now eclipsed not only the lands of the Siberian north-east but also the Transbaikalian territories of Irkutsk *gubernia* which encompassed Kiakhta and its dependencies, the focus of the China trade since 1727.[28]

Only as silver production at Nerchinsk waned after 1774 was the inter-regional balance to some degree redressed and the development of the Siberian north-east and western Transbaikalia resumed. Thereafter, as the frontier was pushed ever further eastward – to the Aleutians, Alaska, and the Pacific coastline of north-western America – during the years 1775–1835, the fur trade came once more to the fore provisioning a rapidly expanding trade to China.[29]

For almost 200 years from *c.* 1635–1835, against a background of steadily extending Russian influence over the lands of east Siberia and beyond, there had emerged an irregular pattern of spatial displacement in economic activity as phases of intense economic activity in the Siberian north-east (in *c.* 1630–50, 1696–1716, and 1775–1835), normally associated with an active fur trade, alternated with equally intense phases of exploitation of the resources of Dauria–Amuria (in *c.* 1650–89 and 1725–75), again initially associated with the fur trade but in the eighteenth century related to the development of the mineral resources of the region. In each area a series of booms and slumps followed each other with an irregular periodicity, but in the eighteenth century at least the intensity of the Daurian mining boom was such that the pattern of fluctuations was superimposed upon a trend towards an increasing concentration of manpower and resources in the Daurian lands where, by the close of the century, the Nerchinsk *nachal'stva* had become the most densely populated region in the whole of east Siberia.[30] It is the story of that *nachal'stva* which may now be told.

Mining activity at Nerchinsk

Following the discovery of the Nerchinsk or Yabolonoi mountains by one Sibiriakov in the opening years of the eighteenth century, the initial focus of mining activity was located in the northernmost part of the mineral tract which was subsequently found to extend along the south-eastern slope of the northern half of the mountain range. This stretched about 120 miles between the longitudinal valley of the Shilka and that of the Arguin, in an area which subsequently was designated the first mining district of the future *nachal'stva*. Here, where the argentiferous lead deposits are mixed with copper, was established in 1704 the shortlived Troitsk mine, on the mountain of that name, which revealed all the basic problems of mining in the region (plan 3.3). Reserves were small, capable of sustaining an annual average output of only 40,000 puds (631 tons) of ore for some six years from 1704 to 1709. The ores raised were, moreover, extremely meagre in their silver content, averaging only about 0.25 zolotniks of silver per pud (2.3 oz per ton) with three-quarters yielding less than the minimum metal content required for refining.[31] In such circumstances, production of both silver and lead was small during the years 1704–9, and it was only with the opening of the larger and, at least temporarily, richer (*c.* 0.7 zolotniks per pud of ore) silver–lead reserves of the Monastir mine in 1709 in association with a reorganization of smelting operations that large-scale production began.

For thirty years, from 1709 to 1739, the Monastir mine (subsequently in 1746 designated the 'Old' or 'First' Monastir) dominated production at Nerchinsk on the basis of its poly-metallic deposits, producing on average 105,750 puds (1,700 tons) of ore per annum which apart from the dominant group of coppers contained 26,800 puds (430 tons) of silver lead with a mean yield of 0.7 zolotniks of silver per pud (6.6 oz per ton). Due to more extensive silver–lead reserves and a greater proportion of high-grade ores the problems which had beset the Troitsk mines had been alleviated but not resolved, for with yields of half of the technologically-determined minimum level for effective refining mining operations at the Monastir workings remained highly precarious. To sustain production, accordingly, the stock of argentiferous lead ores raised each year had to be culled. The better-quality ores, yielding in excess of 1.3 zolotniks per pud (12.5 oz per ton) were set to one side for smelting and refining whilst the lesser ones were stockpiled to be smelted (but not refined) when it was deemed possible to sell the metal produced on the lead market which, as will be shown, had a capacity to absorb only about 6.5 tons of lead a year at constant prices of 1–1.5 roubles per pud.[32] Mining officials' operational strategies in the exploitation of the Nerchinsk silver–lead deposits were thus still highly constrained; stable-state production at prevailing (1:1) ratios of low- and high-grade silver–lead ores amounted to only 6.5 tons of lead and about 1.66 puds of silver. Rarely, however, were officials able to sustain this stable-state situation, and during the years 1709–39 there emerged a distinctive production cycle at the Monastir mine as phases of increasing production, peaking in 1717 and 1722, were followed by acute production crises. During each of these booms, as the production of silver

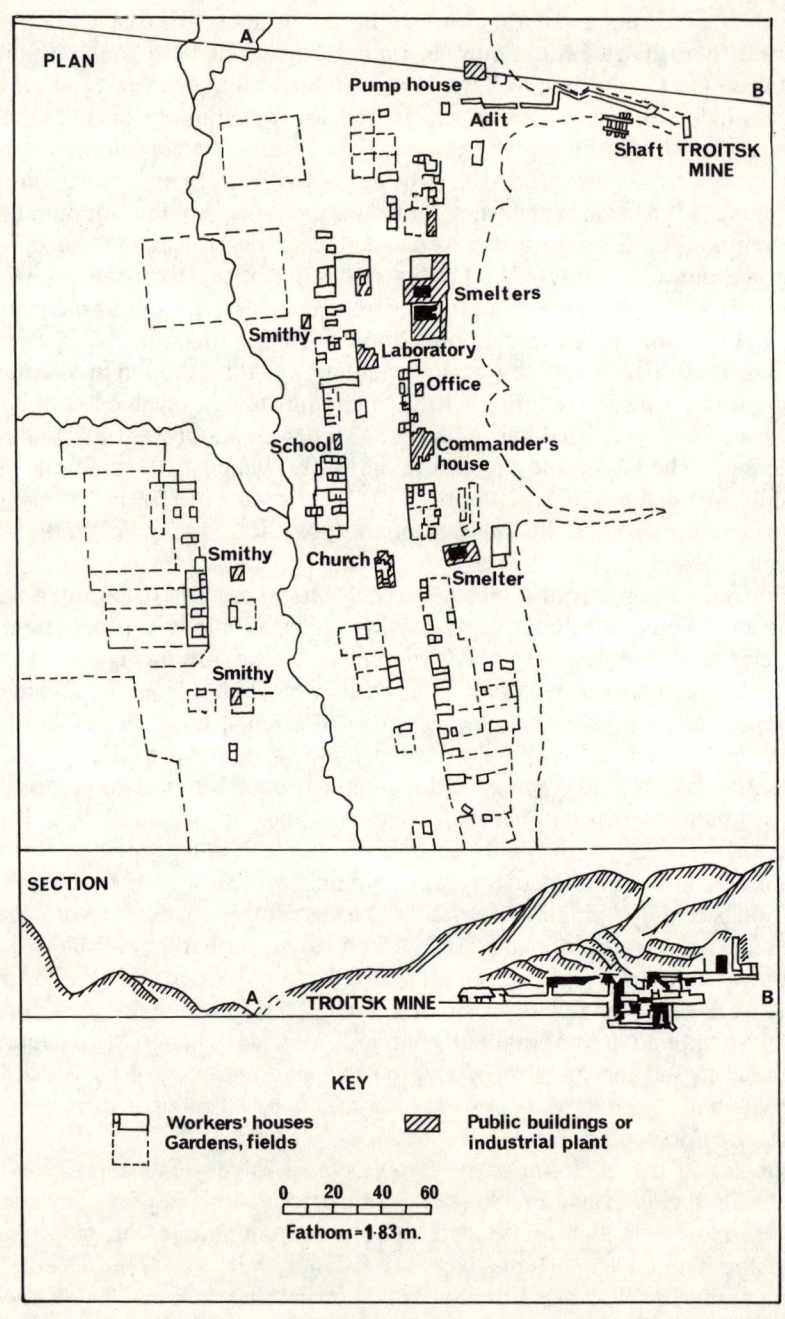

PLAN

Pump house

Adit

Shaft **TROITSK MINE**

Smelters

Smithy

Laboratory

Office

School

Commander's house

Smithy

Church

Smelter

Smithy

SECTION

A **TROITSK MINE** B

KEY

Workers' houses
Gardens, fields

Public buildings or
industrial plant

0   20   40   60

Fathom = 1·83 m.

Plan 3.3   Troitsk mine, Nerchinsk

and 'sterile' lead increased, the lead produced from low-grade ores was gradually displaced from the market. Production of both, however, continued to grow until the lead market was saturated and prices, falling to 72 kopecks per pud, depressed aggregate returns from lead and silver sufficiently to cause a reduction in mining production. Thereafter production fell until existing stocks of low-grade silver–lead ores were cleared and supplies of lead placed on the market once more returned 1–1.15 roubles per pud. During the first fifteen years of mining operations at Monastir, therefore, against a background of sustained production of copper ores, a pattern of five- or six-year production cycles of silver–lead ores emerged at the workings.[33]

When, during the second of these silver–lead production cycles, there was a fundamental shift in the balance of ores raised (2.6:1), however, the periodicity of the cycle was markedly altered as officials were forced to take account of the enormous build-up in stocks of unsmeltable silver–lead ore which imposed such costs upon the enterprise that even the massive rise in silver prices during the years 1725–9 could not stimulate a production response.[34] When these prices fell, their situation was completely untenable and in 1730 production of silver–lead ores ground to a halt leaving officials with about 2,000 tons of ore on their hands. In response this was smelted into 71 tons of lead and a major sales drive was initiated which over the years 1729–34 managed to dispose of 25 tons on the Siberian market and another 16 tons at Moscow, but only at a price (c. 75 kopecks per pud) which rendered any revival in silver–lead production impossible. In conditions of severe resource depletion the nadir of an extended ten- or twelve-year production cycle had been reached. Production of silver–lead ores was halted and with sales of lead produced from existing stocks of low-grade ores proceeding slowly, so that in 1734 officials still had 30 tons in hand, there seemed little prospect of anything more than the most modest recovery. And so it was. With production, predominantly of cupiferous ores, continuing at c. 1,700 tons a year at the Monastir mine, mining of silver–lead ores was resumed during the years 1734–8 as existing stocks of lead were sold off, but output was small rising from 27 tons yielding 11 kg of silver to 133 tons yielding 57 kg. With the extended production cycle continuing its protracted course the time had clearly come to move on.

The year 1739 thus saw the opening of two new mines – the 'First' or 'New Troitsk' located close to the original Nerchinsk mine, and the second, the Zerentuiskii – and the movement of the industry's resource base southward into the zone of silver–lead ores. Production, accordingly, once more increased but otherwise nothing else had changed. The deterioration of the industrial resource base continued. The ratio between the quantities of low- and high-grade ore, which had fallen from 1:1 during 1709–24 to 2.7:1 during 1725–38, continued to deteriorate, falling to 4:1 in 1739–43 and 8:1 in 1744–5. Officials might thus increase silver production, which rose to 15 puds (245 kg) a year, worth *c.* 4,000 roubles in 1745 but this could only be achieved by stockpiling large quantities of lead, which amounted to about 200 tons from current production in that year, which nobody wanted, and which had cost *c.* 4,500 roubles to produce.[35] Plagued by the large and increasing quantities of low-grade, unrefinable ores being raised, and in spite of an effective, if obscure, plan to reduce mining costs, the Nerchinsk mines were bankrupt in 1745/6.

Only the creation of a market for the unwanted lead could resolve the problem and prevent a downswing in mining production at Nerchinsk, and this was precisely what happened in 1746 as events taking place far to the west created just such a market and paved the way for a major production boom at Nerchinsk. The key to the new situation resided in the beginnings of silver production in the Altai utilizing the *Saigerprozess* to process argentiferous copper and other 'dry' ores. This involved the mixing of the silver-rich 'black' copper with argentiferous lead in special low hearths – *Saigerhütten* – where the metal was heated slowly until just above the melting point of lead when the greater part of the silver was taken up by the lead. This was subsequently drawn off and was available for refining. Accordingly the Altai plants required abundant supplies of lead and from 1746 a long-distance trade, involving the transport of the metal some 5,280 miles from Nerchinsk to Kolyvan, was born.[36]

The situation at Nerchinsk was transformed. Officials there in charge of mining and smelting operations were now isolated from the effects of that high and fluctuating level of low-grade ore production which had plagued their predecessors. These ores were now smelted and the unrefined lead produced was despatched to Kolyvan. The ore supplies delivered to the smelters at Nerchinsk for subsequent refining there were henceforth of a uniform and high quality capable of yielding 1–1.5 zolotniks of silver per pud (9–14 oz per ton). Officials, accordingly, now only had to ensure a regular and preferably increasing supply of such ores, a task conditioned by the nature of the mineral deposits and by the skill of prospectors in locating suitable deposits as they were required.[37] In these circumstances the exploitation of the mineral deposits at Nerchinsk evolved in an almost classic manner. The mining frontier extended steadily southward during the years 1739–1829, encompassing first the zinc–lead deposits of the two middle mining districts situated on the Grasimur river and the still richer four southern districts before reaching the most southerly district on the river Uryulungi where the principal veins were of lead, arsenic, and antimony; and each successive decade witnessed a diminution in both the average output and life of the new mines being commissioned (table 3.1). In terms of potentially realizable output of refineable ore prospectors opened up the vista of steadily increasing production until the 1790s when, with the wholesale closure of the older, more productive, mines in an

Table 3.1 Discovery of new deposits at Nerchinsk, 1704–96

	Number of mines discovered	Average life (in years)	Average annual output of refineable ore during operational life (in tons)		
			1704–56	1757–83	1784–96
1704–30	2	37	124	127	—
1731–40	2	34	309	714	—
1741–50	8	29[a]	66	485	—
1751–60	6	27	NIL	403	2,583
1761–70	10	21	—	528	603
1771–80	32	4	—	106	273
1781–90	25	4	—	149	681
1791–96	1	5	—	—	606

Note: a Underestimated subsequently because in successive decades 1,2,2,2,5, and 1 mine was still in operation in 1796.

environment where the new workings were on average both small and shortlived, output would decline. In the event, however, technical difficulties, resulting in the major new workings of the 1770s and 1780 being brought in at below their productive potential, precipitated a downswing in output at that time. The resultant compensatory prospecting boom and the opening of numerous small (under 250 tons a year potential output) mines, moreover, exacerbated the crisis, for the resultant increase in the amounts of low-grade ore raised from the new workings could not be disposed of after smelting, the market at Kolyvan then being subject to its own acute crisis. With only the prospect of further increasing the supplies of such ores and therefore stocks of unwanted lead, exploratory work ceased, and production in the 1790s was concentrated in a declining number of old workings most of which in 1796 already had a long history behind them (table 3.2).[38] The production boom which had begun in the mid-1740s thus aborted in the mid-1770s, the pace of decline thereafter accelerating in the 1790s.

Table 3.2 Number and size of operational mines at Nerchinsk, 1704–96

	Number of mines categorized by average annual output of refineable ore during their operational life (in tons)			
	1000 +	500–1000	100–499	1–99
1704–30			2	
1731–40		1	2	
1741–50	1	2	4	4
1751–60	3	2	5	7
1761–70	4	3	6	14
1771–80	5	3	8	37
1781–90	6	5	15	42
1791–96	4	3	5	4

Plagued throughout its history by endemic problems of high and fluctuating levels (47–88 per cent) of low-grade, unrefineable ore production, output at Nerchinsk only increased rapidly when in 1746 a market was found for the metal smelted from these ores. Thereafter, on the basis of local smelting and refining of high-grade ores and a long-distance – 5,280-mile – trade to Kolyvan for the lead produced from the low-grade ores, mining output increased. Year after year, from 1746 to 1774, new mines were commissioned and production grew, but as on average each successive mine brought into operation had more limited reserves than its predecessor the pace of growth slackened until, beset by technical problems in the late 1770s and early 1780s, output began slowly to decline (figure 3.2). When in the 1790s the effects of this short-term crisis combined with, and exacerbated, more general depletion problems the crisis deepened and the pace of decline in output accelerated.

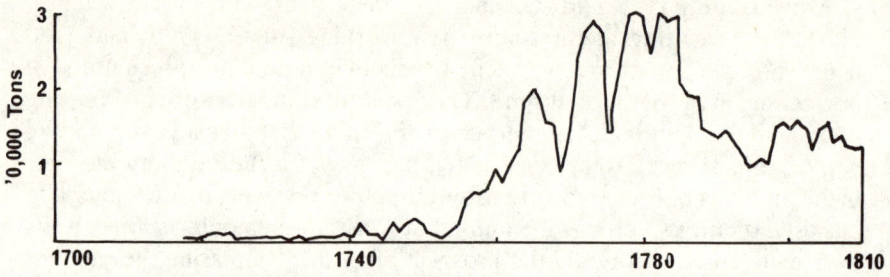

Figure 3.2 Nerchinsk ore production

The labour force at Nerchinsk

After some thirty-five years of intermittent mining activity, therefore, production of silver-bearing ores at Nerchinsk had, during the years *c.* 1745–74, undergone a major mining boom necessitating the creation of a large labour force in the underpopulated and remote wilderness of south-eastern Siberia (table 3.3).[39]

The task facing those charged with recruiting such a labour force was a formidable one but the eighteenth century witnessed not only the creation of a numerically large labour force but one which was also characterized by a high level of productivity. With the initial discovery of the mineral riches of the region and the opening of the diminutive Troitsk mine during the years 1704–9 a small group of convicts and their guards, numbering twelve in all, undertook all mining and smelting operations and were supported by the local Russian population, reinforced in 1708 by a party of 104 peasants despatched thence from the Yenisei district who combined grain production with woodcutting and charcoal-making. Production was small, however,

Table 3.3   The Nerchinsk labour force, 1704–94

		Ascribed peasants		Convict and free labour
1708		104		12
1710		182		12
1725		482		25
1735		n.a.		95
1772		12,900		1,000–1,800
	11,000		1,900[a]	
1782	10,829		1,963	2,000[b]
1794	19,725		n.a.	n.a.

Notes: a Free colonists subsequently ascribed at the Fourth Revision and attached to the mine working. b 'Miners and workers, old, young and staff' (Storch, 1801, II, p. 380).

and the low level of mining activity created few pressures on the local community. Such was not the case with the opening of the Monastir mine in 1709 which ushered in the beginnings of large-scale production at Nerchinsk. The twelve convicts and their guards, who remained the core labour force at the mine until 1720, were numerically quite inadequate to realize the full potential of the new deposits. To supply sufficient manpower, therefore, the previous functional division between convict and peasant labour was broken down as the latter group was pressed into all manner of work. During the 1710s the original party of 104 peasants, who had arrived from the Yenisei district in 1708 and had established farmsteads along the banks of the Shilka, could be found combining the cultivation of an extra half *desyatina* of land in relation to the four *desyatini* of their own to feed the mining population with woodcutting and charcoal-making. They coexisted, moreover, with another group of seventy-eight peasants who had arrived at Nerchinsk in 1710, and who, having established their farmsteads, rendered their quitrents not in grain but in labour at the mine, not only cutting wood and making charcoal but also working alongside the convicts in the bowels of the earth.[40] Mining and smelting personnel after 1710 thus comprised some ninety persons, peasants and convicts, at the Monastir mine who, working at an average annual productivity of 19 tons of ore per man (*c.* 1.4 cwt per man-day), produced about 1,700 tons of silver-bearing ore annually. They were supplied with grain and the works with fuel, moreover, partly by their own endeavours and partly by the efforts of a hundred or so ascribed peasants.[41] Under the control of concessionaries and in conditions of acute labour shortages the burden of increasing output during the 1710s was laid upon the backs of the ascribed peasantry who became jacks-of-all-trades, undertaking a volume and variety of tasks far beyond the limits laid down by imperial edict.

Then, in 1720, with the acquisition of the works by the state and the creation of the mining *nachal'stva*, the whole situation changed as a new order was imposed upon the organization of the labour force at the Monastir mine.[42] The number of mine workers engaged there changed but little, there being some ninety-five miners, workers, and supervisory staff there in 1735 just as there had been twenty-five years earlier. Yet organizationally the structure of this labour force was quite different, as gradually between 1724 and 1735 the number of convicts was increased until in 1735, when they numbered some ninety-five in all, they undertook all mining and smelting operations, producing, like their servile and convict counterparts twenty-five years earlier, some 1,700 tons of ore annually. The corollary of this increased specialization amongst the convict labourers was equally a specialization amongst the ascribed peasants, made possible by the arrival at Nerchinsk in 1725 of some 300 families, drawn from amongst the state peasantry of Ilimsk, Yenisei, and Tomsk districts, which increased the number of peasant 'souls' attached to the works to nearly 500. Greatly enlarged in numbers this group was now also functionally differentiated in ways totally unknown a quarter of a century earlier. Some 238 of their number now supplied the grain needed to feed the mining population, each household yielding up, as their quitrent, some 108 lb of grain, two and a half households providing sufficient to provision one full-time mine worker (125 kg).[43] Another 244, organized into groups of five, cut wood and, employing the new techniques introduced in the 1720s and early 1730s, made enough charcoal to provision the works.[44] In the late 1730s an enlarged, but still basically unstable, level of production was thus sustained by a similarly enlarged labour force, but one now characterized by a highly specialized and productive division of labour, a labour force, moreover, which was sufficiently robust to sustain the pressures engendered by the subsequent production boom of the years 1739–74.

During these years the numbers of workers and production increased rapidly until in 1772 output of ore reached about 2 million puds (34,200 tons) a year, but the organizational forms of 1735 remained intact. Most mining activity remained the preserve of convict labourers, now numbering 1,800, who working at a similar productivity to their predecessors, were able to excavate the greater part (94 per cent) of current output. This group in 1772 was also supported, as forty years earlier, by two functionally specialized groups amongst a greatly enlarged number of ascribed peasants. Some 5,000 'souls' who lived 'some distance from the mines' cultivated 'a certain amount of ground' each and brought 'in winter their produce to the magazines of the foundries'. Another 6,000 were 'employed in cutting and carrying wood, making and carrying charcoal'.[45] Like forty years earlier, they were organized into teams of five for the purposes of rendering their quitrents; three members of the team cut about 40 cubic sazhen (75 m³) of wood to be made into two charcoal beds (*kuchi* or *meiler*) which in laying, heaping, burning, and breaking absorbed the energies of the other two members of the team, who together produced enough fuel to process the ore raised by 0.8 convict-workers.[46] Although greatly enlarged in numbers the members of the Nerchinsk labour force continued to be deployed according to that highly specialized and productive division of labour which had already been established in 1735. Only in one respect was the situation different in the early 1770s to that prevailing forty years before: the principle of division of labour was extended to the mining workforce and a new element evolved amongst the ranks of mineworkers. Most ore production, as has been suggested, was undertaken by convict labour under the supervision of a small group of paramilitary personnel, who were deployed in large mineworking, the greatest number labouring in mines employing a hundred or more workers.[47] The mining boom, however, had witnessed a proliferation in the number of tiny workings each yielding on average less than a hundred tons of ore each year and many producing less than one ton annually. By the 1770s there were thirty-seven such workings, isolated mines located in open country or complex groupings in the immediate vicinity of the great mines, whose exploitation by convict labour posed major problems of supervision. The excavation of ores here thus became the preserve of a new group in the mining workforce: free colonists who, taking advantage of Catherine II's regulations of 1764, gained at least a temporary exemption from the capitation tax and freedom with respect to tillage and woodgathering. For these men mining activity was supplementary to their primary pursuits of tillage and animal husbandry, and their work patterns were accordingly very different from those of the convicts who dominated the mine labour force. Part-time in their work at the mines, which had to fit into the seasonal constraints of their agrarian regimes, their average annual per capita output was but a fraction of that of the convicts, entailing, at equivalent levels of output per man-day, only about 16–18 days a year, or about a quarter of the time expended by the ascribed peasants on work associated with the mines. They were also functionally differentiated in their mining activity from the convicts, playing a central role in the exploratory work at the mine but playing no part in the ongoing exploitation of the main workings. In the 1770s they were thus found either working in new mines which lacked development potential – the isolated pits located in open country where the excavation of the ore occupied two to five peasants annually for a period of one to eight years or the larger groupings of such mines in the vicinity of the greater workings with an equally short life span – or in the major mines possessed of large reserves but only in their initial or terminal stages of development, before or after their exploitation by convict labour. The 1,900 free colonists thus made only a small contribution (*c.* 6 per cent) to total production in the 1770s but played a vital role in exploratory work at the mines. In response to new economic circumstances at the mine, that highly specialized and productive division of labour, which had first been introduced more than forty years earlier, underwent a process of elaboration.

For half a century, *c.* 1724–74, the labour force at Nerchinsk thus maintained a basic organizational stability as production expanded. Then during the years 1779–83, amidst the turmoil of a major crisis and falling production at both Nerchinsk and Kolyvan, that edifice had to be fundamentally restructured. As the management at Kolyvan, in response to the developing crisis, restricted their purchases of Nerchinsk lead during the years 1780–3, the main market for the products of the increasing supplies of low-grade ore, raised in the proliferating number of small mines manned by free colonists, collapsed. With only the prospect of further increasing supplies of such ores and thereby the amounts of unwanted lead, exploratory work ceased and the free colonists lost their *raison d'etre*.[48] As a result they were recruited to the ranks of a new group of workers directly 'attached to the mines and agriculture' who were employed at task work, at a time when, due to Catherine II's labour reform of 1779, their labour was most urgently required. By Catherine's edict task assignments were redefined and the ascribed peasant was excluded from mining activity and coaling, wage rates were set, and labour involvement maxima were established so that the ascribed peasantry could earn the money they required to pay their capitation tax with no more than four weeks' or 28 days' labour a year, the remaining time being free for agricultural pursuits and domestic occupations.[49] With one stroke of her pen Catherine had halved the Nerchinsk peasants' involvement in industrial work, and had posed serious problems for the management of that enterprise which were only resolved by a major reorganization of the labour force. Coaling now became the preserve of the small group of ex-colonists who were henceforth employed as semi-professional workers organized in teams of two, each team working 120 man-days a year making two *kuchi* (= 126 *korobi* of 6.5 chetverts each of charcoal) a year, at task work. They were supported, moreover, by the whole of the remaining 6,000 ascribed peasants involved in charcoal production who were now confined to woodcutting, each felling slightly more than six cubic sazhens (11.8 m^3) in 24–6 days a year.[50] The years after 1779 thus saw a major restructuring of the labour force at Nerchinsk. Work at the mines and smelters remained the sole preserve of the convicts and associated supervisory staff, numbering 2,000 in all, who, confined to a declining number of old workings, were subject to a decline in their productivity; output per man-year fell from 19 tons of ore in 1772 to 14.5 tons in 1794.[51] They were now supported, however, by 1,900 semi-professional task-workers employed at coaling and 10,750 ascribed peasants deployed in either agricultural pursuits or in woodcutting within constraints laid down in 1779 who were distributed in accord with the requirements of the individual foundries.[52]

Expanding ore production at Nerchinsk in the years to 1774 was thus associated with a corresponding increase in the size of the labour force. It was also associated with an increasingly elaborate organization of that labour force, ordered in accord with a highly specialized and productive division of labour. With the decline in production after 1774 and the imposition of Catherine II's reforms of 1779 this whole edifice had to be restructured. Numbers employed at the works slowly declined, and whilst convicts continued to dominate mining and smelting activity as the labour intensity of the ascribed peasantry was reduced, a new group of task-workers, or *urochniki*, was recruited from amongst the free colonists of the region to undertake coaling.

Metal production and markets

The ore raised at Nerchinsk during the eighteenth century largely comprised silver-rich lead ores found initially in combination with copper and after 1739 with zinc. On excavation samples of these ores were sent first to Nerchinsk for testing, and on the basis of these trials the stocks of ore raised were culled. The better-quality ores yielding in excess of 1.3 zolotniks of silver per pud (12.5 oz per ton), were set to one side for smelting and refining and the lesser ones were reserved for smelting but not refining.

The extraction of the silver from the first group of these ores involved the reduction of ore in a simple hearth (*Schmelzofen* or *plavil'naya pech'*) and the production of an argentiferous lead described as 'fertile' (i.e., *viplavleno iz serebryanikh rud svintsa, nazivaemogo sirogo, is kotorogo otdelyetcya serebro*). This was then placed in a cupelation hearth (*Treibofen* or *obzhigatel'naya pech'*) and subjected to an oxidizing blast which converted the greater part of the lead into lithage, leaving in the hearth a residual deposit of hearth lead (*Herdblei* or *gert*) and the 'unclean' silver (blicksilver) still containing traces of other metals (*ostravshago of razdeleniya serebra, gletu i gertu, soderazhaushago v sebe svinets*) and requiring further refining. The amount of silver obtained at this stage of operations depended almost entirely on the original silver content of the ore and the amount of concentration required to produce a refineable lead. In the processing of 'soft' unenriched lead, produced from high-grade ore, the operation worked simply with a minimum of loss. The silver content of the 'fertile' lead was obtained with the minimum of loss. The lead was converted into lithage, and on resmelting yielded its metal content as 'sterile' or de-silvered lead (*hartes Blei*). Overall lead losses amounted to no more than 10 per cent. The more the 'fertile' lead had been enriched to obtain satisfactory levels of silver concentration, however, the more it resisted conversion, the losses in refining lead produced from the 12.5 oz ores found at Nerchinsk being particularly dramatic. About 20 per cent of the lead remained in the hearth containing almost two-thirds of the silver in a compound which was almost totally resistant to both secondary or tertiary refining. Each component element of the technology deployed at Nerchinsk thus worked best in the processing of high-grade ores containing about 100 oz per ton and when applied to the ores raised at the eastern Siberian mines it was stretched to its very limits. Only the very highest grades of ore raised there, containing more than 12.5 oz of silver per ton, were worth refining, and even these could be made to yield up only 45 per cent of their silver content and then only at a cost of losing some 87 per cent of the ore's extractable lead.[53] Nerchinsk was thus a highly marginal silver production centre. In normal circumstances only 22–53 per cent of the ores excavated there contained more than the technologically-determined minimum level (12.5 oz of silver per ton) necessary for effective refining. In smelting and refining these ores, moreover, losses were colossal. Only 13 per cent of the lead content of the ore was converted into sterile lead after smelting and refining, and only 45 per cent of the silver contained in the ore was obtained, and this still contained many impurities.

In spite of these constraints, however, the Russians persisted in their exploitation of the meagre Nerchinsk deposits of refineable ore, and after some thirty-five years (1704–39) a major production boom, in two phases (1739–54 and 1764–84), pushed output of blicksilver up to 10,129 kg a year in 1774 before a prolonged decline set in (figure 3.3).[54] Year after year a caravan was despatched from Nerchinsk, normally at the end of December, carrying large quantities of blicksilver to St Petersburg. There the metal passed out of the jurisdiction of the mining authorities and was delivered to the imperial laboratory under the control of the procurator general for the separation of the component elements contained in the blicksilver, including the gold (amounting up to 3 per cent) which was recovered with the use of nitric acid. Finally the refined products, both gold and silver, were delivered to the mint.

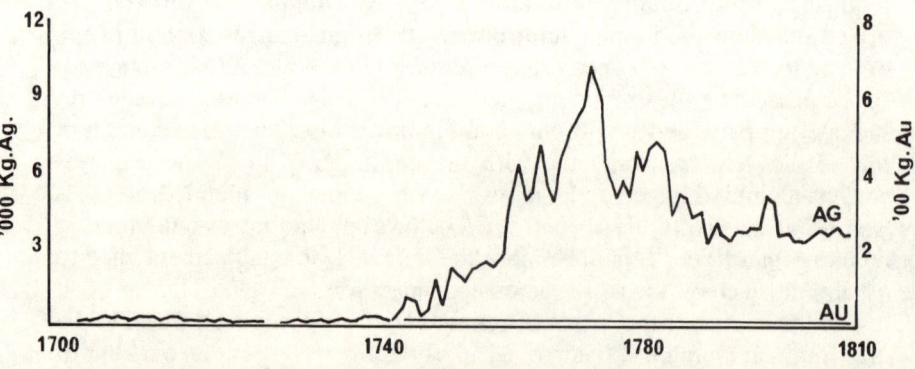

Figure 3.3 Nerchinsk silver and gold production

Whilst the Russians, in a hostile environment and in conditions of acute technological difficulties, thus sustained a major silver production boom at Nerchinsk during the years 1745–74, the course of production there was determined neither by geological factors, involving changes in the silver content of the refineable ore, nor by alterations in the production processes described above. The key to changes in silver production lay in the management's ability to dispose of the low-grade ores raised at the mines. Such ores, containing less than 12.5 oz of silver per ton, could not, within the framework of the existing technology, be made to yield up their precious metal content, refining merely increasing lead losses to impossible levels whilst leaving 30 per cent of the metal as a hearth lead, containing all of the silver, which was totally resistant to secondary or tertiary refining. These ores, which had been won by the deployment of so much manpower and resources, could, accordingly, only be smelted and disposed of in a 'soft', unconcentrated state – no mean task in the remote environment of east Siberia, as was revealed during the first long production cycle of the years 1704–45. At that time, against a background of increasing silver production which rose from less than one to sixteen puds a year and a long cycle in the proportion of low-grade ores raised which declined from 75 per cent during the operation of the Troitsk mine to 50 per cent during the opening years of the Monastir workings before again increasing to 72 per cent in 1723–39, 80 per cent in 1739–43, and 89 per cent in 1744–5, the ingenuity of the management in devising ways to dispose of the lead produced was stretched to its limits – and beyond. The potential amount of the product available for sale steadily increased, rising from about 15 to 200 tons a year, but new outlets for its disposal were hard to find and at a price of one rouble per pud, which was sufficient to cover production costs, they found themselves able to sell *c.* 6.5 tons annually. At first, from 1704 to 1725, they had been able to exceed this figure, apart from small quantities used at the works, selling 3.67 tons a year at the factory for local use and another 10.67 tons at Tobolsk and other Siberian towns and *ostrogs* from whence much of the metal found its way, illicitly, across the border to China. With the reorganization of the China trade during the 1720s and the enforcement of the export prohibition on lead, however, the management found itself able to dispose of only 4.5 tons a year. Its valiant, if unrealistic, attempt to sell 16 tons on the Moscow market in 1729 proved no more successful and the experiment was never repeated.[55] Production and sales of lead had existed during the years 1704–25 in an uneasy equilibrium, short-term production cycles of five or six years serving to clear stocks about an equilibrium level of *c.* 15 tons a year. But thereafter they followed divergent courses. Between 1725 and 1745 the amounts of low-grade ore smelted steadily increased whilst, thanks to government intervention, the market for the lead produced collapsed. In response to this situation the management curtailed current production of ore during the years 1725–39 in the hope of clearing existing stocks. As a result silver production declined but, in spite of a major sales drive, stocks were only marginally reduced. In 1739, accordingly, the policy was reversed. Silver production once more increased but so too did stocks of unwanted lead, until in 1745 the net increment to stocks from current output amounting to 200 tons cost more to produce – c. 4,500 roubles – than the total value of current silver production which amounted to no more than 4,000 roubles.[56] Plagued by high and fluctuating levels of low-grade ore production since its inception in 1704, silver production at Nerchinsk was thus highly marginal, the profitability of the enterprise being entirely conditioned by the ability of the management to dispose of the lead produced from the low-grade ore – a task rendered almost impossible by the restricted nature of the Siberian market and by government interference with export markets.

In these circumstances the *sustainable* level of silver production at Nerchinsk was no more than one or two puds (16–32 kg) a year. Producers might exceed this amount, but as long as average lead sales amounted to no more than 6–7 tons a year they merely saturated the market for that metal, pushed down prices, built up stocks of unwanted lead, and rendered the whole enterprise unprofitable. Only the creation of a market for the unwanted lead produced could free Nerchinsk silver producers from their difficulties. But as later writers were to observe, there was not much possibility, within the marketing strategies of the year 1704–45, of achieving this objective. The China market was closed to them, European Russian markets were too distant for them to be competitive against British imports, and, for domestic use, the Siberian market was highly restricted. Yet in 1746, as has been suggested above, just such a market emerged, thanks to the beginning of silver production at Kolyvan and the introduction of the *Saigerprozess* there which gave birth to a new long-distance trade involving the carriage of Nerchinsk lead some 5,280 miles, a two-month journey, to Kolyvan.[57]

Henceforth lead sales, and accordingly silver production, at Nerchinsk, became entirely dependent upon the fortunes of Kolyvan silver producers – and upon those Kolyvan silver producers' ability to obtain alternative supplies of the base metal (figure 3.4).[58] Throughout the years 1746–71, except the period 1751–9 when the initial discovery of the lead-rich Nicholaev mine made Kolyvan self-sufficient in the metal, the Nerchinsk management organized caravans to carry lead to Kolyvan. The amount of metal despatched thence rose to 160 tons in 1748, 320 tons in 1762, and 530 tons in 1768, and short-term supply fluctuations, induced by either production or commercial factors, were eliminated by the existence of stocks first established at Nerchinsk, Yakutsk, and Krasnoyarsk during the years 1744–6.[59] From 1771 to 1786 the situation was transformed. With the discovery of new, lead-rich deposits at the Nicholaev and Beresov mines of the Kolyvan–Voskressensk complex that centre once more, during the

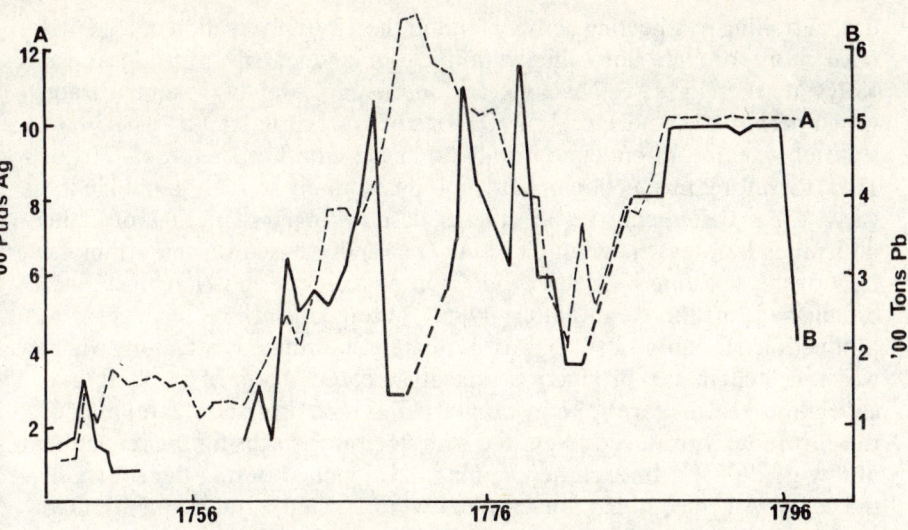

Figure 3.4   Nerchinsk lead sales

A   Kolyvan silver output
B   Nerchinsk lead sales

years 1771–5, became self-sufficient in lead, causing the market for the Nerchinsk metal temporarily to collapse. When recovery occurred, moreover, it was short-lived, and after a brief revival in the Nerchinsk–Kolyvan trade between 1775 and 1780 it once more declined, this time as a result of declining silver production at Kolyvan. During the years 1771–86, therefore, the market at Kolyvan for Nerchinsk lead was subject to two major crises, 1771–5 and 1780–6. At the same time, as has been shown above, supplies of low-grade ore rapidly increased, due to technical difficulties at the major new Nerchinsk workings and a compensatory prospecting boom there. As a result stocks of unwanted lead increased at Nerchinsk. It was reported that in 1771–3 'only 14–18,000 puds' were 'required for the separation of the silver from the copper at the foundries of Kolyvan' and that 'many millions of puds (tens of thousands of tons) remain useless on the spot'; and in 1789–92 that 'a million puds' of ore 'is unwrought', whilst the commercial organization created for its transport remained unused.[60] In response to this situation the Nerchinsk management closed down its transport department in 1773 and henceforth contracted out the work of transporting lead to Kolyvan. From 1773 to 1777 the Tobolsk carriers Semen and Vladimir Shevirin undertook its transport, and thereafter a variety of carriers including Ivan Telnikh of Yenisei, Piotr and Stepan Shumilov of Tomsk, and Ivan Purmov of Biisk.[61] The management also, more slowly, set about reducing supplies of unwanted low-grade ore, curtailing prospecting activity until in the 1780s production of lead was once more brought into line with the demand of silver producers at Kolyvan, thereby re-establishing the *status quo ante* 1771, and a trade which permitted the disposal of 500 tons of Nerchinsk lead a year.[62]

After a period of endemic instability at Nerchinsk mines from 1704 to 1745, therefore, production of lead and silver rapidly increased, until in the early 1770s the mining boom collapsed, ushering in a period of acute difficulties before a new equilibrium, once more based on the symbiotic relationship existing between production at Kolyvan and Nerchinsk, was established in the years after 1786. Throughout these changes the production of lead and silver was intimately interlinked, a feature which was reflected in the productive capacity created. Pairs of smelting and cupelation hearths for the reduction of high-grade ores and the refining of the 'fertile' lead produced co-existed with smelting hearths for the reduction of low-grade ores. In each new mining field opened during the course of the century such combinations of plant were created, establishing by 1796 an extensive complex of works, each forming the focus for a major concentration of men and capital equipment (table 3.4).

Table 3.4 The Nerchinsk works in 1796

Foundries	Date	Pairs of smelting-cupelation hearths	Smelting hearths	Associated mines	Convicts	Ascribed peasants
Nerchinsk	1747	4	8'[a]	Blagodatsk	180	
Dutarsk	1776	3	9	Voskresensk	100	
				Pavlovsk	88	
				Rezanovsk	48 } 500	2,623
				Vozdvizhensk	29	
				Novaya otrad ⎫ Ekaterino } Blagodatsk ⎭	55	
Kutomarsk	1776	4	14[b]	Kaidansk	647 } 878	4,768
Ekaterinsk		closed		Klichkinsk	231	
Grasimirsk	1778	1	3	Kutogorsk	103	544
Alexandrovsk		1	3	Ekaterinsk	288	1,513
Shilkinsk	1767	2	6	Grazimur-Voskressensk	36	189
Petrovsk	1791	Iron works		Baletinsk		200

Notes: a   Plus a proving hearth for testing ores, b   Only eight hearths of the eighteen operational at Kutomarsk whilst Ekaterinsk was said to have 'stood cold for some considerable time'.

Kolyvan–Voskressensk (maps 3.4–3.5)

The ability of the Russians in the 1660s to secure the line of the middle Ob establishing regular communications between Surgut and Narim, as has been suggested above, initiated a new phase in the settlement of east Siberia. No longer confined to the long, circuitous northern route via the estuary of the Ob in their passage to the east, increasing numbers of Russians arrived at Yeniseisk which became a springboard for the penetration of Dauria–Amuria and the Siberian north-east.[63] The effects of these changes were, moreover, no less profound on west Siberia.[64]

Prior to the 1660s, the Russians had been confined to the periphery of the west Siberian plain, settlements assuming the form of an arch with its pedestals located at Tobolsk-Tara in the west and Tomsk-Kuznetsk in the east and with the estuaries of the Ob and Taz, on which were established Obdorsk and Mangazeya, forming the creststone. The Siberian plain, within the arch, was the preserve of the Khirghiz, and the Russian or native aboriginal venturing there would have been foolhardy indeed. Indeed, even within the walls of their fortresses they enjoyed little security, the main centres of Russian settlement, such as Tobolsk, being subject to repeated attacks by the nomads. Beyond those walls they ventured at their peril.[65] In such circumstances the position of the incoming Russians was highly precarious. Unable to cultivate land in the vicinity of their forts, for both climatic and security reasons, they remained throughout the years 1580–1660 dependent upon supplies of grain from European Russia which they were forced to transport along a communications network which was confined to the riverine system of the lower Ob, Taz, and Yenisei – a system which, to say the least, provided only the most tenuous of links. Even in favourable conditions passage from Verkhotorie to Yeniseisk took a year to complete, and rarely were conditions favourable. 'Great weather' in the Ob Gulf often destroyed vessels, personnel, and cargoes whilst passage on the Yenisei was hampered by swift currents and many rapids. As late as 1666–7 the authorities at Tobolsk could report that vessels went carelessly between that city and Mangazeya, and consequently were often wrecked at sea with the loss of men and supplies.[66] The difficulties were not confined to those engendered by weather and poor seamanship as a report of one Dimitri Cherkasov, who made the journey between Tobolsk and Mangazeya in 1643–4, reveals. He left Tobolsk in a convoy of four boats in mid-July, and initially at least all went well, the convoy arriving safely at Beresov at the beginning of August. Then, however, their difficulties began. Putting out to sea once more they encountered contrary winds in the Ob estuary and beat slowly towards the Taz, but on approaching the entry to that river the wind began to lash the boats, causing them to capsize and scattering both government and traders' stores, whilst the people floated and struggled to shore as best they could. At this point, having gathered the survivors together, it was decided to split the party. One group comprising two Tobolsk servicemen and five merchants were sent back to Tobolsk to report the disaster, another group of fifteen remained at the site of the shipwreck over the winter to guard their goods and those of the traders, whilst the main party, numbering some seventy persons in all, having repaired one of their boats, resumed

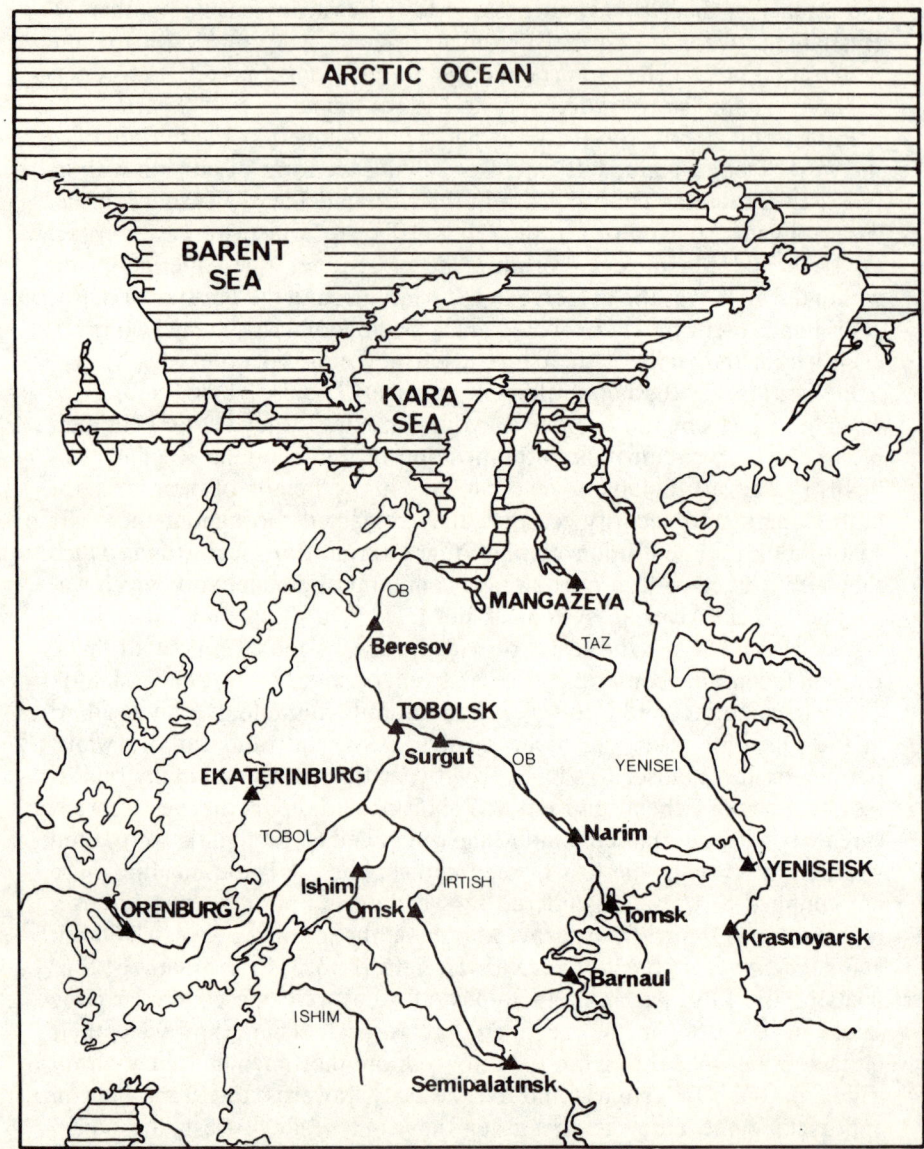

Map 3.4 West Siberia

their journey. Once again, however, after a single day's passage they were wrecked, only to be besieged by hostile Samoyed on the foreshore for eight weeks. Only in November were they able to resume their journey, this time by sled, and subject to continual attack by the Samoyeds who not only inflicted numerous injuries but in delaying the Russians' passage caused them to run out of supplies. For two months they travelled from one *zimov'ye* to another, their numbers being depleted as men and women alike died of wounds or hunger, until in mid-January 1644 they came to Mangazeya, only twenty remaining of all the serving men and traders who had set out from Tobolsk. It had been a nightmare journey, clearly enough; but its hardships were not unique, and probably were not even exceptional.[67] Yet the Russians persisted in their use of this hazardous route for it was their only lifeline to supply their settlements which were at best precariously perched on the edge of the Tartar-controlled Siberian plain; when that supply system failed, as it so often did, they virtually ceased to exist, the inhabitants suffering 'every want and impecunity' and being reduced to eating 'pine and larch bark'.

The securing of the passage of the middle Ob and the establishment of regular communications between Surgut and Narim completely transformed this situation. By preventing nomadic incursions beyond the river, the Russians brought security to the northern sector of the Siberian plain and the mid-seventeenth century, accordingly, witnessed a southward movement of the aboriginal peoples, Ostyak, Samoyed, and Ket all making the long trek to new hunting grounds and drawing a stream of Russian *promyshlenniki* in their wake. Russian settlers were not slow to take advantage of the new communications system and the security afforded them from Kirghiz attack, and the Russian population of the province increased rapidly, attaining 70,000 in 1660 and 229,000 in 1709 when for the first time the incomers outnumbered the aboriginal population.[68] The whole balance of population within western Siberia was subjected to profound change. As the main transit route across Siberia shifted southward so the settlements along the old and now disused northern passage decayed. Beresov and Obdorsk survived but in a depleted state, so that when in 1734 a visitor arrived at the latter settlement seeking a guide to the waters of the estuary he found it decayed and with no one who knew enough to act as a guide.[69] Obdorsk was now at the limit of Russian influence on the lower Ob and beyond lay *terra incognita* – a land abandoned to the Samoyed in which the traces of former Russian settlement at Nadim and Mangazeya had disappeared. Mangazeya was once a great town which for half a century after its foundation in 1601 had been the centrepiece of the old route, in spite of being 'a desolate fort deep in the frozen tundra, almost on the Arctic circle itself, amidst the warring tribes of "bloodthirsty Samoyed" and other "unpeaceful natives", cut off from Rus and even from Siberia by the storms of the Mangazeya Sea'. It had contained, it was reported, 500 dwelling houses, together with a midsummer 'rendezvous' and fair. In 1660 it had still been a major centre but by 1672 the garrison was withdrawn and soon after the city was replaced by a small *ostrog*. By the end of the century it too had disappeared.[70] The Arctic north of western Siberia had become a complete backwater within the short space of half a century. The focus of settlement now lay far to the south, Russian peasants gathering at the nodes of the new trans-Siberian route along the Ob – on the Tobol-Irtysh in the west and the Tom-upper Yenesei in the east – where they established an agrarian regime which, as early as 1685, ensured that the Siberian deliveries could be stopped 'because now they till grain in the Siberian towns and much grain is grown in Siberia'.[71] The arch of Russian settlements had been reformed but now, far to the south, the pedestals of the new system were formed by fortresses at Ishim-Petropavlovsk in the west and Semipalatinsk-Biisk in the east, the crest by the course of the middle Ob, and the lands between still remained the preserve of the nomads whose raids maintained a state of insecurity in the Russian settlements of the periphery. For a century, c. 1660–1760, it was on this frontier that Russian settlement in west Siberia evolved, the balance between nomad and incomer gradually swinging in favour of the latter as their numbers were rapidly augmented, rising from 229,000 men in 1709 to 297,000 in 1737 and 420,000 in 1763.[72] Russians were flooding into the southern section of the west Siberian plain, and with the implantation of villages of free colonists, combining military and agricultural functions, in the lands of the Tara and Barabinskaya steppe after 1763, the region was made secure from nomad raids. This allowed the main transit route across west Siberia to shift southward once more, directly linking Irbitz with Irkutsk by way of Ishim, Tomsk, and Krasnoyarsk.[73]

Subsequent development of western Siberia, during the last quarter of the eighteenth century, took place within the context of these changes. Settlers now poured into the lands between the new road and the Ob, again augmenting the population of the province which reached 576,000 in 1796/7 and slightly more than 1,100,000 in 1815, and the previously isolated nodal settlements on the Tobol–Irtysh and Tom-upper Yenisei gradually merged into the general agrarian regime of the region.[74]

Mining activity and metallurgical production

Discovery and settlement   It was against this background of the settlement of west Siberia that mining activity began in the Altai, giving birth to the Kolyvan–Voskressensk complex which grew to such importance in the eighteenth century that it acquired a grandiose popular title – the Russian Potosi. The story of this important, if little known, mining and metallurgical complex began in 1724 when a group of Kuznetsk peasants, hoping to gain the lands of the upper Ob, engaged the Kirghiz and Dzunghir Kalmuck and fought their way to the river. There they discovered the Blue Mountains and the ancient workings which contained incredibly rich deposits (25 per cent metal content) of blue copper, in a total wilderness.[75] Amongst their number was a workman who had been previously employed at the Demidovs' Ekaterinburg works and realized the significance of the discovery. He immediately rushed back to his ex-master with samples,

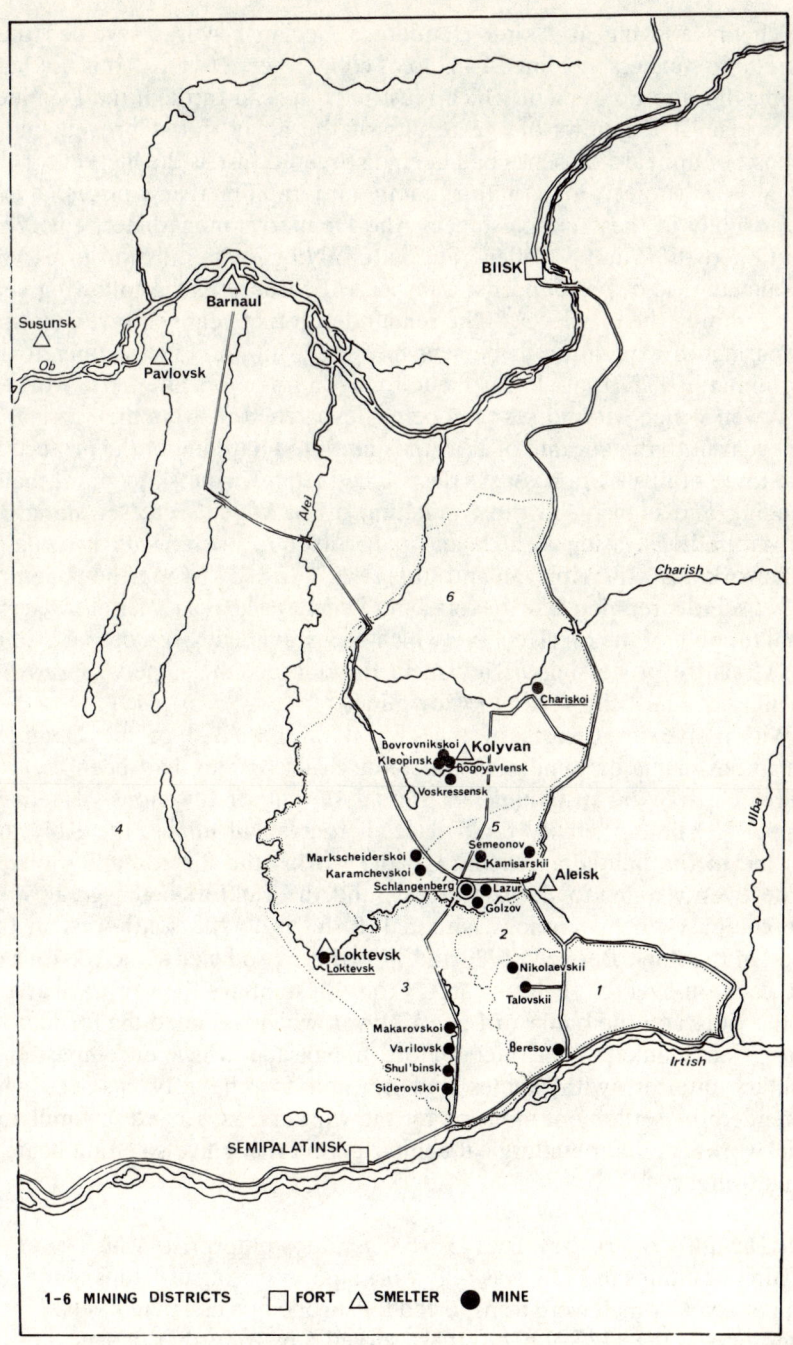

Map 3.5 Kolyvan–Voskressensk

which on assaying at Akinfi Demidov's Neviansk works revealed their incredible richness yielding 21–53 lb of copper per zentner.[76] It is not hard to imagine the excitement which must have spread through the Demidov enterprise at the news of the results of the assay, for it proved that a deposit of unheard of riches had been discovered just as the market for the metal was subject to a major boom and rapidly rising prices.[77] Not surprisingly in these circumstances, the Demidovs immediately (in 1725) petitioned the Mining College at Ekaterinburg for permission to exploit the newly discovered deposits, and on 26 February in the following year were granted the privilege.[78] The remainder of that year was taken up with preparing an expedition to be sent into the Altai, and it was only at the beginning of 1727 that all was made ready, a force of some thirty workers and twenty men with horses then being despatched.[79] From their arrival in that year until the autumn of 1729 they laboured hard under the protection of a force of fifty Tara cossacks despatched from Kuznetsk to guard them, building and blowing in three smelting ovens which were set alongside other buildings inside a ditch and palisade, and opening up two major mineworkings – the Kolyvan and Loktevsk.[80] In 1730 two further smelting hearths and a refining hearth were built which in the next year allowed the local refining of the black copper which had previously been despatched to the Urals for processing at Neviansk. In addition, the same year saw the opening of the Pichtov and Golzov mines.[82]

Within five brief years of the arrival in the Altai of the Demidov expedition a major mining and metallurgical complex had been created there, capable of sustaining an annual output of fifty tons of refined copper.[83] Mining activity took place in four small mines: the Kolyvan, located in the immediate vicinity of the works; the Pichtov and Golzov, some twenty miles to the south-east; and the Loktevsk, the greatest of the complex's copper mines, situated 40 miles to the south-west in the bend of the Alei. Between 1729 and 1731 these produced some 100 tons of ore each on average, every year.[84] From these mines the ore was carried to the works on the banks of Lake Kolyvan which assumed the form of an *ostrog* surrounded by a ditch before a palisade which encompassed a military quarter, with stables and housing for the fifty cossacks who defended the settlement, housing for the workers, associated saw-mill and brickworks, and a metallurgical complex comprising five smelting hearths and a refinery.[85]

The Demidov enterprise   By 1731 the Demidov enterprise, which had first begun operations in 1729, was fully working, producing fifty tons of refined copper a year which were transported to European Russian markets. In the aftermath of the 1725–9 boom these served now to further depress prices, aggravating a crisis which was already causing a rationalization of the Urals industry. Before the full impact of the Siberian product could be felt, however, the promising experiment of the Demidovs came to an end in a series of disasters, as in 1731 the Golzov mine was gutted by fire and then in 1732 the whole Kolyvan–Voskressensk works was burnt to the ground, 'as a result' of which 'all work in the factory ceased and . . . all the working people were dispersed in neighbouring towns, field and villages'.[86]

In spite of unfavourable market conditions and the destruction of the works, however, the Demidovs persisted. In 1733 a smelter was brought from Ekaterinburg to test local ores and to prepare a plan for a reconstruction of the enterprise. On the basis of his deliberations it was decided to undertake the exploitation of ten mines; to recruit a major labour force comprising fifteen German miners together with 176 day workers and 419 peasants guarded by a military force of one sergeant, seven troopers, and 100 cossacks, and to construct a major complex of works comprising not only a reconstructed Kolyvan but also new plants at Barnaul and Shul'binsk.[87] It was on the basis of this plan that work, accordingly, began but it was to be another ten years before it was to be finally realized.

For the first quinquennium after the fires (1734–1739) reconstruction merely served to restore the *status quo ante* 1731. Mining activity remained confined to the Kolyvan–Voskressensk, Loktevsk, Pichtov, and Golzov mines. Smelting capacity was reconstructed only at Kolyvan where at the time of Gmelin's visit in August 1739 there was once again a major metallurgical complex enclosed within a fortress or *ostrog* and comprising five wards. The first contained five copper smelting hearths and a copper hammer, the second two kilns and associated crushing equipment as well as two refining hearths, the third stocks of lead and copper, whilst the fourth contained housing and stables and the fifth the residences and workshops of blacksmiths and other craftsmen. Everything was much as it had been in 1731, and of developments beyond the walls of Kolyvan, envisaged in 1733, there were as yet (in 1739) few signs. A few of the Demidovs' peasants had constructed settlements on the banks of the rivers Charish, Loktev, and Barnaul but these contained only forty or fifty persons in comparison with the 800 at Kolyvan, and on his visit to these settlements Gmelin found little which would cause him to anticipate their future importance.[88]

Within the next quinquennium (1739–43), however, the whole of the 1733 plan was realized. Prospecting now resulted in the discovery of lead ores at Voskressensk in 1741 and the opening up of new copper mines at Mursinsk in 1739, Chuporshnevsk, and Chariskoi in 1741, and Bogoyavlensk and Yurkensk in 1742. By 1743 the full complement of ten mines, whose exploitation had been envisaged in 1733, were in operation and although the new mines like their earlier counterparts were small, employing no more than five or six miners and each mine producing about 100 tons of ore annually, supplies delivered to the metallurgical establishments doubled.[89]

To cope with these increased supplies, moreover, new capacity was brought into existence, again leading to a realization of the 1733 plan. In 1739/40 the Barnaul works was commissioned with four refining hearths and two smelting hearths for the reduction of Mursinsk and Loktevsk ores transported to the new works by way of the Alei and Ob. In 1743 the Kolyvan works was extended, with two new smelting hearths and, perhaps more significantly, a silver production complex comprising two *Saigerofen*, one *Daarofen*, and two cupelation hearths, or *Treibofen*, whose construction marked the beginning of the end of the family's control of the enterprise.[90]

In the space of little more than a quarter of a century, 1727–44, the Demidovs had built a major metallurgical complex in the remote wilderness of the Altai comprising three factories – Kolyvan–Voskressensk, Barnaul, and Shul'binsk – protected by an *ostrog* whose garrison was amply equipped with sixteen cannon and 187 muskets as well as massive stocks of mines and shells. The largest of these works remained Kolyvan, whose workers, salaried staff, and soldiers occupied some 325 houses. It was also equipped with most of the capital equipment: seven copper smelting hearths and a hammer with housing and waterwheel; three kilns and water-powered crushing stamps as well as nine refining hearths, all employed for copper production, and two *Saigerofen*, a *Daarofen*, and two *Treibofen* for silver production; a cornmill, saw-mill, and brickworks; and two blacksmith's shops equipped with six smithies, hearths, and anvils. The other two works – Barnaul built in 1739–40 and Shul'binsk in 1743 – which were complementary in their functions to each other were much smaller. Shul'binsk was nothing more than a mining camp, with thirty-nine houses providing accommodation for the peasantry attached to the works and the small group of workers – four preparing the *Rohstein* and sixteen excavating ore at the Loktevsk and Mursinsk mines. It also had a blacksmith's shop, containing three smithies with hearths and anvils for repairing their tools. Here the major activity was mining, at the end of the Demidov epoch, some 2–300 tons of ore being delivered annually by way of the Alei and Ob to Barnaul. At Barnaul there was no mining activity, only the reduction and refining of the ores brought from Shul'binsk, the capital equipment at the works reflecting its specialized function: six water-powered crushing stamps, six smelting hearths, and five refining hearths (*Garofen* or *garmakherskikh pech'*).[91]

From its inception in 1729 the Demidovs' enterprise in the Altai had evolved, however, in a quite different manner from the state works at Nerchinsk. Initially, from 1729 to 1731/2, the works, located on the banks of Lake Kolyvan, were equipped only with smelting and refining hearths producing respectively black and refined copper. Nor did this situation change with the rebuilding of the plant after the disastrous fires of 1731/2. As new mines were opened up and a new smelting works established at Barnaul in 1739/40, the primary focus of activity remained the production of copper. Smelting and refining capacity at Kolyvan remained at five and two hearths respectively from 1729–31/2 and 1734–9 before there was an extension of capacity to seven smelting and nine refining hearths at Kolyvan and six smelting and five refining hearths at Barnaul during the years 1740–6. Throughout the the Demidov period (1726–46), this plant was largely employed in processing high-grade copper ores capable of yielding between 8 and 53 per cent unrefined, 7–8 per cent refined copper, production of the latter metal steadily increasing from the blowing in of the first furnaces in 1729 to 1746. Ever expanding, the course of refined copper production was not a stable one, however, for prior to the completion of the first refining hearth in 1730 and again in 1733–4 and 1738/9 much of the black (unrefined) copper smelted at the works was transported to the Urals and refined at the Demidovs' Viisk, Nizhnii Tagil, and Neviansk works.[92] Yet whether refined in the Urals or the Altai production of copper increased, rising from c. 32 tons a year in the early 1730s to 64 tons in the latter part of that decade and c. 100 tons in the early 1740s and dominating activity at Kolyvan–Voskressensk – and sales of copper on European Russian markets in the 1740s.[93]

The Fermor–Reiser (1735) and Beyer (1743) commissions Only on two occasions during the Demidov era was the supremacy of copper production at Kolyvan–Voskressensk called into question, with disastrous results on current production. In both 1735 and 1743, on the occasion of state seizures of the works, current production ground to a halt as stocks of ore were simply stockpiled awaiting the outcome of current investigations. On the first occasion, when the enterprise was subject to an investigation by captain-of-engineers Fermor and assessor Reiser as part of the general inquiry into the Demidovs' affairs, current output was estimated as being worth c. 4,000 roubles (= 43 tons), an amount which was held by the state against the payment of the family's tax arrears.[94] On the second occasion, in 1743, the enterprise was subject to investigations by Brigadier Beyer, and existing stocks of copper from that year's production, amounting to nearly 30 tons of black copper and 75 tons of refined and bloom copper, as well as existing stocks of ore – 56 tons – and all new production between 1743 and 1745 – 675 tons – were again stockpiled, only some 209 tons being smelted.[95]

On each of these occasions the avowed reason for bringing production to a halt and stockpiling ore until it could be tested lay in a belief in government circles that the Demidovs were extracting large quantities of silver from the base metal. Indeed such a claim figured prominently amongst the list of indictments levelled at the family by the Chancellor, Graf. Kapustin, in 1733 when the general inquiry into the Demidovs' affairs was ordered. It was declared with respect to the Kolyvan–Voskressensk works that the management had not only unlawfully supplied arms to the Kalmuck and Tartars but had also 'founded in their factory silver ore, which should not by law be smelted'.[96] The investigation of this claim accordingly figured large in the work of the Fermor–Reiser commission despatched as part of this general inquiry and whose work resulted in the seizure of the works in 1735. Nor were the avowed reasons for the seizure of 1743 very different. On this occasion government action was triggered by the arrival in St Petersburg of a disaffected Demidov employee, one Trager, who brought with him samples of ore from the Voskressensk and Chuporshnevsk mines for proving. In an attempt to forestall him, Akinfi Demidov presented the Empress Elizabeth with some 27 funt 8 zolotniks of silver, possibly the product of the lead ore samples that they had sent from the Chuporshnevsk mine to Neviansk in the previous year. If Akinfi considered that this gift, which was coupled with a request that he be allowed to send knowledgeable men to explore the possibility of silver production, would allay imperial suspicion, however, he was sadly mistaken, for that year saw the appointment of the Beyer commission whose investigations dominated activity at the works for the next three years (1743–5).[97] On and off for a decade from 1733 to 1743 the government had thus intervened into the Demidov's affairs at Kolyvan–Voskressensk, avowedly in the belief that the family were extracting large amounts of silver from the copper produced at the works. The basis of this belief lay in rumours which circulated both within and outside the works of the riches of the Voskressensk ores which though poor in copper had been found in 1733 to be rich in silver, capable of yielding up 1.5 zolotniks of silver per pud (14.5 oz per ton).[98] Yet in spite of this knowledge of the argentiferous nature of the copper ores, which circulated widely after 1733, there is no evidence whatsoever that the Demidovs made any attempt to exploit the silver-bearing characteristics of these ores. They remained in 1743, as they had been in 1729, copper producers and it is in relation to this activity that the motives for state intervention must be sought.

In 1735 the state already had a long history of intervention into the affairs of the copper industry behind it, arising largely from a concern over the unprofitability of its Urals copper works during the copper-market crisis of 1730–1734/5. Successive commissions had deliberated during the years 1730–2 and 1733–5 and had revealed that, at prevailing prices, the state works were indeed unprofitable, partly as a result of internal inefficiencies and partly because of competition, including that of the Demidovs' Altai works whose products in the years 1733–5 were once again making inroads into European Russian markets. In these circumstances accordingly it was decided to 'privatize' the state works but not before their capital value had been enhanced.[99] This involved the government in a reform of its works and perhaps also a suppression of competition, for what could be easier with the Fermor–Reiser and other commissions undertaking their investigations into the Demidovs' affairs than to use the alleged silver smelting as an excuse to seize the Kolyvan works and suppress production, thereby enhancing prices and raising the value of their rather dubious assets. Although market forces in the course of 1734/5 resolved their problems, by autonomously raising prices, it is unlikely that the state's resolution to seize the works weakened. With the Demidovs owing the government 800,000 roubles in back taxes the stocks of copper built up in 1735 and that produced in 1736–8 were at least a realizable asset. These years accordingly were ones of ruthless exploitation by the state, with output being pushed up to an unheard of 80 tons in 1737, with the result that when the works were handed back to the Demidovs in the following year they were a wreck, posing no major threat to the state enterprises of the Urals.[100] As has been suggested above, however, this situation did not last long, and with the rebuilding and extension of the works after 1739 Altai copper once more flooded European Russian markets, causing prices to fall, private entrepreneurs to rationalize production, and the state works yet again to make endemic losses.[101] Perhaps not surprisingly, therefore, in 1743 the state, with alleged evidence of silver smelting at Kolyvan before it, once again took the works in hand whilst the Beyer commission carried out its investigations into the allegations. As in 1735 the avowed reason for its actions was the alleged silver-smelting activities of the Demidovs but if this was the case then the Beyer commission found singularly little evidence of such activity and certainly not enough to justify a formal seizure of the works in 1743. On arrival they discovered the unfinished fabric of a silver extraction complex whose construction may have been begun in 1742. This they completed in 1743 to process the ores whose silver-bearing nature they wished to investigate. Thereafter for three years, from 1743 to 1746, they pursued their course but with sadly unrewarding results. Of 42,042 puds (= 675 tons) of silver-bearing ores raised only 3,000 puds (= 48 tons) were deemed worthy of smelting and even this only yielded 46 funt 84 zolotniks of silver (14 oz per ton).[102] If the government's objective was the seizure of a major silver-producing plant then, on the basis of the Beyer commission's investigations, it was to be badly disillusioned. If its objective was to suppress copper production at Kolyvan it was much more successful. Output of refined copper which had been running at 96 tons in 1741, the penultimate year of Demidov control, averaged 9 tons a year during the period of the Beyer commission, 1743–6.[103] It thus seems probable that throughout the Demidov era from 1726–46 both the family and the government viewed the Kolyvan–Voskressensk works in the context of copper *not* silver production.

This situation only changed in 1745. During that year the commission continued its unrewarding work testing ore from the Voskressensk mine, but on this occasion with somewhat better than usual results. A sample load of 87 tons of ore derived from the workings on smelting yielded up as usual only 17.6 cwt (one per cent) black copper, but on refining this produced 250 funts of silver containing 3 funt of gold (40 oz of blicksilver per ton). Even this rich find was soon eclipsed, however, when miners working in the shafts of the Schlangenberg, which had been briefly exploited by the Demidovs in 1736 and by the state in 1737, produced a similar quantity of ore which on smelting yielded up 33 puds, 37 funt, 33 zolotniks of silver – 224 oz per ton![104] Beyer and the Captain Bulyakov, in charge of the Schlangenberg party, were stunned. After three years of fruitless work, which might have served the primary purpose of keeping Kolyvan copper off the market but which had yielded little silver, they had hit the mother-lode. The silver, containing 0.6 per cent gold, was immediately sent back to St Petersburg with a recommendation that the works be seized, a recommendation eagerly welcomed and implemented in 1746.[105]

As a result of a chance find in an abandoned mine-working which had been picked over and deserted by both Demidov and Crown workers almost a decade before, the course of the history of the Kolyvan works was changed as a new management was installed and silver production began.

'The Russian Potosi': the Schlangenberg or Zmeinogorsk mine in the eighteenth century (plan 3.6)   As a result of the discoveries of 1745 and the subsequent seizure of the Kolyvan–Voskressensk works by the state in the following year a new age dawned in the mining and metallurgical history of the Altai – and indeed in that of Europe as a whole. The organization of production was totally reformed, giving birth to the Kolyvan *nachal'stva* in 1747, and whilst mining activity continued in the traditions of the Demidov era, with a proliferation in the number of small mines producing less than one hundred tons of ore a year, total output during the years *c.* 1745–85 became dominated by a single working – the Schlangenberg or Zmeinogorsk mine – popularly known in the eighteenth century as 'the Russian Potosi'.[106]

In part the new mining ordinances issued on 1 May 1747 merely effected a transfer of the existing Demidov smelters, mines, and workers to the organization newly created by the Crown for the administration of the works – the Kolyvan *nachal'stva* or *Oberbergamt* – which was headed by Beyer, who was now promoted to the rank of general, and which was recruited both from amongst the members of the Beyer commission and from the staff of the Ekaterinburg works.[107] They also, however, laid the foundations for a complete reorientation of production within the works. Henceforth as new mines were opened up the primary objective of production was to be silver, which in the form of blicksilver was to be sent to St Petersburg for the separation of the silver and gold at the imperial laboratory, and not copper, which for the next twenty years was simply stockpiled. In the reduction of this ore cognizance was also taken of the depletion of timber reserves at Kolyvan and in the refining of the metal

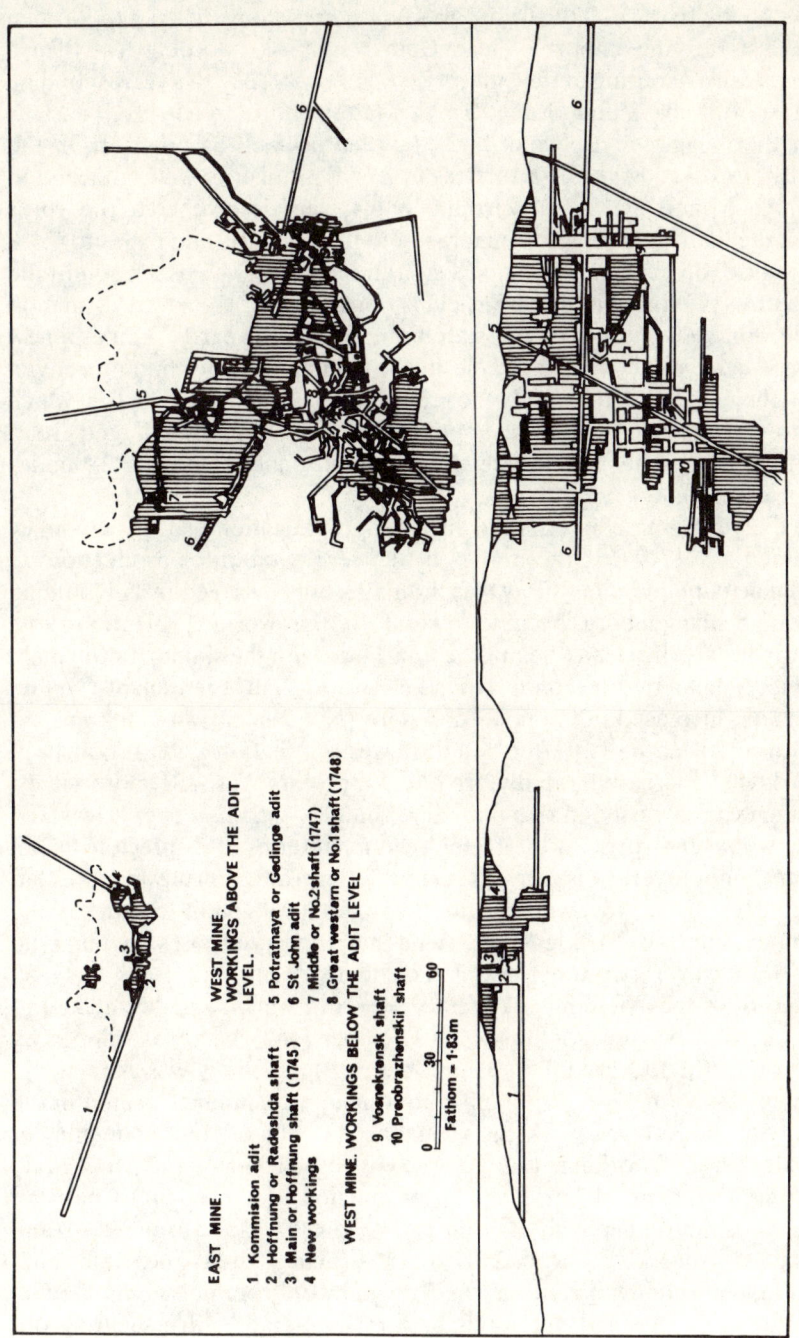

EAST MINE.

1 Kommision adit
2 Hoffnung or Radeshda shaft
3 Main or Hoffnung shaft (1745)
4 New workings

WEST MINE.
WORKINGS ABOVE THE ADIT LEVEL.

5 Potratnaya or Gedinge adit
6 St John adit
7 Middle or No.2 shaft (1747)
8 Great western or No.1 shaft (1748)

WEST MINE. WORKINGS BELOW THE ADIT LEVEL

9 Vosnekransk shaft
10 Preobrazhenskii shaft

0   30   60
Fathom = 1·83m

Plan 3.6   Schlangenberg mine, Kolyvan

produced the total inadequacy of local lead supplies for the introduction of the *Saigerprozess* was also noted, leading to plans for the creation of a new works on the Irtysh and the transport of most ores along the Alei to Barnaul as well as for the import of lead from Nerchinsk. Nothing less than a complete restructuring of the enterprise was envisaged with a corresponding expansion of the labour force. To this latter end, accordingly, peasants from the villages of Belyarsk and Malyshev as well as others from the Berdsk *ostrog* and the Biisk fortress in Kuznetsk district were attached to the mine. Miners, particularly from Olonets, were despatched to the works whilst the management was encouraged to put colonists and exiles to work and to hire, on contract, foreign (Saxon) miners. In 1747, as a result of the ordinances published in that year, plans were laid for a total transformation of the works taken over by the state in the previous year.[108] Under a new management, directly accountable to the Mining College, mining activity was to be reorientated towards the requirements of silver production which was to take place in new plant supplied with charcoal by an enlarged group of attached peasants and with lead by a long-distance trade of 5,280 miles with Nerchinsk.

In the years immediately following the promulgation of the new policy document, *c.* 1747–51, the impact of the reorientation of production on mining activity and metallurgy was dramatic. As envisaged in 1747 mining on the Schlangenberg became pivotal to the working of the whole enterprise. Work already begun there in 1745, with the sinking of the main (middle) shaft by the then Brigadier Beyer and Lieutenant-Captain Bulyakov, increased in tempo. An eastern (No. 2) shaft was sunk in 1747 and was followed shortly after by a third western shaft in 1748. Production, accordingly, increased rapidly. In the four years 1747–50 more silver-bearing ore was raised in this one mine complex than had been excavated in the whole enterprise in the Demidov era. In terms of its precious-metal content, moreover, this ore was in a totally different league to that previously raised, for whereas in the main production centre of the Demidov era the Voskressensk mine had produced ores yielding up 12.5 oz of silver per ton the Schlangenberg during the years 1747–50 produced 48 tons yielding 187–710 oz per ton, which was despatched to Kolyvan, and 804 tons yielding 112–31 oz per ton, which was shipped by way of the Alei to Barnaul.[109] The whole focus of activity was changing in ways envisaged in 1747. Production continued in the mines located in the vicinity of the Kolyvan works, now enhanced in number by the opening of a small group of workings (Kleopinsk, Gustokashchinsk, and Medvedev) near the Voskressensk mine and in production by the mighty Chariskoi mine whose output increased from *c.* 130 tons a year between 1741 and 1745 to 490 tons a year from 1747 to 1750 (plan 3.7). At the same time, smelting and refining activity at the Kolyvan works became dominated by the processing of small quantities of very high-grade, silver-bearing ore

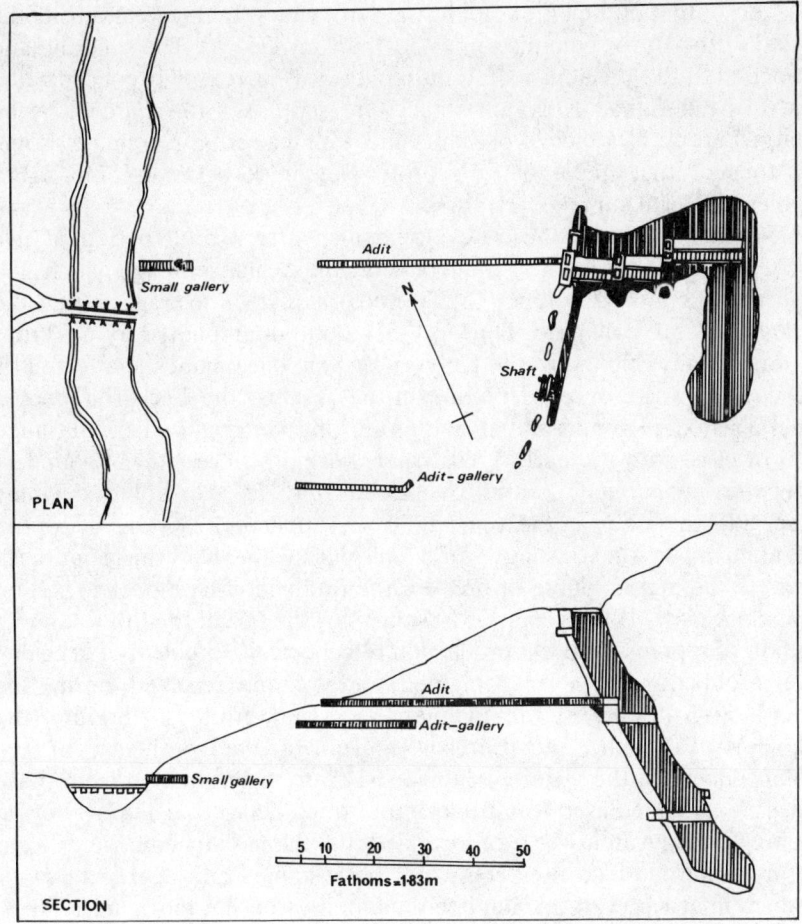

PLAN

SECTION

5 10   20   30   40   50
Fathoms =1·83m

Plan 3.7 Chariskoi mine, Kolyvan

derived from the Schlangenberg workings.[110] That mine also now supplied large quantities of silver-rich ore, by way of the Alei, to Barnaul. Nor was that proposal to open up production in the Loktevsk–Shul'binsk forest forgotten. Even as work was set in hand for the construction of the new Shul'binsk smelting works on the Irtysh, production of ore continued at the Loktevsk mine and the diminutive Shul'binsk mine, opened in 1743, was extended northward, the new Shemaraichinsk workings yielding up not only copper ores but also silver-bearing lead spas of 10 per cent plumbiferous content.[111] During the years 1747–51, therefore, the new development programme for the Kolyvan–Voskressensk mining enterprise was fully implemented. Production, increasingly concentrated on the new Schlangenberg workings, underwent that major reorientation from copper to silver, output of the latter metal rising from a few funts a year during the period of the Beyer commission to 367 puds in 1751.[112] The smelting and extraction of the precious metal, moreover, was increasingly concentrated in works other than Kolyvan which were supplied with charcoal by an enlarged group of attached peasants and with lead from Nerchinsk, some 520 tons of that metal being sent to Nerchinsk between 1746–51 to supplement local supplies.[113]

1751, however, saw these developments, after a brief period of five years, brought to naught. Production on the Schlangenberg, which had been running at *c*. 200 tons (13,250 puds) a year on average during the period 1747–50, collapsed. During 1751–8 output amounted to *c*. 80 tons of ore annually. Nor were the Kolyvan or Shul'binsk mines unaffected by the developing resource depletion crisis. Against the background of a general retreat in mining activity within existing workings (which continued in most cases into the early 1760s when workings were either flooded or otherwise abandoned) output increased only at the Loktevsk mine (plan 3.8) which during the years 1750–67 supplied *c*. 250 tons of copper ore annually to the Barnaul works, displacing the Kolyvan–Shul'binsk mines as the major source of ores for the production of copper to sell on local markets.[114] If the years 1751–8 thus saw the rise of the Loktevsk to a position of supremacy in the production of copper, the problem of declining silver production, however, remained, and was only resolved, during the years 1750–5 at least, by the opening up of a new group of workings (the Nikolaev, Talovskii, and Beresov mines) to the south-east of the Schlangenberg in the vicinity of the Ulba river.[115] Production from these mines steadily increased to a peak output of *c*. 478 tons in 1757–8 but on smelting at Kolyvan it was soon revealed that the initially high yields from the argentiferous lead ores could not be sustained and aggregate silver production at Kolyvan, having been maintained during the years 1750–5, thereafter declined. The 1750s thus saw a brief displacement of the Schlangenberg and the emergence of new mining complexes within the Kolyvan *nachal'stva*. The Loktevsk mine rose to a position of supremacy in the production of argentiferous copper ores. The Ulba mines came to dominate silver production and also lead production, displacing supplies of that metal from Nerchinsk and thereby precipitating a decline in silver production there.[116] In spite of this restructuring of production, however, the new situation at Kolyvan was an unstable one and when in 1755–6 silver production there began to decline, as it had at Nerchinsk some years before, there emerged during the years 1757–60 a general specie production crisis.

Even as the first symptoms of the impending crisis began to appear, however, changes were taking place – on the Schlangenberg – which were to herald the dawning of a new era, not only in the mining history of the Altai but in that of Europe as a whole. The crisis in that mine and in others

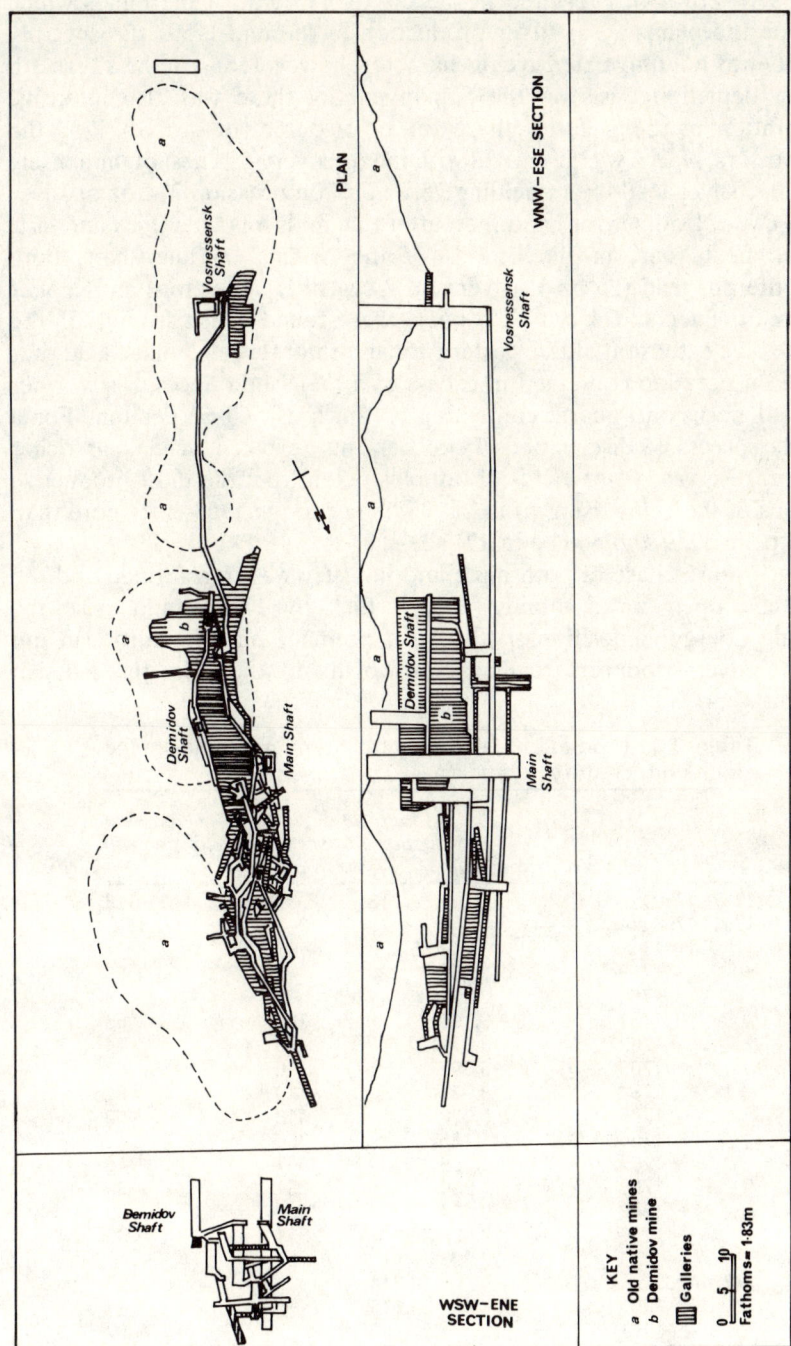

Plan 3.8 Loktevsk mine, Kolyvan

PLAN

WNW–ESE SECTION

WSW–ENE SECTION

KEY
a Old native mines
b Demidov mine
▦ Galleries

0 5 10
Fathoms=1·83m

Vosnessensk Shaft

Demidov Shaft

Main Shaft

Demidov Shaft

Main Shaft

in its vicinity in 1750/1 had set men searching for new deposits of ore. As has been shown this resulted in the discovery of the Ulba mines which became the mainstay of silver production in the mid-1750s. Prospecting activity was not only extensive in character, however, and in the aftermath of the depletion crisis on the Schlangenberg there was also intensive exploration of the galleries there for new sources of ore. In 1753 the prospectors' search was rewarded with the discovery of a nest of immensely rich, fat, silver-lead ores (yielding 75 oz and on occasion 224 oz of silver and 8 cwt of lead per ton) and thereafter rich finds followed thick and fast. In the next year, in the bowels of the original middle shaft, both argentiferous lead (25 oz of silver and 2 cwt of lead per ton) and copper (75 oz of silver and 4 cwt of copper) were found whilst from 1753 the progressive extension of the western Johann shaft revealed black lead spas which on occasion contained in excess of a thousand ounces of silver per ton and hornstone quartz containing as much as 560 oz per ton. For a decade successive discoveries of ore deposits revealed richer and richer mineral reserves, silver yields of samples taken from the most productive sections of the veins rising from 25–75 oz per ton in 1753–4 to more than 1,000 oz in 1759 and 560 oz in 1761/2.[117]

The resource base of both mine and industry was transformed and the path laid open for a mining boom which by 1771 would raise the Schlangenberg to a position of absolute supremacy amongst European and Asiatic silver producers rendering it worthy of the title 'the Russian Potosi'.[118]

Table 3.5 Production of the Schlangenberg mine in the eighteenth century

	Annual ore production (tons)		Silver yield (oz per ton)
	primary ore	wash ore	
1747–1750 av.	213		c 116
1751–1758 av.	78		
1759–1771 av.	8,252		
1759	2,053	42	69
1760	3,271	88	46
1761	3,165	109	53
1762	9,877	43	44
1763	6,397	26	47
1764	8,077	119	51
1765	9,223	418	53
1766	11,125	756	62
1767	9,985	567	46
1768	11,320	231	30
1769	10,624	508	50
1770	17,593	1,351	48
1771	18,203	1,521	
1772–1783 av.	18,314		23
1787	24,115		14
1791	21,130		22

From 1759 to 1771 whilst most workings in the Altai were subject to a process of decay – the mines being either abandoned or flooded – activity on the Schlangenberg intensified. The meteoric rise in output brought into existence a formidable production complex, and a whole new community was born on the banks of the Smeevska brook containing 362 houses in 1771 for its 2,449 inhabitants who worked in the mine.[119] It was protected, moreover, by the Smeevskaya fortress – an earthwork and ditch structure created in 1759 which was almost a mile in circumference, garrisoned by a company of troops from the Barnaul mining battalion equipped with twelve cannon and containing amongst other buildings the mine office and laboratory and warehouses for the storage of provisions. This was home for the mineworkers. Their places of work were elsewhere. About 2,000 of them, divided equally between the three main workings (the Kommissions, Middle, and Great Western shafts) spent the mining season (May–October) each year labouring in the galleries of the mine, each worker raising three puds (c. 1 cwt) of ore a day or nine tons a year and their collective output amounting to more than a million puds (c. 16,000 tons) of ore annually.[120] Supporting these workers there were another 400 ascribed peasants sorting and grading ore and working in the wash-works constructed to process low-grade ores. Progressively since the foundation of the mine the number of such washes had grown. Initially they had been located on the Smeevska brook, the first complex completed during the years 1748–52 comprising nine stamping presses and twenty washes, but capacity was extended as the boom got under way, two new washing complexes similar in form to the original one being created on the waters of the upper Korbolicha in 1763/4 and yet another complex being built further downstream in 1769.[121] In the early 1770s the mountain was the scene of intense activity, some three quarters of all the miners in the jurisdiction being concentrated in this one mining complex.

As in the late 1740s, however, the Schlangenberg mine, which once more assumed a pivotal position in the operation of the whole enterprise, provided only one element in a bilateral system involving the transport of the ore some 150–160 miles north-westward along the Alei to smelting plants located on the tributaries of the Ob in lands rich in timber stands. Initially production here remained concentrated in the Barnaul works, originally founded in 1730 but completely rebuilt in 1749 on its elevation to the position of principal mining office, and still in 1771 a formidable production complex. At that time the settlement, which had assumed the status of mining town (*Bergstadt*), was dominated by the palisaded fortress (*kreml'*) with its guardhouse, school, and magazine, whose garrison of 368 soldiers and 12 cannon guarded the community. It also comprised the *Hüttenhof* and township. The former was situated by the side of the stream against the massive dam (532 yards long by 28 yards wide with a 20-foot fall of water) constructed in 1745, and contained most of the enterprise's productive capacity. This comprised a smelting hut with twelve smelting hearths (*Krumofen*) built in 1749, thirteen cupelation hearths (*Treibofen*), a refining hearth (*Garofen*), and an ore crushing mill (*Pochwerk*); another smelting hut with eighteen smelting hearths and a hammer built in 1750, a roasting hut, constructed by Mr Kolleigienrath Hans in 1769, with twelve roasting ovens; the laboratory; a corn-and saw-mill; a smithy and bellows-manufactury; a large storehouse for tools and provisions; a small stone treasure house; and an enormous open space for the storage of charcoal, ore, and other items. The *Hüttenhof* was the productive centre of the settlement, its plant being capable of processing *c.* 700,000 puds (11,250 tons) of ore annually and of providing employment for 200 'smelting people' who lived in the adjacent town. The town proper was the residential heart of the settlement, embellished with office buildings and two churches built of stone. It also possessed a market, warehouse, and apothecary's shop, as well as a brewery, brick-and glassworks, and an indoor riding school all for the use of the population, some 5,448 in all, who were provided with 949 houses for their accommodation. The community at Barnaul was thus a large and varied one. Apart from 4,300 peasants who made up the bulk of the population, there were eighteen mining officials, 200 smelting workers, soldiers, and some 168 merchants and 107 craftsmen who lived in part by their trade and in part by ore-carrying. All were under the authority of the mine chancery which was responsible for their welfare, regulating markets, policing the town, and providing fire cover with a shiny brass English fire engine which was the responsibility of one Captain Polsonov. It was a benign administration, ensuring a supply of provisions which allowed workers earning *c.* 15 roubles a year to satisfy their gargantuan appetites whilst expending no more than about 30 per cent of their income and leaving them *c.* 10 roubles to spend on the wide variety of Chinese and European wares acquired each year at the Irbitz fair by two officers despatched from Barnaul to provision the market there. This, then, was the Barnaul works, a major production complex whose denizens enjoyed a remarkably rich lifestyle given the remoteness of their situation.[122]

Given the magnitude of the mining boom on the Schlangenberg, however, even this mighty complex lacked the capacity to process all the ore which poured forth from the silver mountain, and as early as 1761 plans were laid for the creation of a new works to be supplied by water along the Alei and Charish.[123] As a result in 1763 the Novo Pavlovsk works was established, some 34 miles from Barnaul on the river Kasmala 8 miles above its confluence with the Ob, where capacity was steadily increased from six smelting hearths in 1763 to twelve in 1769 and twenty-four in 1771.[124] Capacity had been enhanced by 80 per cent and a new smelting complex of no mean dimensions created. As at Barnaul, plant was concentrated in the *Hüttenhof* which was again situated by the stream next to the dam which in winter provided an 18-foot fall of water but which in summer was dry. Here was a whole complex of blockwork huts and houses, the longest containing the smelting hearths and others containing the laboratory, a hospital and a school, a bellows-manufactory, a saw-mill, a crushing-mill and a large warehouse for the storage of copper stone (*rohstein*) before its despatch to Barnaul for refining. Beyond was the large mining village (*sloboda*) with a fine wooden church, mining office, and market hall equipped with stalls for the merchants to display their wares, as well as 264 houses laid out in straight rows to accommodate the 1,900 inhabitants, Orthodox and Old Believers, who gained their livelihood either by trade or by work at the smelters.[125] At the height of the mining boom in the early 1770s, therefore, the intense mining activity taking place on the Schlangenberg was counterbalanced by equally intense activity 150 miles to the north-west where a major smelting complex, comprising the Barnaul and Pavlovsk works, was created to process the Schlangenberg ores.

Yet thanks to events taking place in both European Russia and Siberia this bilateral system, which established Russian supremacy amongst European and Asiatic specie producers, remained only one element in that metallurgical complex which was the Barnaul mining district. Against the background of a severe crisis in the Urals copper industry which occasioned a marked shortfall in mint supplies of that metal, the decision was made in 1763 to institute an autonomous Siberian copper coinage, reserving Ekaterinburg mint output for European Russia and bringing to an end that twenty-year period when Siberian copper had been kept off the market and largely stockpiled at the works.[126] To implement this new policy, accordingly, the construction of a new mint was authorized and opened in May 1764, forming part of a new plant – the Susunsk works – which was completed the following summer and henceforth smelted low-grade Schlangenberg ores and those derived from the argentiferous-copper deposits discovered in the Dzunghir mountains by an expedition despatched thither under Major Petrulin in 1761.[127] A further bilateral system was thus superimposed upon the first, providing the basis for a secondary boom – in copper production – which provided the basis for an autonomous Siberian currency.

Central to the operations of this new system was production at a group of mines (New Lazur–Gusov, Hausensk, and Ivanovsk) opened up as a result of Petrulin's discoveries in 1761 which yielded up rich brown and green copper ores containing up to 10 cwt of copper and 12.5 oz of silver per ton and lead ores containing 2.75–4.5 cwt of metal per ton. Production here, particularly at the New Lazur mine, increased rapidly over the years 1762–5 and, although output thereafter collapsed, during the first phase of operations enormous quantities of ore were despatched to Susunsk.[128] Together with some 5,540 tons of low-grade argentiferous copper ores

Table 3.6 Production at the New Lazur mine

	tons
1762	160
1763	95
1764	353
1765	887
Total	1,495

from the Schlangenberg these New Lazur ores provided the basis for operations undertaken during the years 1762–71 at the new Susunsk works – a plant of no mean dimensions.[129]

Situated in two settlements above and below the stream which flowed into the Ob some 80 miles west of Barnaul, the Susunsk works was a worthy counterpart to the neighbouring Barnaul and Pavlovsk works. Above the stream in Nizhne–Susunsk was situated the smelting and refining plant, under the supervision of three officers and employing 260 workers (*Hütten-* and *Meisterleute*). This comprised: an ore crushing mill with three stamps; a warehouse for the storage of the copper ore; a roasting hut containing four ovens (*Kaltsenirofen* or *obzhigatel'naya pech'*); a 'windhouse' and smelting hut containing twelve smelting hearths (*Krumofen*) producing copper stone (*Rohstein*) which then passed to the roasting hut for conversion into black copper before returning for refining in three reverberatory furnaces (*Spleisofen*) and three refining hearths (*Garofen*). Below, beyond the 230-yard-long dam and its pond, was the mint and warehouse, for the storage of the refined copper, under the control of a manager (*Oberhüttenverwalter*) and his assistant. This employed a hundred workers and, together with the laboratory, smithy, bellows-manufactory, saw-mill, office, hospital, and school formed the industrial quarter of the lower settlement of Verkhne–Susunsk which also contained a large residential village (*sloboda*) of some 2,285 people who were provided with a church for their enlightenment and 308 houses for their accommodation.[130] This then was the Susunsk works, which during the boom years of the late 1760s annually produced some 5,500 puds (*c*. 88 tons) of copper which was converted into small-denomination coins worth 137,500 roubles for circulation in Siberia.[131]

On the basis of two interlocking mineral supply systems, rooted in the production of the mighty Schlangenberg and diminutive Lazur mines whose ores were processed in an enormous metallurgical complex situated 200 miles to the north-west, the Barnaul mining district rose, during the years 1759–71, to a position of international prominence. An extensive local and regional supply system afforded its denizens with an ample lifestyle and the enterprise's now extensive requirements for construction materials to make and repair equipment engendered not only the creation of a large number of saw-milling enterprises but also the construction of a major iron and steel works. Indeed, even as the boom reached its height in the years 1770–1, an old plant at Tomsk was being replaced by a new complex, comprising a blast furnace with three beam hammers and three smaller hammers (weighing 1.6 cwt each), a steel works and hammer, and a wire hut, capable of producing some 30–40,000 puds (482–643 tons) of ferrous metal a year.[132] Yet another element was thus added to the great metallurgical empire of the Altai, which now at its height in 1771 was not only self-sufficient in iron, but also produced enough copper coins to meet the monetary requirements of the Siberian population and poured forth such quantities of gold and silver as to be unrivalled in either Europe or Asia and certainly in sufficient amounts to justify its popular name – the Russian Potosi.

Even as the boom on the Schlangenberg, which rendered Russia supreme amongst European and Asiatic specie producers, reached its peak, however, the signs of decay were already becoming apparent. Amidst general dilapidation amongst the older mines and smelting plants, production at the New Lazur workings had already collapsed in 1765. Now yields of the Schlangenberg ores showed signs of instability. Having varied between 5 and 6 zolotniks of auriferous silver per pud (46–56 oz per ton) until 1766, the next quinquennium saw average ore yields fall to 4.8 zolotniks per pud (44 oz per ton), and thereafter the decline was relentless. During the administration of Lt General von Irmann (1770–80) they amounted to no more than 4 zolotniks (37 oz per ton) and in the last year of his successor's regime, 1785, 2.5 zolotniks (23 oz per ton). The nadir was reached two years later in 1787 at 1.3 zolotniks (14 oz per ton). Subsequent years saw a brief recovery but the golden days of the 1760s were past and as the century drew to its close the silver ores delivered from the Schlangenberg to the smelters contained about 2.5 zolotniks of the precious metal and the argentiferous coppers 1.5–2.[133] To compensate for this disastrous fall in the metal content of the ores production on the Schlangenberg was increased to 1.3–1.5 million puds (20,900–24,100 tons), an increasing volume of minerals passing along the Alei and capacity at the great Barnaul–Pavlovsk–Susunsk metallurgical complex being fully occupied in its processing. Yet in spite of this increased activity the silver production of the works collapsed, falling from 1,232 puds (20,172 kg or 97 per cent of total output) in 1771 to 604 puds (9,892 kg or 57 per cent of total output) in 1791.[134]

For half a century, c. 1745–95, the Schlangenberg dominated production in the Altai. After a false start during the years 1745–50, when average annual production of ore rose to some 200 tons of ore yielding some 32 puds (526 kg) of auriferous silver, and the resultant crisis of 1751–8, the discovery of new rich deposits laid the foundations for a major production boom. From 1759 to 1771 activity on the mountain was intense, a meteoric rise in output bringing into existence a formidable production complex which with an output in excess of a million puds (18,200 tons of ore) yielding 1,232 puds (20,172 kg) of silver reigned supreme amongst European and Asiatic production centres. Already in the early 1770s, however, the first signs of instability were beginning to appear, and whilst thereafter production of ore continued to grow falling yields ensured that the amount of silver produced from Schlangenberg ores steadily declined both absolutely and relative to other centres which now rose to prominence.

Crisis and renovation, 1771–86 Even as the first signs of instability began to appear on the Schlangenberg, 1768 witnessing a 40 per cent fall from normative levels of metallic content of the ore raised, men began to cast around for new deposits. Old abandoned workings were re-opened (e.g. Nikolaev mine). Recently discovered workings which had been neglected during the boom (e.g. the Semeonov mine) were revitalized. New mines (e.g. Cherepanov) were discovered. Within fifteen years a new production complex was created alongside the Schlangenberg whose salient features were contemporaneously described by Hermann who knew the industry well.[135] As he noted, the principal mine in 1786 remained the Schlangenberg which produced three-quarters of total output – 1.5 million of 2 million puds of ore. Now, however, others challenged its position. Amongst these the best groove was said to be the Semeonov.[136]

Initially discovered in 1732 the Semeonov deposits, situated 20 miles south-west of the Schlangenberg on the Korbolicha, were only exploited in 1763 when, as a result of a short-term production downswing at the main workings, a shaft was driven and silver-rich ore containing up to 375 oz of silver and 8 cwt of lead per ton was discovered in the gallery opened up. For three years activity was intense but with recovery on the Schlangenberg in 1766 the labour force was withdrawn. In response to the crisis of 1768 recovery at the Semeonov works began but production concentrated on previously discarded wash ores processed in a new works opened in 1769 on the lower Korbolicha and it was only in 1771, when production from Schlangenberg ore declined, that mining activity once again commenced.[137] The initial boom, however, was shortlived and only in the late 1780s with the rebuilding of the workings did the Semeonov mine rise to a position justifying Hermann's description of it as the best groove on the Schlangenberg.

In the interim it was older workings, eclipsed since the 1750s, which dominated mining activity outside of the Schlangenberg. Amongst these the group of workings situated about the Nikolaev mine in the Brobovskii mountains was of primary importance. Production here had been of some importance in the period 1751–9 but thereafter, as the Schlangenberg rose in importance, output had declined, the ores raised at the Nikolaev mine in 1771, containing 18 oz of silver and 2 cwt of lead per ton, being hardly

Table 3.7 Production of ore at the Semeonov mine in the eighteenth century

	Annual ore production (in tons)		
	primary ore		wash ore
1763–1771 av.		305	
1763	288		nil
1764	524		nil
1765	127		nil
1766		Abandoned	
1767	520		nil
1768	No miners		nil
1769	No miners		66
1770	No miners		107
1771	1057[a]		58
1772–1783 av.		479	
1791		3,062	

Note: a to October.

worthy of exploitation.[138] The breakthrough at these workings only came during the years 1775–8 as production at the Semeonov mine declined. In 1775 output at the Nikolaev mine rose to 10,000 puds (160 tons) and in 1778 at the Beresov mine it increased to 40,000 puds (640 tons). The boom, as before, was shortlived, however, production at the Nikolaev mine after the initial increase declining to 1783 and, in spite of the sinking of a new shaft 2 miles from the river Ulba at the Beresov mine in 1781 which sustained production there in that year at 41,400 puds (665 tons), also at the Beresov mine where output amounted in 1783 to 9,701 puds (156 tons) – yielding 10.3 kg of silver and 16 cwt of lead.[139] Neither of the great mines – the Semeonov and Nikolaev – noted by Hermann as important components in the new production complex had before 1783 realized their full potential.

Production initially concentrated in the Semeonov groove, amounting during the years 1771–5 to about 1,000 tons of ore a year yielding 4,770 kg of silver and 450 tons of lead, had thereafter declined whilst production at the Nikolaev group of mines had increased. Total ore output in the new silver-production complex was thereby maintained, but as yields on average fell every effort had to be made to maximize returns. Accordingly in 1778 a new smelting complex was constructed – the Novoaleisk works – some 12 miles from the Semeonov mine and 26 from the Ulba workings.[140] Here were built three smelting furnaces and an ore-crushing works, operated by two mining officers and three men 2–3 miles from the outpost whose peasants and cossacks cultivated various grains as well as melons, gherkins, and other garden fruits. Its function was to smelt the lead ore and silver slag obtained by smelting the Semeonov ores which were not transported to Barnaul, a task occupying 4–6 weeks a year in the processing of 2,000–3,500 puds (32–56 tons) of wash ores. In addition the ores from the Ulba workings were also reduced here, in 1778 some 700–1,000 puds (11–16 tons) taking three weeks to process. In all, during its first year in operation, work at the new plant involved the smelting of 1.6 tons of roasted leadstone and a similar quantity of silver slag containing 3.5 ounces of silver and about 40 pounds of lead per ton.[141] In the following year, after the delivery of 40,000 puds of ore from the Beresov mine yielding 64 tons of leadstone, similarly containing 4.5 ounces of silver and 42 pounds of lead per ton, output increased but its complementary relationship to the Barnaul works remained. The new works processed the low-grade ores whose low precious-metal content did not justify the costs of transport to Barnaul. The old works continued to process the high-grade minerals. Together, however, they contributed only *c.* 1,850 kg of silver to total production and only 170 tons to lead output. The new production complex, which had contributed a third of total Kolyvan production of silver in the first half of the decade, now in 1778 provided only 12 per cent. Similarly, lead output, which had completely displaced Nerchinsk supplies during the years 1770–5, by 1778 held only a third of the market and lost even this position in silver and lead markets by 1783, leaving the Schlangenberg once more supreme but at much reduced production levels.

Only during the years 1783/4–86 with the rebuilding of these workings and the discovery and exploitation in 1784 of the Cherepanov deposits, 7.25 miles to the north-west of the Schlangenberg,[142] did the new complex re-establish its position, contributing a quarter or 500,000 puds (8,308 tons) of total ore output before declining to 400,000 puds (6,388 tons) in 1791.[143] By this time, however, even as Hermann eulogized the productive capacity of the new (Semeonov, Nikolaev, Cherepanov) complex which was in 1786 sustaining total output at a level 40 per cent below that of 1771/2 but 50 per cent above 1783/4, new developments were taking place which heralded the birth of the nineteenth-century industry.

Towards the nineteenth-century industry: the Salairsk, Ziryanov, and renovated Loktevsk mines For twenty years, during the period 1771–90, as production at the Schlangenberg mine steadily declined, a new production complex, initially comprising the Semeonov and Nikolaev mines and later the newly discovered Cherepanov deposits, intermittently (1771–8 and 1783–6) augmented supplies of silver-bearing ores, creating a completely restructured industry by the latter date, when Hermann on the occasion of his first tour of duty at the enterprise described the complex.[144] As has been indicated, the principal mine remained the Schlangenberg but its output of 1.5 million puds (24,116 tons) of ore, far larger than in 1771/2, now comprised only three-quarters of total industry output. The residual output derived from the Semeonov mine, which produced lead ores and gravels as well as zinc, the Cherepanov, with hornblend and ochre in quartz, and the Nikolaev mine, from which was raised silver-bearing lead ochre. Out of these grooves came some 500,000 puds of mixed ores, which, with the 1.5 million puds from the Schlangenberg, were only slightly crushed and smelted into *Rohstein* which was then seigered with lead producing *c*. 780 puds (12,764 kg) of blicksilver. Whilst the production of ore had thus doubled, the decline in yields was such that production of blicksilver was reduced by some 40 per cent in comparison with 1771/2. At best the 'new' mines, upon which Hermann set so much store, were only a temporary stop-gap, their output on occasion (1772–6 and 1783–6) retarding but not halting the process of decay and short-term crises therein (1779–83/4 and 1787–91) precipitating acute dislocations in the industry. What might be described as the eighteenth-century Altai silver industry was, during the years 1771–91, firmly set on the path of decline, a path which led to its ultimate extinction in the early nineteenth century.

Even as Hermann described the industry, however, and unchronicled by him, new deposits were being discovered which would allow the industry to weather the downswing of the late 1780s and to establish the production base of a new complex which, during the years 1791–1868, maintained a high output level, on occasion equalling that of 1771/2 and consistently producing 1,000–1,100 puds (*c*. 16,000–18,000 kg) of blicksilver a year or *c*. 80–90 per cent of eighteenth-century peak output levels.[145] Of these new deposits the earliest to be discovered lay about 100 miles to the north-east of Barnaul where in 1782–3 the Salairsk mine was opened, shortly followed by the Tishovsk mine in 1788 and the Gerikhovsk mine in 1789–90, creating a compact complex of three mines each situated no more than 0.75–1.5 miles from each other. The minerals here were extremely low-grade argentiferous lead ores containing between 0.75–1 zolotniks of silver per pud (7–9.3 oz per ton) which, excavated in vast quantities, rising from *c*. 0.3 million puds (*c*. 5,000 tons) in 1792 to 1 million (*c*. 16,000 tons) a year in the period 1840–50, provided a firm base for an industry experiencing the decline of the great eighteenth-century workings (table 3.8). The main dynamic in the nineteenth-century industry was provided, however, by another group of mines – the Ziryanov–Riddersk workings – located far to the south of the Schlangenberg no more than 50 miles from the Chinese border. The poly-metallic ores of these deposits were of quite a different order to those of Salairsk. Containing 4–8 zolotniks of silver, 6–10 funts of lead, and 2 funts of copper (37–74 oz of silver, 3–5 cwt of lead, and 1 cwt of copper per ton), they allowed the Ziryanov mines to assume the position in the early-nineteenth-century Altai industry that the Schlangenberg had enjoyed in the eighteenth century. The story of these workings, which begins in the case of the Salairsk mines in the late 1780s and with regard to the Riddersk–Ziryanov group in the early 1790s, thus more truly belongs to the nineteenth-century history of the Altai industry, but as the first group – the Salairsk mines – had begun to make a major impact on total ore production before 1800 it may perhaps be considered here.[146]

Table 3.8 Production of ore at the Salairsk group of mines, 1782–95

| | Salairsk mine | | Tishovsk mine | | Gerikhovsk mine | |
	output (tons	yield (zol./pud)	output (tons)	yield (zol./pud)	output (tons)	yield (zol./pud)
1782–3	1,125	1.25–3.25	—	—	—	—
	210	0.5 –1.00				
1784	286	1.25–3.0	—	—	—	—
	276	0.5 –1.00				
1785	333	1.25–1.5	—	—	—	—
	1,105	1.00				
1786[a]	707	1.25–1.5	—	—	—	—
1787	643	1.25–1.5	—	—	—	—
1788	1,423	1.25–3.0	702	1.6[b]		
	117	1.00				
1789	3,084	1.25–4.5			112	1.25–3.5
	186	1.0				
1790	7,310	1.25–5.0			78	1.5 –2.5
1791	7,264	1.25–5.00			70	1.5 –1.5
	408	1.00				
1792	5,371	1.25–2.5	959	0.5 –1.0[c]	65	1.5 –4.5
	781	1.00				
1793	5,649	1.25–3.5			7	0.75–4.5
	710	1.00				
1794	6,798	1.25–7.0			103	2.0 –5.5
	1,353	1.00				
1795	8,075	1.25–4.0	NA	NA	NA	NA
	2,845	1.00				

Notes: a plus, during the years 1786–7, 38.5 tons of copper ore containing 3–10 funts per pud from the Chechyulkhinsk mine. b Total 1788–91. c Total 1792–4.

Created during the crisis years 1782–5 and sustaining an annual average output of 830 tons at that time, production at the Salairsk mine subsequently declined, falling to 707 tons in 1786 and 643 tons in 1787, as production in the secondary (Semeonov, Nikolaev, and Cherepanov) complex briefly recovered. When decline once more set in at the older workings, however, production at the Salairsk mine and the new workings opened in 1788–9 rapidly increased, rising to 1,715 tons in 1788, 7,563 tons in 1790, 8,573 tons in 1794, and in excess of 10,920 tons in 1795 when the workings were already well set on that path which would ultimately cause output to exceed the 1-million-pud (16,000-ton) mark in the 1840s. Within a decade of its foundation a major mining community, comprising 628 houses, had been created at the new workings encompassing in 1791 some 575 mine operatives and *c*. 10 troops (table 3.9) and this was only the beginning. By 1795 their numbers had increased to 760 and in 1809 to 931, and this marked only a stage in the expansion of the labour force to 1,811 in 1850.[147] From 1789 more and more men were employed at the Salairsk mining complex excavating a rapidly increasing quantity of ore, which from 1795 underwent primary smelting at a new works – the Gavrilovsk plant – built 95 miles to the north-east of Barnaul and about 3 miles from the mine

on the Talmovaya brook.[148] The Salairsk mining camp was thus comple-
mented by a small smelting works from 1795, comprising eight smelting
furnaces, two smithies, and a saw-mill, whose function, like the Novoaleisk
or Korbolicha/Zmeev works created in 1778, was to convert the low-grade
ores into concentrated *Rohstein* for despatch to Barnaul.[149]

As the eighteenth century drew to its close, therefore, yet another major
mining and metallurgical complex had been created within the Kolyvan
mining *okrug* which already in 1791 contributed 17 per cent of the
2,329,898 puds (37,458 tons) of ore then being produced, an amount
sufficient to raise silver production within the enterprise to its new
equilibrium level of 1,000–1,100 puds a year. Operations within that
bilateral system of silver production were, accordingly, both extended and
transformed, heralding the creation of a new silver-production complex in
the Altai based on the processing of vast quantities of low-grade ore from
the Salairsk and Riddersk groups of mines and the exploitation of the rich
poly-metallic reserves of the Ziryanov mines.

A similar process, moreover, may be discerned in that other bilateral

Table 3.9 Labour force at the Salairsk mine (1791) and at the Gavrilovsk
smelting works (1795)

Salairsk mine	
Chief mine officers	2
Deputies	3
Trainees	4
Miners	
Class 2	4
3	37
4	284
Mineworkers	15
Invalids	145
Ore collectors	27
Ore analysts	13
Carpenters, machines, and mines	33
Blacksmiths	2
Barber/surgeons	2
Stable hands and sled drivers	2
Foresters	3
Gavrilovsk smelting works	
Smelting master	1
Machine apprentice	1
Smelters	16
Servers at factory	37
reserves	13
Sawyers–woodcutters	1
Charcoal-makers	56
master	1
look-outs	2
apprentice	1
Woodcutters	40

system of copper production which had been created in the boom years of the 1760s, although in this instance the impetus to expansion after 1783 came from the renovation of the old Loktevsk (plan 3.8) and New Lazur mines.[150]

Both mines had been abandoned in the late 1760s as low-grade argentiferous-copper ores from the Schlangenberg became the mainstay of production at the Susunsk smelting works and mint. Only as production of cupriferous ores from the Schlangenberg declined during the years 1772–83 was there a revival of activity at the old workings. Mining was restarted at the Loktevsk mine in 1770, some 450 tons of ore containing 4 oz of silver and 7.5 cwt of copper per ton being raised in 1770–1 and about 600 tons a year on average from 1772–83 or a level which was almost 2.5 times greater than that attained during the previous boom of 1751–66.[151] During these years the fortunes of the Loktevsk mine were in the ascendant and by 1782/3 it had completely displaced the Schlangenberg as the principal source of copper ore within the mining *okrug*, production of the refined metal, as a result, actually increasing from *c.* 5,500 puds (88 tons) in the late 1760s to 18,800 puds (302 tons) in 1782 when the Siberian coinage was called down to the European Russian standard to maintain monetary stability (table 3.10).[152] Subsequently production declined as the workings were flooded but, with the installation of a handle-pump driven by three horse-whims, the main Nosvishinskoi shaft was drained and extensive reserves of argentiferous copper, first discovered in 1782, were opened up to exploitation. The resource base of this sector of the industry was transformed and production at the Loktevsk mine rapidly increased, reattaining antediluvial levels of output in 1788 and thereafter increasing to in excess of 1,500 tons a year in 1791 and 1793 when, with the output of the Golzov and Lazur mines, reopened after 1783, total production of copper ores amounted to *c.* 3,000 tons.[153] Within little more than a decade, from 1781–2 to 1791, output of copper ores increased three and a half times in relation to mining activity, reaffirming the importance of at least one element in the bilateral system of copper production.

Yet, paradoxically, even as this reaffirmation was made with respect to the mining element within the system the other smelting element lost its separate identity, as the motivation of those directing operations during the mining boom of 1781/2–91 totally changed. During these years copper no longer figured greatly in their calculations, and as yields fell so too did production of refined copper from *c.* 300 tons in 1782 to *c.* 200 tons a year in the period 1792–6 and *c.* 83 tons a year in 1800–3.[154] The ores were now prized for their silver rather than their copper content, and as such became part of the resource base of the silver-production complex, further augmenting poly-metallic mineral supplies processed therein and contributing to a complete reorganization of its smelting capacity. By 1791 the copper ores of the Loktevsk, Lazur, and Golzov mines were merely one element

Table 3.10 Production at the Loktevsk mine in the period 1767–94

	Annual ore production (tons)	Silver yield (oz per ton)
1767	21	—
1768–1769	nil	—
1770	22	—
1771	430	—
1772	346	—
1773	320	—
1774	67	—
1775	227	—
1776	511	—
1777	791	—
1778	584	—
1779	195	—
1780	288	—
1781	885	—
1782	795	9–70
1783	1,327	21–100
1784	1,254	9–79
1785	666	14–93
1786	1,684	12–119
1787	1,106	9–107
1788	1,338	9–125
1789	1,200	18–112
1790	1,440	9–128
1791	1,523	9–98
1792	775	16–56
1793	1,554	9–124
1794	1,041	14–168

in a massive quantity of high-grade (in excess of 1.5 zolotniks per pud/14 oz per ton) argentiferous poly-metallic ore, amounting to 1,638,000 puds (26,334 tons), which passed to the smelting works of the Ob complex. This had been totally modernized in the early 1780s to process the greatly increased quantities of relatively high-yield, silver-bearing ore, a programme which involved the transformation of the Pavlovsk and Susunsk works. Whilst smelting capacity remained unchanged at the Barnaul works, which had thirty hearths, at Pavlovsk the number of smelting hearths was increased to twenty-two and at Susunsk to nineteen for the initial reduction of the argentiferous polymetallic ores and their conversion into *Rohstein* and *Kupferstein*. Each, moreover, was now equipped with two or three roasting ovens, to concentrate and eliminate the impurities in the primary metal, and two or three leading hearths, where the concentrated 'fertile' stone was mixed with lead which on heating took up its silver content, leaving a 'sterile' multi-metallic compound comprising, for example, copper and lead. By 1791 these works thus merely served to undertake the primary and secondary processing of argentiferous ores of which copper was only one element, normally amounting to 7–15 per cent of the total amount of ore reduced.[155] Subsequently, the 'fertile' lead obtained from the saigering of the *rohstein* or black copper was despatched to Barnaul where it was processed in three cupelation hearths to extract the blicksilver which was the primary objective of operations. The copper–lead compounds were also despatched to Susunsk for refining, but as the amounts of such compounds declined so too did refining capacity which was reduced to three *spleisofen* in the 1790s, a mere adjunct to a works whose primary function had become the processing of argentiferous ores.[156]

If the high-grade cupriferous ores, yielding in excess of 14 oz of silver per ton, thus merged into the general supplies of high-grade, silver-bearing ore delivered to the remodelled Barnaul–Pavlovsk–Susunsk smelting complex, the low-grade copper ores also were assimilated into the system evolved for the processing of such ores, involving, as in the case of the Aleisk and Gavrilovsk plants, the construction of a works close to the mine. In 1781 yet another subsidiary plant – the Loktevsk works – was built, to which the copper ores of the neighbouring mines were brought for the production of *Kupferstein*. Like the other subsidiary plants it was a simply structured complex comprising a crushing works and eight smelting hearths, together with a hospital and officers' quarters of wood with stone foundations and guardhouse; a workshop for bellows-making and smithy; and a warehouse for materials and provisions for the miners and landsmen who were said 'to live in the greatest freedom' with magnificent herds of cattle and bountiful garden grounds.[157] Initially, the supplies of low-grade copper ore for processing here were small, and in 1783, even with the smelting of Schlangenberg ores, containing 11 oz of silver per ton, and Nikolaev ores, containing 14 oz of silver and 0.75 cwt of lead per ton, only four smelting hearths were operational. But as production at the Loktevsk mine rapidly increased existing capacity was soon fully occupied and before 1792 was increased to ten smelting hearths with two roasting ovens.[158] As the production of copper ores thus rapidly increased, reaffirming the importance of this element in the operations of the enterprise, the previously autonomous bilateral system of copper production lost its independent identity, merging into an all-embracing unitary silver-production complex and establishing a new industrial order which would set the pattern of operations for the nineteenth century.

Kolyvan–Voskressensk at the close of the eighteenth century Anyone observing the new production complex which had emerged within the Kolyvan mining *okrug* during the closing years of the eighteenth century had laid out before them the whole history of the enterprise. Mining operations were extensive and involved the exploitation of mines spanning three generations (table 3.11).[159]

The original Kolyvan mine, opened in 1729, remained operational as did a number of the other mines – Golzov, Loktevsk, and Murzintsev – which

Table 3.11 Mining operations at Kolyvan–Voskressensk, 1793

Mine	Output of ore (tons)	silver (oz/ton)	Metal content lead (cwt/ton)	copper (cwt/ton)
Schlangenberg	21,130	14–47	—	—
Semeneov	3,062	12–28	1–3.5	—
Nikolaev	2,862	12–28	0.5	—
Cherepanov	436	21–44	—	—
Petrov	1,839	12–56	—	—
Riddersk	700	12–448	1.5–7.25	2–2.5
Golsov	500	2	0.5	0.5–3.5
Lazur	1,019	—	—	0.5
Loktevsk	1,554	9–124	—	1.0–5.1
Mine No. 8	79	—	1	0.6–3.5
Beresov	30	—	—	1.0–1.5
Murzintsev	24	—	—	0.6–6.75
Kleopinsk	2	—	—	—
Akimov	40	—	—	0.9–2.5
Kolyvan	90	—	—	2–3.5
Salairsk	5,645	12–47	—	—

had formed part of the Demidov copper-production enterprise, producing during the years 1729–45 beween *c.* 75 and 235 tons of ore a year (figure 3.5). Only the Golzov and Loktevsk mines were now of any

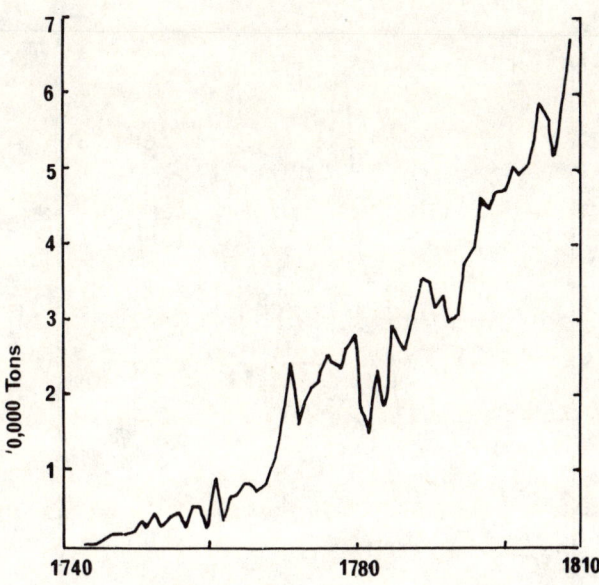

Figure 3.5 Kolyvan ore production

Table 3.12 Ore reduction at Kolyvan–Voskressensk, 1792

Primary works		Secondary works	
1 Barnaul		(i) Aleisk	
Ore processed		Ore processed	
Silver, 21–23 oz/ton,	11,254 tons	Silver	12–19 oz/ton
Copper	1,186 tons	Silver–lead	12–19 oz/ton
Rohstein and Werkblei,	386 tons	1431 tons	2.75 cwt/ton Pb
		(ii) Kolyvan	
		Ore processed	
		Silver	12–21 oz/ton
		1608 tons	Silver lead	9–14 oz/ton
		1.5 cwt/ton Pb
		(iii) Gavrilovsk,	uncompleted
2 Pavlovsk		(iv) Loktevsk	
Ore processed		Ore processed	
Silver, 21–23 oz/ton,	7,556 tons	Silver	21–3 oz/ton
Copper	511 tons	Silver–lead,	12–19 oz/ton
Rohstein and Werkblei	370 tons	3778 tons	2.9 cwt/ton Pb
3 Susunsk			
Ore processed			
Silver 21–23 oz/ton,	6,805 tons		
Copper	1,447 tons		
Total ore processed			
Silver, 21–23 oz/ton	25,615 tons	Silver and	6,817 tons
Copper, yields n.a.	3,144 tons	silver–lead	12–23 oz/ton,
			1.5–2.9 cwt/ton Pb.

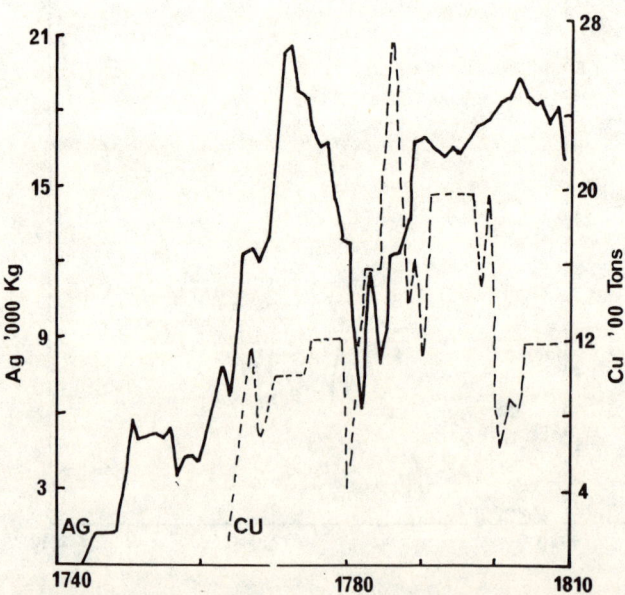

Figure 3.6 Kolyvan 'silver' and copper production

Table 3.13 Industrial capacity at Kolyvan–Voskressensk 1746–95

Works	1746	1771	1795
Kolyvan (1729)			
Smelting hearths	7	8	6
Roasting hearths	—	—	1
Refining hearths	10	—	—
Saiger hearths	2	1	—
Cupelation hearths	2	—	—
Barnaul (1739/40)			
Smelting hearths	6	30	30
Roasting hearths	—	1	3
Refining hearths	5	1	—
Saiger hearths	—	—	3
Cupelation hearths	—	13	3
Pavlovsk (1763)			
Smelting hearths		24	22
Roasting hearths		—	2
Saiger hearths		—	2
Cupelation hearths		—	2
Susunsk (1764)			
Smelting hearths		12	19
Roasting huts		4	3
Refining hearths		6	3
Saiger hearths		—	2
Aleisk (1778)			
Smelting hearths			8
Roasting hearths			1
Loktevsk (1781)			
Smelting hearths			10
Roasting hearths			2
Gavrilovsk (1795)			
Smelting hearths			8

significance, however, having been redeveloped during the copper booms of 1751–66, 1772–81, and 1783–91, and, with the Lazur and Kleopinsk mines, first opened in the years 1751–66 and subsequently redeveloped in the period 1783–91, comprising the major elements of that cupriferous ore complex, whose output had steadily grown from 554 tons a year in the period 1762–71 to 3,200 tons in 1793. Similarly the Schlangenberg mine, which during the years 1745–51 and 1758–71 had pushed aggregate production of silver and silver–lead ores up to 6,875 kg and 24,284 tons a year respectively, still dominated the industry in 1793 although now with an output of 21,130 tons, only contributing 59 per cent to an aggregate production of 35,750 tons. The residual output in 1793 came from the new third-generation mines, the Semeonov, Petrov, and Cherepanov in the vicinity of the Schlangenberg and the Nikolaev to the south-east which during the years 1772–83 and 1786–87 had served to support ore production at levels equal to, or slightly above those of 1771, and the Salairsk and Riddersk mines which, with the newly discovered Ziryanov mines, pushed production up beyond the 2-million-pud mark and laid the foundations of the nineteenth-century Altai industry. Within fifty years of the seizure of the Demidov's Kolyvan works, namely between 1745 and 1794, ore production, from successive generations of mines, had increased from *c.* 200–250 tons a year to *c.* 39,000 tons or almost 2.5 million puds (figure 3.5).

To process this ore new capacity had been brought into existence, most of which still remained operational in 1792 (table 3.12).[160] The original Kolyvan works, founded in 1729, remained in operation but had been reduced to secondary status after a protracted phase of decline, the number of smelting hearths falling from seven in 1746 and eight in 1771 to six in 1795 (table 3.13).[161] In contrast, the other great Demidov plant, the Barnaul works, went from strength to strength in the Crown epoch, completely dominating silver production, which on the basis of the rich Schlangenberg ores rose to 20,754 kg of unrefined metal in 1771, forming the centrepiece of a silver-production complex now augmented by Pavlovsk and a copper-production complex centred on Susunsk with an output of *c.* 88 tons (figure 3.6).[162] The subsequent decline in silver yields resulted in a diffusion in the smelting of high-grade ores amongst the principal works and the creation of new plant – Aleisk, Loktevsk, and Gavrilovsk – to process low-grade ores, but still, as the century drew to its close, the Barnaul works reigned supreme in an enterprise which in processing *c.* 39,000 tons of ore was able to maintain silver production at a consistent level which was only slightly below the peak levels of the early 1770s.

The Kolyvan labour force

To man this enormous enterprise a vast labour force had to be recruited during the years 1747–96 in lands which only a few years before had been a wilderness (table 3.14).[163]

Within a mere half-century a group of workers and their dependents had been assembled at the Kolyvan–Voskressensk enterprise comprising 12

Table 3.14 The Kolyvan labour force, 1747–96

	Master workers and mine workers		Ascribed peasants	'Revision souls'
	Total	At work		
1747	1,758	736	3,121	4,879
1763	1,944	n.a.	38,064	40,008[a]
1771	n.a.	4,775	n.a.	n.a.
1782	13,151	5,573	54,750	67,901[b]
1796/7	14,306	7,453	55,306	69,612[c]

Notes: a–c Third to Fifth Revisions.

per cent of the Russian population and 7.5 per cent of the total population of Siberia.[164]

The Demidov era   Even in 1746/7, as the enterprise passed into the Crown's hands heralding this great expansion of the labour force, however, the presence of the Russian operatives was not without significance in the local economy of the Altai. Already there were some 736 full-time workers, a vastly greater number than the original party of fifty who had arrived twenty years before to construct the new works.[165] Nor were most of these additional workers drawn from the mining camps of the Urals or European Russia. A few workers had been despatched from Neviansk and other Demidov works in the Urals, and in 1733 a party of fifteen Saxon miners had been recruited on the recommendation of an Ekaterinburg smelter sent to the Altai to prove the Kolyvan ores, but even at that time such miners were a minority amongst a labour force which already numbered 419 persons.[166] The main source of recruits to the Kolyvan labour force was the local population who, in the Kuznetsk district alone, numbered 2,874 'souls' in 1722–4 and 6,347 in 1738 together with 11,346 in Tomsk district.[167] Larger numbers of fugitive peasants as well as others with passports were flooding into the Altai and by decrees issued on 31 March 1726 and 9 August 1727 the Demidovs were empowered to recruit from amongst their ranks, by the latter decree being ordered to select for their Altai factory 'fugitives, both living in the forest and amongst the Kalmuck and roaming about the fields and monasteries and amongst the trading people and peasants without paying capitation tax'.[168] Numbers of such recruits employed at the works rapidly increased during the years 1727–38, the great majority being fugitive serfs from Arkhangel town, Moscow, Kazan, Nizhnegorod province, and the lands about Ustiug and Vyatka, who by an ordinance of 13 November 1736 were permanently attached to the enterprise, forming by far the largest element in a labour force which had grown from 419 workers in 1733 to 482 in 1738, and with their dependents the largest element in a population numbering in 1738 some 900 'souls'.[169]

Thus was created by 1738 the group of master workers, free workers, or peasants attached by the ordinance of 1736 to the works and maintained at the cost of the Demidovs. They were paid a salary (in accordance with the wage schedule of 1723 revised in 1737) from which was deducted the cost of provisions drawn from the company store, and henceforth were to form a permanent element in the Kolyvan labour force. Together with the operatives brought in from other mining camps they were described by Gmelin on the occasion of his visit to the works in the summer of 1739.[170] At that time the principal focus of activity within the enterprise was Kolyvan, which was encompassed within the walls of a fortress or *ostrog*, wherein there was housing for the great majority of workers and their dependents, 800 in all. Here were the workers who had been brought from Ekaterinburg and Neviansk as well as most of the locally-recruited master workers, who were provided with factory houses but spent most of their summers working in the mines. Typical of this group were the thirty men working in the Ploskogorsk mine, opened in 1734, who at the time of Gmelin's visit laboured in the 56-feet-deep workings, carried the ore to Kolyvan – and fought a running battle with the Kirghiz.[171] Outside the walls of the *ostrog* life was, to say the least, precarious, and numbers settled there were small. Only forty or fifty people lived in each of two groups of settlements on the Charish/Barnaul and Loktevsk rivers under the protection of a military force, whose numbers had been steadily augmented from the fifty Tara cossacks, who had been despatched in 1727 to protect the original party of workers from the incursions of Kirghiz and Dzunghir raiders, to the 100 cossacks who with a sergeant and seven troopers defended the works in 1738.[172] In this beleaguered state thus lived the Kolyvan labour force in 1738, as yet a tiny element (2.5 per cent) of the Altai population.

Within the decade 1738–47, however, whilst the insecurity remained an endemic feature of frontier life, the situation was transformed as the numbers attached to the works rapidly increased to some 4,879 persons. Almost all of the increment were new migrants to the Altai, some 3,979 having arrived after the 'First' Revision of 1737/8 and being described in 1747 as 'new comers and of unknown kin'.[173] In part these newcomers were recruited to the ranks of the master workers bringing the total number of such workers and their dependents to 1,758, of whom 756 were employed at the works, distributed between the three main settlements – Kolyvan, Barnaul, and Shul'binsk.[174] By far the largest of these settlements remained Kolyvan, protected by an *ostrog* below which was a street containing 325 houses for the workers, salaried staff, and soldiers. In comparison the other settlements were tiny. Barnaul, whilst containing a not inconsiderable amount of the enterprise's capital equipment, in 1747 was purely a smelting establishment with only seventy-nine houses for its workers, whilst the Shul'binsk mining camp had but thirty-nine houses and a smithy. Together these settlements encompassed 756 workers and their dependents or some 1,758 persons, a small but not insignificant element in the population of the Altai.

Now, however, this group of full-time workers was numerically dwarfed in both the Kolyvan labour force and the overall population of the region by a new group of part-time peasant operatives who in 1747 numbered some 3,120 'souls'.[175] As in the case of the master workers these peasants were drawn from the ranks of the migrants to the Altai. From 1740 an increasing number of the incomers were attached to the works in accord with the ordinance of 23 March 1734 which allowed entrepreneurs to claim fifty households for each thousand puds of refined copper they produced.[176] In 1740 the first such grant was made: a decree of the Siberian governor's chancery assigned 200 households to the Kolyvan works, in accord with its six furnaces of 4,000 puds (64 tons) productive capacity. These households were in the village of Berdsk, the Malyshev *sloboda*, and the Belyarsk *ostrog*, situated 100–350 miles from the works, numbering in 1745 some 1,500 'souls' of whom 1,002 were said to be at work each year. Two years later, in 1742, a further allocation was made, this time to the Barnaul works which possessing six furnaces also received 200 households – 100 in Berdsk, Malyshev, and Belyarsk in Kuznetsk district and another 100 in the village of Chaus and the Sosnov *ostrog* in Tomsk district, 240 miles from the works, numbering together 1,620 'souls' of whom 1,002 were at work each year.[177] In 1746/7 some 3,120 'souls' were thus attached to the works in accordance with the ordinance of 1734 and formed a new group of ascribed peasants.

These individuals, living 100–350 miles from the works, were engaged, if they lacked a horse, in stockpiling charcoal, ore, timber, and other materials and, if they possessed a horse, carrying these materials to the works, thereby displacing the master worker in the support of mining and metallurgical activity. In undertaking this service function, however, they laboured within a highly organized framework of rules laid down in their ordinances of ascription. Production norms were set for their labour and wages were fixed in accord with the pay schedule of 1724 as revised in 1737. Those engaged in charcoal production for the purpose of rendering their capitation tax either spent some sixty days a year making a 20-cubic-sazhen charcoal bed (*kucha* or *meiler*) which in laying, heaping, burning, and breaking fully absorbed all their energies for a payment of 1.20 roubles, or cut wood, expending a similar amount of time, felling some 12 cubic sazhens at 25 kopecks per 'official' cubic sazhen to earn the same amount.[178] Whether cutting and coaling or carrying charcoal or ore, however, each peasant worked some sixty days or two months a year at an 'official' day rate of 5–6 kopecks and an actual rate of 2 kopecks to earn

Table 3.15 Wages at Kolyvan–Voskressensk, 1745 (kopecks per 'official' day)

Category	Official	Unofficial
Peasant without horse		
winter	4	7
summer	5	$\left\{\begin{array}{l} 7^a \\ 6 \end{array}\right.$
Peasant with horse		
winter	6	10
summer	10	12

Note: 1   7 kopecks 'urgent smelting and mining work', 6 kopecks for the rest.

enough to pay their poll-tax and obrok of 1.10 roubles. With an additional unpaid twenty days spent in transit to their work, moreover, their total commitment to industrial work rose to about one and a half days a week. In 1747, therefore, 756 full-time workers, engaged in mining and smelting, were supported by 3,120 peasants who, within the framework of the prevailing labour legislation, spent on average one and a half days a week stockpiling and transporting raw materials.

In spite of this massive augmentation of the labour force, however, those employed remained insufficient in numbers to perform all the work of the enterprise.[179] The Demidovs, accordingly, were forced to pay over the odds for the labour they required.[180]

They offered an additional 2–4 kopecks a day over the official rate to peasants to induce them to undertake 'urgent mining and smelting work' and other tasks, thereby further enhancing the peasant's commitment to industrial work. As at Nerchinsk prior to 1720, therefore, in conditions of acute labour shortage, the burden of sustaining production at Kolyvan in 1745 was laid on the backs of the ascribed peasantry who became jacks-of-all-trades undertaking a volume and variety of tasks beyond the limits laid down by imperial edict and earning in the process not inconsiderable sums.

The Crown enterprise: administrative staff With the seizure of the Kolyvan works by the state in 1746/7 a new administration, the mining *nachal'stva* or *Oberbergamt*, was created directly accountable to the Mining College and subject to the administrative ordering of that body, officials henceforth being appointed to functionally differentiated posts in accordance with the official table of ranks (table 3.16).[181]

Table 3.16 Table of ranks of mining officers in the eighteenth century

Title[a]	Line officer's rank[b]	Administrative grade
Unter-Shichtmeister	Ensign	13–14
Schichtmeister	Lieutenant	
Berggeschvören	Lieutenant	12
Probirstchik	Lieutenant	
Hüttenverwalter	Captain	
Forster	Captain	10
Markscheider	Captain	
Bergmeister	Major	8
Oberhüttenverwalter	Major	
Oberbergmeister	Lieut-Colonel	7
Berghauptmann	Colonel	6
Oberghauptmann	General[c]	3–5

Notes: a On the duties of these officers see below. b As artillery or engineering officers each enjoyed one higher line rank. The officers at Nerchinsk had their rank doubled 'because of the distance and penury of their posting'. c Or equivalent civil rank.

The borders and internal ordering of the jurisdiction were first delineated in the original mining ordinances of 1747 and thereafter extended in 1759/60 and 1785 until, incorporating the districts of Kolyvan, Biisk, Kuznetsk, Semipalatinsk, and Krasnoyarsk, the mining jurisdiction covered 390,000 square versts – an area which was as large as such western European countries as England or the Netherlands.[182] This vast territory was administered from a central mining office, which was initially located at Kolyvan but in 1749 transferred to Barnaul, to which was subordinated six district offices.

In 1747, when the total administrative staff numbered no more than fourteen or fifteen officers, the authority of the central mining office, which extended beyond simple administrative duties to the exercise of full civil and criminal law, was represented at district level by a number of junior ranks.[183] First, there were the chiefs of mining operations (*Unter-* and *Oberschichtmeister*), six in number, the senior at Kolyvan and the remainder distributed amongst the other five district offices. These were in charge of all materials, were expected to verify the presence of all operatives at their place of work and to assess the quantities of ore raised a well as fixing wages, and to despatch at three-monthly intervals reports on mining operations to the central office. Second, there was a master smelter (*Hüttenverwalter*), represented only at Kolyvan and Barnaul works, who was in charge of all smelting operations in the district and who in collaboration with an ore-driver and assayer (*probirstchik*) supervised this activity and despatched reports on their work to the principal office of the mining region.

Here the chief administration was located under the supreme authority of a commander (*nachal'nik*), enjoying the rank of general or councillor, whose staff at Kolyvan comprised a chief director of mines (*Oberbergmeister*) with general technical and managerial responsibilities and a number of specialists. In relation to mining operations there was a director of mines (*Bergmeister*), and responsibility for smelting resided with a director of furnaces (*Oberhüttenverwalter*). Assisting each of these officers was a surveyor (*Markscheider*) whose duties involved prospecting, making technical drawings of mines and furnaces, and undertaking surveys of lands and forests. These then were the executive officers of the jurisdiction, supported by a small group of Chancery clerks, to whom the junior officers at district level were responsible.

In spite of the small numbers involved the creation of this skeletal administration had been no mean task. Supreme authority was vested in Beyer who was promoted from brigadier to general and who recruited the junior levels of his administration locally from amongst two groups. First, a lieutenant and five soldiers of the local garrison were promoted to the position of *Schichtmeister*, manning the district offices and assuming responsibility for local mine administration. Second, and far more important as a source of recruits, was the group of Saxon miners brought to the works by the Demidovs. These filled the ranks of junior officers in charge of smelting operations: Junghans and Schmit became chiefs of smelting (*Hüttenverwalter*) at Kolyvan and Barnaul respectively and Gillicher became head of the laboratory (*probirstchik*) at the former centre. One of their number even rose to supreme authority in mining matters becoming *Bergmeister*. His fellows in middle management were, however, all drawn from amongst the ranks of the Beyer commission: Bulyakov, the discoverer of the Schlangenberg, was raised to the rank of guards-captain and put in charge of the Surveyor's office (*Markscheider*), and Beyer's second-in-command, Ulrich, became head of smelting operations (*Oberhüttenverwalter*). The basic administration was thus recruited locally but was reinforced by incomers brought from Ekaterinburg. To assist Beyer, an experienced administrator, Poroshin, who had considerable knowledge of Swedish mining operations, was appointed chief director of mines (*Oberbergmeister*), assuming the role of heir apparent. With him from Ekaterinburg also came a number of cadets, who were assigned as trainees to the middle management (*Bergmeister*, *Markscheider*, and *Oberhüttenverwalter*). Thus was brought into existence the basic administration of the jurisdiction (*nachal'stva* or *Bergamt*).

Subsequent years saw it grow rapidly in size, the original fourteen or fifteen officers of the period 1747–61 thereafter increasing to thirty-eight in 1771, fifty-six in 1785, and eighty-five in 1795.[184] In part this was a result of a simple multiplication in the number of incumbents to existing posts. In 1771 the single *Schichtmeister* with his five *Unter-Schichtmeistern* of the 1740s had been replaced by four principal officers and fourteen subordinates and a similar process may be observed in smelting operations where there were nine *Hüttenverwalter* and two assayers. Throughout the administrative hierarchy increased manning was the order of the day, twenty-six officers occupying posts held by nine a quarter of a century before. Increases in the scale of operations, however, led not only to enhanced manning of existing posts but also to the creation of new ones as the administration became more elaborate and new functions were added. As senior management proliferated in numbers during the years 1747–71 the Chancery supporting it also expanded, some fifty-five clerks and copyists being employed at the latter date under the control of two secretaries and an office manager. Its function also changed. It ceased to be simply an office of record, receiving from the district offices and storing the accounts recording the operations at each mine and smelting establishment. By the 1770s it was also handling a large volume of documentation generated by the executive action of the chief administration. At that time the administration was no longer dependent upon information from subordinate officers for the running of the concern. It had created its own team of field officers (*Berggeschvören*) who transmitted periodic reports on the state of the enterprise based upon their personal inspections rather than upon the testimony of subordinates. The administration had thus not only become larger but also more elaborate.

It had also assumed responsibility for the provision of a new services, schools for the education of master workers' and soldiers' children, a fire service, and a hospital service. In 1771 these services were as yet embryonic, the schools employing only four teachers and the fire service being only the part-time responsibility of one of the *Hüttenverwalter*, one Captain Polsonov, whilst the hospital at Barnaul employed, apart from a few paramedics, only two surgeons and an apothecary. The pattern for the future, however, was clear.

A quarter of a century later, in 1795, all aspects of administration had grown enormously.[185] The eighty-five staff-officers at that time were supported by a Chancery employing ninety-three staff. Service provision was also greatly enlarged. In the provision of medical services alone each minor establishment, like the Salairsk mine, had its team of non-commissioned paramedics, and the larger workings, like the Schlangenberg, were equipped with a proper hospital employing eighteen staff surgeons and orderlies, the total number of officers employed in the medical service rising to thirty-six. So in the half-century since the creation of the new mining jurisdiction in 1747 the rapidly increasing number of mining officers was supported by a burgeoning bureaucracy: 138 persons in 1771, 258 in 1785, and 350 in 1795.

The recruitment mechanisms established in 1747 proved perfectly adequate in supplying numbers of suitably-trained personnel to fill the ranks of the officer elite. At steadily diminishing intervals surviving officers were promoted to senior posts in their own or other enterprises. Their posts were then filled by newcomers to the officer elite. Junior officers continued to be recruited from amongst the ranks of master workers. Mining and smelting people, like the soldier's and master's sons I.I. Polzynov and K.D. Frolov who began work as miners in 1742 and 1744, were raised to the rank of *Schichtmeister* after some twenty years or so. For their counterparts in the 1790s their stay in the ranks before promotion to the officer class was only about ten years.[186] Recruits to senior and middle management were brought in from outside. In 1761, on the occasion of the first round of promotions, twelve cadets from the Land and Sea Corps and from Moscow University were despatched to Kolyvan and apprenticed to the senior mine officials in preparation for promotion to the officer elite, and thereafter the number of such apprentices steadily increased. In 1771 there were thirty allocated in twos and threes to the senior officials. In 1795 there were 165.[187] The educational institutions of European Russia thus poured forth a steady stream of young hopefuls for training at Kolyvan before they entered the promotional ladder of the mining Corps which, at intervals shortening from *c.* twenty to *c.* ten years during the period 1746–

1806, steadily elevated the successful in rank. By a policy of promotion from the ranks of the master workers or recruitment from the educational institutions of European Russia numbers entering the officer corps of the Kolyvan *Bergamt* thus steadily increased over the years 1747–95, creating a rapidly expanding pool from which the expanding numbers required for the administration of the enterprise could be drawn through a process of promotion which took place at steadily shortening intervals.

The Crown enterprise: master workers   Subordinated to, and a major source of recruits for the lower echelons of the officer corps, were the master workers – free workers or peasants attached to the works in perpetuity by the ordinance of 1736. They were maintained at the cost of the management, being paid a salary (in accordance with the wage schedule of 1723 revised in 1737, 1769, and 1783) from which was deducted the cost of provisions drawn from the company store, and had their own hierarchical ordering of ranks and order of advancement similar to that of non-commissioned officers in the army. Discipline, wages, and punishment, indeed, were all very nearly military. An industrious group, they each also possessed, however, a house and small vegetable garden and kept livestock which was in-wintered, supplementing family labour in the management of this holding with the time made available to them by the numerous holidays of the Russian calendar when there was no public work.

Table 3.17[188]   Wages of master workers, 1723–96

Category	Wage (roubles per annum)		
	1723–68	1769–83	1783–96
Master worker[a]	30	36	32–40
Under-master	15	24	24
Qualified worker	12	15	22
Apprentice and supervisee	6–12	n.a.	18–20

Note: a   Miners grade 1–4, grade 1 = colour sergeant of engineers.

In 1747 their numbers, amounting to some 736 in all, had been insufficient to undertake all mining and smelting activity, emergency cover being provided by members of the ascribed peasantry who were paid 2–5 kopecks over the official rate to undertake the work. With the mining crisis of 1751, however, this situation was transformed as a slightly enhanced labour force, augmented by fifty miners brought in from Olonets, completely displaced the part-time work of the peasantry; this secured for the master workers an exclusive domain in the fields of smelting and mining activity. Nor did this situation change as the mining boom of the years 1758–71 began and an alternating pattern of labour flows between the ranks of the master workers and ascribed peasantry was superimposed upon patterns of recruitment in the industry. In spite of an only slightly rising level of nominal and real wages, the pious wishes, expressed in the mining ordinances of 1761, to recruit additional master workers were more than amply fulfilled, enhancing their numbers to 4,704 in 1771 or almost six times those of a decade earlier. From a vast pool of ascribed peasantry attached to the works, who in spite of the mining boom could not be employed in sufficient numbers for all to be allotted work to meet their fiscal obligations, recruits flowed thick and fast into the ranks of the master workers, creating by 1771 a major element in the labour force and with their dependents a major element in the population of the Altai (table 3.18).[189]

Table 3.18 Master workers at the Kolyvan *Bergamt*, 1771

Occupation	Number	Occupation	Number
Administrative personnel		*Metallurgical workers*	
Officers' apprentices	30	(a) *Iron and steel*	
Chancery clerks and copyists	58	Furnacemen	3
Draftsmen	3	Steelmakers	8
Barber-surgeons	4	Mouldmakers	3
Schoolmasters	4	Wiredrawers	2
Others	28	(b) *Silver and copper*	
Craftsmen		Crushers, foremen/masters	19
Glassmakers	10	Smelters	391
Smiths	23	Roasters	24
Carpenters and bellows makers	54	Hammersmiths	34
Ropemakers	6	(c) *Mint*	
Mineworkers		Minters	42
Deputy directors	14	Assistants	72
Deputy foremen	22	Charcoal makers	36
Miners	2,915	Dam builders	4
Children, at work/in school	566		
Washers	257		
Stable hands	72	Total	4,704

In 1771 amongst these 4,704 workers by far the largest element (two-thirds) were mineworkers, and amongst the miners by far the largest number worked in one group of workings – the Schlangenberg or Zmeinogorsk mine – which in that year engrossed almost three-quarters of the total mine labour force.[190] In the short space of a decade the meteoric rise in production on the mountain had brought into existence a formidable production complex and mining community containing 362 houses on the banks of the Smeevska brook for its 2,449 inhabitants. By 1771 this was no mean settlement. Apart from the houses laid out along the road, it also contained an administrative complex, with the mine office and surveyor's office, a laboratory, hospital, school, specie store and provisions store, guard-house, church and watchtower, and the mighty Smeevskaya fortress – an earthwork and ditch structure created in 1759 – which was almost a mile in circumference and was garrisoned by a company of the Barnaul mining battalion equipped with twelve cannon. In comparison the other mining camps, containing collectively about 1,000 workers, were tiny.

The settlements associated with the smelting works, though containing a smaller number of master workers than the residual mineworkers, were, however, substantial communities.[191] Even Kolyvan, though smelting activity there had decayed due to a shortage of timber, was a substantial village (*sloboda*) which, with the smelting plant on the banks of the Belaya, was surrounded by an earth wall and ditch, and contained a church, materials store, and 127 rough-hewn wooden houses for the 963 inhabitants. The denizens here were, however, all peasants, who apart from cultivating their vegetable gardens and raising a few stock earned their living by ore-carrying. Elsewhere, master workers were more prominent in the smelting communities (Pavlovsk, Susunsk, and Barnaul), which were located 30–80 miles from each other on tributaries of the Ob, and which now dominated the processing of the ore raised in the enterprise. Amongst these by far the largest was Barnaul, which by 1771 had assumed the status of mine town (*Bergstadt*). It was a settlement of no mean dimensions. Dominated by a palisaded fortress (*kreml' or Hüttenhof*), located by the side of the stream in the town and containing a guardhouse, school, and magazine in a palisade as well as the smelting plant, the town proper was both elaborate and extensive. Its population, some 5,448 in all, was large and all were well provisioned with both a regular supply of cheap consumables and a wide range of exotic wares brought from the Irbits fair and sold at a mark up of one-third (table 3.19).[192]

This single community, which encompassed half of all the master workers outside the mining sector, was thus not only the principal administrative, manufacturing, and smelting centre of the *nachal'stva* but also the principal distribution centre for local and foreign wares in the region. In comparison the other smelting works supported only small communities. Pavlovsk had, besides the *Hüttenhof*, a large village (*sloboda*) with a fine wooden church, mining office, and market hall containing stalls for the merchants to display their wares as well as 264 houses laid out in straight rows to accommodate the 1,900 inhabitants, including the 160 master workers employed at the plant. Of similar size was the large village at Verkne–Susunsk of some 2,285 persons, including 360 master workers employed at the works and mint, provided with a church for their enlightenment and 308 houses for their accommodation. Much smaller was the settlement near Tomsk where the sixteen master workers employed at the ironworks lived with their dependents amongst some seventy peasants in the small village or *dorf*.

Within the space of little more than a decade a large group of master workers, numbering 4,704 persons, had been recruited largely from amongst the ranks of the ascribed peasantry. Two-thirds were employed in the mines, particularly the mighty Schlangenberg, where a large community

Table 3.19 Goods sold within the mining jurisdiction of Kolyvan–Voskressensk in 1771

Commodity		Price
Provisions		
Barley grits	(pud)	15–25k
Wheatmeal	(pud)	20–25k
Malt	(pud)	25k
Peas	(pud)	10–15k
Onions	(pud)	15–25k
Melons	(each)	2–5k
Beef	(pud)	23–30k
Mutton	(pud)	60–70k
Butter	(funt)	3–4k
Tallow	(pud)	60k
Fertilizer		
Bonemeal	(pud)	9–10k
Livestock		
Horse	(each)	4–10r
Calf (4–5 weeks)	(each)	30–40k
Cow	(each)	3–4r
Hens	(pair)	6–10k
Clothing		
Shoes	(pair)	30–50k
Foreign goods		
Chinese silks		'are cheap'
Woollen cloth	(ell)	3–5r
White silk stockings	(pair)	3r
Writing paper	(quire)	1.60r
French wine	(maas)	80k
French brandy	(maas)	3r
Raisins	(funt)	12k
Rice	(funt)	12–15k
Coffee	(funt)	40k
Sugar	(funt)	30–35k
Tea	(funt)	1.20r
Almonds	(funt)	15–20k

grew up mainly of single men, trainees, and qualified workers earning 12 roubles a year, who now outnumbered the married old-timers, the masters and under-masters earning 15–30 roubles a year.[193] Thus at one focal point of those interlocking, bilateral systems of silver and copper production there emerged a large community of miners – the Schlangenberg mining camp – which like its tiny satellites overwhelmingly comprised master workers. In contrast at the opposite node of the systems, in the smelting communities (Barnaul, Pavlovsk, and Susunsk) far to the north-west, the numbers of master workers were much smaller. In these townships, whose populations numbered 2–5,000 persons, they and their families rarely comprised more than 20 per cent of the total. Yet whether resident in the mining camps, whose populations were made up overwhelmingly of master workers, or in the smelting towns and villages, where they formed a minority group in a predominantly peasant population, all were engrossed within a distribution system organized from Barnaul which, by allowing them to augment the produce of their holdings at low cost, ensured a remarkably rich lifestyle. By expending no more than 30–50 per cent of their salaries all the food they required was theirs, and with the residual 10 roubles left at their disposal they could buy a wide variety of Chinese and European wares brought each year from the Irbits fair to the works. A remarkably rich and numerically large group of workers had thus before 1771 emerged within the jurisdiction.

Its rapid growth in numbers had been largely a result of a massive inflow of recruits from the ranks of the ascribed peasantry, however, and with Catherine II's reforms of 1779, which led to a marked improvement in the condition of this latter group, recruitment to the ranks of the master workers became a much more difficult matter. Numbers henceforth grew slowly, rising only by some 14 per cent between 1771 and 1785 to c. 5,500.[194] The source of this growth, moreover, was quite different from that of the 1760s for, as one contemporary explained, 'as the work is very prejudicial to the health and likely to shorten life one is barely able to replace losses' by recruitment, 'the children of the miners now providing true augmentation'. Others agreed. As one remarked, 'the main source of workers is from amongst the children'.[195] Over these years, 1771–85, children became a more and more important element in the ranks of the master workers, after undergoing free schooling, from the age of 10 (subsequently lowered to 6), sorting and washing ore and then entering the ranks of the master workers at the age of 14. In 1771 they had numbered 554 (366 at school and 188 at work) against an adult labour force of 4,083.[196] Now in 1785 there were 1,029 against 4,186 adult workers.[197] Truly, as the number of peasant recruits to the ranks of the master workers declined and the proportion of married men therein increased, the latter's offspring became the principal source of augmentation of a labour force which was undergoing a major structural transformation.

Inevitably this phase could not last, for as the population which had given birth to the child-workers of the 1780s aged so its reproductive capabilities diminished, and the numbers of child-workers entering the labour force in the 1790s fell to a mere 161. But now the children of the previous decade had entered the ranks of the master workers and some at least were having children of their own who, though too young (less than 6 years of age) to enter the labour force, were poised to form the next generation of workers.[198] Even as the eighteenth century closed, judging by the experience of the Salairsk labour force, the pattern for the future was becoming evident as self-reproduction became the dominant characteristic of the master worker group. In 1795 at that mine the cohorts of workers recruited from amongst the peasantry prior to 1779 who were now aged 34 and more still comprised a significant element (23 per cent) of the labour force, although now more than a quarter of their number were classified as invalids and assigned only light work. Most of the workers, however, had been recruited since Catherine II's reform and in this group recruits from amongst the peasantry were of steadily declining importance. In the cohort entering the ranks of the master workers during the years 1775–84 they still remained dominant, but in the next cohort of 1785–94 they comprised only 62 per cent of the total, the remaining 38 per cent being masters' and soldiers' sons born after 1771 who, with the current generation born between 1780 and 1789, now comprised 25 per cent of the total labour force. Already in the workforce of the late 1790s members born of that group comprised 25 per cent of the whole and thereafter this proportion steadily increased to 40 per cent in 1820/31 and 62 per cent in 1841.[199]

There were not only changes in the patterns of recruitment to the group of master workers, however, but also in their distribution amongst the various activities at the works. As numbers continued to grow – from 5,215 in 1785 to 7,453 in 1795 – more and more workers were deployed to the non-mining sectors of the enterprise, numbers engaged in work at the mines falling from 3,217 (or two-thirds of the total) in 1771 to c. 2,700 (or c. a third) in 1795, the numbers employed in administration and work at the proliferating number of smelting plants correspondingly increasing.[200] The years 1771–95 had thus seen major changes in the structure of the master-worker group as it began to become self-reproducing and was deployed away from mining activity, changes which, like those of the years 1747–71, can only be understood in the context of parallel changes in the force of ascribed peasants.

The Crown enterprise: ascribed peasants   As has already been indicated, some 3,121 peasants in Kuznetsk and Tomsk districts had already been attached to the enterprise under the Demidovs. Now, with its transfer to the Crown in 1746/7, their numbers were rapidly augmented until in 1759/ 61 the entire peasant populations of the two districts were encompassed within a remodelled and territorially extended jurisdiction.

The initial plan for peasant attachment under the new regime was outlined in the original mine ordinance of 1747, when numbers required were set at 10,935, to be drawn from Beloyarsk, Malyshev, Biisk, and Berdsk – the sources in Kuznetsk district of the original attachments of 1740–2. Implementation of this scheme was not as easy as the planners envisaged, however, and it was to be another decade before the manpower quotas were fully assigned (table 3.20).[201]

Reserves of manpower within the original settlements of assignment had already in 1751 been exhausted and henceforth the growth of the labour force depended on natural increase, leaving the management with a 3,073

Table 3.20   Assignment of peasants to the Kolyvan–Voskressensk enterprise, 1751–7

Settlements	1751	1757
Beloyarsk *sloboda* and adjacent villages	2,119	2,302
Berdsk *ostrog*	3,068	3,115
Malyshev *sloboda*	1,300	1,900
Biisk *krepost*	242	545
Sosonov *ostrog*	418	430
Chaus *ostrog*	333	327
Subtotal	7,480	8,619
Kolyvan works with Charish cossack squadron		360
Kolyvan works, newcomers settled thereabouts		110
Settlements about Kolyvan and Barnaul		667
Barnaul works		94
Shul'binsk works		94
Subtotal		1,325
Total		9,944

shortfall to its requirements in 1757. Quotas were accordingly only fulfilled by an extension of the net into Tomsk district, where a further 757 'souls' were attached in Sosonov and Chaus *ostrogs*, and by the incorporation into the ranks of the assigned peasantry of the cossacks at Kolyvan and of the newcomers who in the 1750s began to settle about the works. Within the original area of assignment the population was fully engaged by 1757, comprising some 14,306 'souls' of whom 7,859 were said to be labouring each year at the works.[202] Thus when, against the background of the contemporary mining boom, it was decided to further enlarge the numbers of ascribed peasants it proved necessary to cast the net even wider. By a Senate decree of the 22 July 1759, accordingly, *all* the peasantry of Tomsk and Kuznetsk districts were attached to the works, bringing a further 23,758 male 'souls' into the fold, enhancing total numbers at the time of the 'Third Revision' in 1763 to 40,008, and closing the frontier of attachment.[203]

Henceforth the net could be cast no further and the subsequent increase in the number of male 'revision souls' attached to the works (excluding master workers) was slow, rising from 38,064 in 1763 to 54,750 in 1782 and 63,093 in 1796. As a result of emancipation and non-incorporation of newcomers into their ranks, moreover, as the century drew to its close the proportion of peasants attached to the works in the total Altai population had already begun to decline (table 3.21).[204]

Even as the mining boom of the years 1757/8–82 began to gain momentum, therefore, from 1761/3 the possibility of making new attachments to the works came to an end; whilst between 1761/3 and 1782 production increased 2.8-fold the number of ascribed peasants attached to the works increased only 1.4-fold. Henceforth, accordingly, instead of attaching more peasants to the works, as had been the case from 1740 to 1761, the existing ones, from 1761 to 1779/83, were made to work harder, thereby providing all of the manpower required by the industry as it rose to a position of supremacy amongst Old World producers.

This was achieved in the Altai not by adjustments to the peasants' rewards and obligations, a major feature of labour intensification in the Urals, but by a process of alterations in task assignments. Indeed, an examination of the former aspects of the official terms of ascription would suggest that the Kolyvan peasants were not only lightly burdened but that even amongst the peasantry attached to metallurgical works they were a particularly favoured group (table 3.22).[205]

Under the Demidovs, as the inquiries of the Beyer commission revealed, the ascribed peasantry of the Altai paid a slightly lower capitation tax and obrok than their counterparts in the Urals but were rewarded in the same measure; their prescribed labour was remunerated in accordance with the pay schedule of 1723, their non-obligatory labour, which in both regions was much in demand because of acute shortages of labour, being paid for at even higher rates than in the Urals.[206] Nor did this situation change with their transfer to the Crown in 1746/7. Even though the enhancement in the capitation tax and obrok in 1763 resulted in labour intensification in both regions, allowing the abandonment of the use of non-obligatory labour, the impact of this new imposition was very different in the Urals and the Altai. In the former district it resulted in a 42 per cent increase in labour intensity. In the latter only those involved in woodcutting and coaling were affected and even then, due to a shift to payments at summer rates, the increase was limited to 24 per cent. In the case of carriers there was no increase at all; indeed in this case the number of 'official' man-days they were forced to undertake actually fell by 8 per cent.[207] Subsequently, moreover, as labour intensity in the Urals continued to increase, until in

Table 3.21 Attached peasantry at Kolyvan, 1782–96/7

Okrug (uezd)	1782 attached	1796/7 attached	1796/7 unattached
Kolyvan gubernia			
Kolyvan	16,936	19,763	802
Biisk	15,188	18,025	1,406
Kuznetsk	10,287	12,136	575
Semipalatinsk	1,369	1,560	4,865
Krasnoyarsk	295		
Total	44,075	51,484	7,652
Tomsk oblast			
Tomsk	8,672		
Kainsk	1,852	n.a.	n.a.
Achinsk	151		
Total	10,675	11,609[a]	n.a.
Total Kolyvan–Tomsk	54,750	63,093	n.a.

Note: a Of whom 7,787 peasants were freed from factory work in Tomsk, Kainsk and Achinsk by an edict of 3 March 1797 (*PSZ*, 1st series, XXIV, no. 17862) leaving 3,822 attached workers in the oblast.

Table 3.22 Rewards and obligations of the ascribed peasantry: the Urals and the Altai, 1724–96

Category of peasants	wage rate (kopecks per day)				
	1724–62	1763–68	1769–78	1779–83	1784–96
Urals					
Peasant with horse					
Winter	6	6	8	12	12
Summer	10	10	12	20	20
Peasant without horse					
Winter	4	4	5	8	8
Summer	5	5	6	10	10
Altai					
Peasant with horse					
Winter	6*	6	6	6	6
Summer	10	10	10	10	10
Peasant without horse					
Winter	4*	4	4	4*	4*
Summer	5	5*	5*	5	5
Region	Capitation tax and obrok (roubles)				
Urals	1.20	1.70	2.70	2.70	3.70
Altai	1.10	1.70	1.70	1.70	1.70

Notes: Asterisked figures indicate the rates actually paid as reported by contemporary writers. In the case of the capitation tax and obrok paid in 1763–79 in the Altai this was raised to 2.70 roubles but some sixteen years earlier than in the Urals only 1.70 roubles had to be rendered in labour at official rates. Finally, on the post-1799 day-rate for woodcutting see table 3.23 below.

the period 1769–78 it was 70–88 per cent higher than in 1740–62, in the Altai the peasantry continued, in terms of their rewards and obligations, as before. At a time (1762–79) when there was a massive increase in the demand for labour and when the Altai management dispensed with the use of high-cost, non-obligatory labour, therefore, the peasantry of the region at best were obliged to increase the number of 'official' man-days they worked per year by some 24 per cent and in the case of carriers actually worked less 'official' man-days than before, seemingly, relative to their Urals counterparts, the Altai peasantry were a very privileged group indeed.

The reality was very different, however, for in calculating their workload not only was the 'official' day-rate taken into account but also the task specification of the work in hand which was also laid down in the articles of ascription. In the Urals changes in task specification had little or no effect on the levels of 'real' wages or labour intensity – at least in the state works. Task rates, set for woodcutting, coaling, or carrying were continually adjusted in line with official day-wage rates, thereby remaining in conformity with the task specifications laid down in 1737.[208] Throughout the period under consideration men cutting wood were paid at a rate commensurate with their ability to fell one 'cubic sazhen' of 9.5 chetverts in six days (= 1.3 cubic sazhen of 6 chetverts in five days).[209] Similarly in the making of a standard 20-cubic-sazhen charcoal bed or *kucha* the task specification first set down in 1737 remained the norm at sixty-eight man-days throughout the period under consideration as did the task specification for carrying work of 30 versts per load (of 250 puds) per day.[210] With such constant work assignments labour intensity thus varied with changes in the peasants' rewards and obligations. Such was not the case in the Altai. Here the task to be undertaken per 'official' man-day was steadily enhanced in the years to 1779.

In the case of woodcutting the official production norms were set at the rate of 1 (actual) cubic sazhen to be cut in six days at a rate of 25 kopecks per sazhen or about 4 kopecks a day, the peasant fulfilling his tax obligations in twenty-six 'official' man-days in the period 1740–62. Actual work assigned, however, differed greatly from the official norm (table 3.23).[211] As the actual cubic sazhen remained an administrative and accounting fiction, even in the period 1740–62 when the 9.5-chetvert cubic sazhen was in use, workloads were some two and a half times above the 'official' rate, the peasant working slightly more than two months a year cutting wood or twice as hard as his counterpart in the Urals.[212] Nor did this differential significantly change as the latter's labour increased during the years 1763–78. By introducing the 11-chetvert measure as its work standard the Kolyvan management yet again increased labour intensity at a time when it was already rising, by some 24 per cent, as a result of fiscal pressures, so that the peasant during these years, 1763–78, increased his workload by c. 44 per cent, closing the differential slightly in relation to his Urals counterpart but still working almost twice as hard at some 100 days a year.

Table 3.23 Woodcutting in the Altai, 1740–96

Cutting rate[a]	Payment	Capitation tax
Official 1 sazhen³ = 6 days	25k = 4k/day	1.10r = 26 days
1740–62 2.5 sazhen³ = 15 days	25k = 1.6k/day	1.10r = 68 days
1762–78 2.9 sazhen³ = 17.5 days	30k = 1.7k/day	1.70r = 100 days
1779–96 1.8 sazhen³ = 11 days	44k = 4k/day	1.70r = 43 days
	44k = 6k/day	1.70r = 28 days[b]

Notes: a Whilst the 1 actual cubic sazhen was used for accounting purposes throughout the period under consideration the actual amount of wood cut in the period 1740–62 was the 9.5-chetvert cubic sazhen (= 2.5-actual cubic sazhen); 1763–78 the 11-chetvert cubic sazhen (= 2.9-actual cubic sazhen). b From 1779 official calculations were made on the basis of an enhanced daily productivity involving the cutting of a 6-chetvert cubic sazhen of 1.8 actual cubic sazhen in 7.5 days at c. 6 kopecks a day.

Nor was this pattern of increased work-intensification through increased task assignments exceptional in the Altai. In the case of coaling there was a similar divergence between 'official' and actual production norms (table 3.24).[213] Official production norms, as has been shown, assigned some sixty-eight man-days for the making of charcoal beds or *kucha* which in the

Table 3.24 Coaling in the Altai, 1740–96

Coaling rate days per kucha		Payment kop/day roub/kucha	'Real wage' kop/day	Capitation tax days	roubles
Official	68	5 = 3.40	5	22	1.10
1740–62	24	5 = 1.20	1.7	66	1.10
1763–78	14	5 = 0.85	1.25	136	1.70
1778–96[a]	68	7 = 4.76	7	24	1.70

Note: a After 1778 this was a 'free-market' rate.

Urals was paid for at an official rate of 3.40 roubles or 5 kopecks a day. In the Altai on the other hand the time assigned for the task was considerably shortened. During the period 1740–62 the peasant there was allowed only twenty-four days at the official wage of 5 kopecks a day (24 × 5 kopecks = 1.20 roubles per kucha) to accomplish a task which took sixty-eight days, thereby reducing his 'real' wage to 1.7 kopecks a day and forcing him to work slightly more than two months a year to render his taxes. In the period 1763–78 the divergence became even more acute. Each peasant undertook to make two charcoal beds in lieu of his capitation tax and obrok, being allowed only fourteen days at an official day rate of 6 kopecks for the making of each bed, thereby reducing the 'real' day-rate to 1.25 kopecks and doubling his labour intensity to in excess of four months a year. As in the case of woodcutting so also with coaling the Altai peasant's involvement, already high in comparison with that of his Urals counterparts in the Demidov era, steadily increased until he committed three or four months' labour a year to this work.

Finally turning to the question of carrying services, once again in the Demidov era those peasants employed with their horses in the winter season carrying ore from the operational mines to Kolyvan (Bogoyavlensk and Kolyvan mine, 2.6 miles; Korbolicha and Zmeev, 24 miles; and Medvev and Ploskogorsk, 55 miles) were set assignments far greater than the official norms. In the period 1740–62 they were about 4 times greater, reducing the 'real' day wage to 1.5 kopecks and increasing the peasant's involvement to c. 73 days a year. Nor did this situation change in the subsequent period 1763–78 when, at prevailing 'real' day rates, fiscal pressures enhanced the peasants' involvement to c. 102 days a year. At this time, however, such peasant-workers played only a small role in ore-carrying, solely transporting minerals from the Schlangenberg and Semeonov mines to the new Aleisk smelter. As envisaged in the mine ordinances of 1747, most ore was now carried by peasants drawn from the Kolyvan, Biisk, and Semipalatinsk districts who each summer in late June and July shipped the minerals along the Alei some 130–185 miles north-west from the mines to the new smelting complex (Barnaul, Pavlovsk, Susunsk works) on the tributaries of the Ob. The impact of this change in transport mode was dramatic. The official day-rates rose, the peasants being paid at the summer tariff of 10 kopecks a day, but by using water transport their productivity was increased by 3.5 relative to transporting ore by pack-horse, and unit transport costs were greatly reduced. By a combination of organizational innovation and increased labour intensity, similar to that imposed upon the peasants involved in woodcutting and coaling, therefore, the management of the Altai works increased the volume of carrying.[214] The Altai peasantry, already working at twice the intensity of their counterparts in the Urals when in 1746/7 the enterprise was transferred from the Demidovs to the Crown, however, had to work harder to achieve this enhancement of output, in woodcutting, coaling, and carrying, their involvement in industrial work rising to three or four months a year.

During the mining boom of 1759–79, therefore, a process of extension of the network of attachment, bringing the Tomsk peasantry into the enterprise's labour force, increased the number of attached peasants 2.6-fold from 14,306 to 38,064. The act of ascription embodied in the edict of 1759, which attached all of the peasantry of Tomsk and Kuznetsk districts to the works, marked the end of the phase of extensive growth of the labour force, however, and henceforth the numbers of 'revision souls' attached to the enterprise grew slowly, by about 50 per cent between 1762 and 1779, largely as a result of natural increase. Far more important in realizing the labour potential of the peasant population during these years was an intensification of their labour as changes in task assignments resulted in a doubling of their involvement in industrial work. As far as the ascribed peasantry were concerned, however, such intensification was a double-edged weapon. As their 'real' earnings fell and those of the master workers increased, members of the administration attempting to recruit to the latter group experienced no difficulties at all. Peasants flocked to join the ranks of the master workers, and their numbers, which had been growing slowly up to 1762, thereafter grew about six-fold over the years 1763–79, thereby reducing the rate of increase in the number of ascribed peasants to about 20 per cent.[215] Whilst, during the boom years 1759–79, the number of master workers thus increased six-fold, largely as a result of recruitment from the peasantry after 1762, the number of ascribed peasants increased only about three-fold. Intensification of the latter's work through changes in their task assignments, however, ensured that the labour available from this group also increased six-fold, permitting a balanced growth within the labour force and a parallel six-fold increase in mine output.

The Altai mining boom of the years 1759–79 thus rested on an increasing exploitation of the peasantry. Those fleeing the increased workloads imposed upon them filled the ranks of the master workers pushing up the numbers six-fold. Those who remained contributed an equivalent six-fold increase in the amount of labour available for auxiliary work – cutting wood, coaling, and carrying.

In 1779 Catherine II brought this situation to an end. By her edict issued on 25 May in that year she attempted to regulate the condition of the ascribed peasantry 'of a manner just and conforming to humanity'.[216]

Henceforth the peasantry could not be employed in mining work or coaling, and in the tasks that remained to them, woodcutting and carrying, they were only allowed to work in accordance with fixed task assignments and wages which limited their labour to four weeks a year to obtain the money to pay their capitation tax, leaving the remaining time free for agricultural pursuits and domestic occupations. At one fell swoop labour intensity levels amongst the ascribed peasantry were reduced, in the Urals to half their former levels and in the Altai to a quarter or a fifth. The impact on the metallurgical enterprises of the Urals and Siberia was disastrous as managements cast around for new sources of labour. One contemporary, writing more than twenty years later, illustrates the confusion that reigned in the Urals as a result of the edict. Of labour market conditions in its aftermath he wrote:

> free workers were few in number and if they were employed in the mines the furnaces languished. However, in the copper and iron mines of the Urals they are usually employed in the transport of ore because the attached peasants are only numerous enough to cut wood and carry charcoal. As most crown peasants in the neighbourhood of the mine are attached it is nearly impossible for those without serfs to work a mine because it is very difficult, even if one pays a very considerable price, to have enough free workers and a great amount of capital is required, which is probably why so few mines have been opened since [the edict of 1779 and] the manifesto of 1782 where the privileges of the mines were extended and assured.[217]

The reverberations of the imperial edicts of 1779–82 thus echoed through the Urals works long after their promulgation. If conditions here were bad in the aftermath of the reforms, however, they were nothing short of catastrophic in the Altai where the peasantry's workloads had been double those of the Urals counterparts prior to 1779. Excluded from mine work and coaling, the Altai peasantry now worked solely at woodcutting and carrying and even in this restricted field of activity, under the new labour legislation, they were found numerically wanting. To meet the enterprise's growing requirements for cut wood which could be made into charcoal the number of ascribed peasants assigned to the task of woodcutting had to be increased massively (table 3.25).[218]

Table 3.25 Peasants engaged in woodcutting at Kolyvan–Voskressensk, 1771–96

	Number	Output (Actual sazhen^3)
1771	4,156	80–82,000
1785	15,882	108,000
1796	19,002	129,000

From constituting one-eighth of the total workforce before the reforms, those engaged in woodcutting increased thereafter to a quarter of the working population.[219] The remaining three-quarters in the post-reform years engaged in ore-carrying, forming a group whose numbers had similarly been augmented in order to compensate for the decline in labour intensity resulting from the edicts of 1779–82 (table 3.26).[220]

Table 3.26 Peasants and free workers engaged in ore-carrying at Kolyvan–Voskressensk, 1771–97

	Ascribed peasants	Free workers
1771	24,735	
1783		c. 13,157
1789	36,098	c. 8,864
1790–97	46,504	c. 5,534

In this instance, however, the numerical increase in the size of the work-group was totally inadequate, leaving the management with insufficient manpower to undertake the work required. As early as 1783 it was estimated that, with almost 2.5 million puds (c. 40,000 tons) of ore to transport, it would require, under the new labour legislation, an extra 74,000 peasants to undertake the work, enhancing total numbers involved in ore-carrying to 86,800 and the total wage bill to 147,554 roubles plus 61,989-roubles passage money, or approximately a five-fold increase over the outlays of the pre-reform years. In these circumstances, accordingly, the management resolved to enter the free market to secure the labour it required, finding that at a day-rate of 6.5 kopecks or a half-kopeck over the official rate it could secure from the available peasants, numbering 13,157, all the labour it required, each individual being willing to work some 185 days a year for a reward of 12 roubles (equivalent to the earnings of a grade-4 miner in the ranks of the master workers) and their collective wage bill amounting to 157,891 roubles or more than 50,000 roubles a year less than estimated costs.[221] At prevailing wage rates during the years 1779–83 the ascribed peasantry thus considerably improved their lot, earning a sum which after the payment of taxes left them with as much spare cash as the master workers and working only three-quarters of the time expended by the latter group. Management similarly had benefited as it had 'emancipated' those ascribed peasants engaged in ore-carrying from the shackles of Catherine II's edicts, obtaining all the labour it required at below statutory cost. As wages rose, however, increasing to 8 kopecks a day or 15 roubles a year in 1789, so the peasant priced himself out of the market causing the management to once again enforce its statutory rights in relation to the peasant's labour.[222] From 1783–97 the numbers of peasants employed in ore-carrying under the terms of the imperial edicts of 1779–82 steadily increased until during the 1790s they numbered 46,500. To maintain ore-carrying capacity at the levels first set in 1783, however, it still remained necessary to employ 5,534 peasants in the capacity of 'free workers'. When fully engaged, therefore, engrossing three-quarters of the total labour force, the ascribed peasantry employed in ore-carrying were still, if deployed in accordance with the edicts of 1779–82, numerically insufficient to maintain existing production levels. Only by the enrolment of some 5,534 of their number in the ranks of the 'free workers', where they earned 12–15 roubles a year and contributed some 44 per cent of output, could the production targets be achieved.[223]

As in the Urals, therefore, Catherine II's labour legislation of 1779–82 wrought havoc in the Altai. As a result of her edicts the multi-functional group of ascribed peasants had been rationalized into a bi-functional grouping excluded from minework and coaling and engaged only in woodcutting and carrying services. Yet even in this restricted field of activity, at the low labour-intensity levels imposed by the new legislation, only in the area of woodcutting were there sufficient men on hand to undertake the work required. In ore-carrying, even to maintain existing levels of production, part of the available workforce had to be deployed in expensive 'free labour' and existing levels of transport services were by the 1790s insufficient to meet existing demand, particularly in relation to the new mining fields situated far to the south. In spite of a reorganization of the ascribed peasantry the Kolyvan management thus still found itself, in the aftermath of Catherine II's reforms, acutely short of labour. Nor in its hour of need could it turn to the master workers for succour. Enhanced 'real' earnings amongst the ascribed peasantry and 'free workers' ensured that few from this group, in the years after 1779, would seek employment in that quarter and, as has been shown, the numbers of master workers grew slowly during the last twenty years of the century and the proportion of this group engaged in mining activity steadily declined. If the growth of mine output was to be sustained during these years, therefore, a new source of labour had to be found.

The Crown enterprise: urochniki   For the Kolyvan management its most pressing problem, in the aftermath of the reforms of 1779–82, resided in the areas of proscribed peasant labour, coaling, and mine work, and during the years 1779/83–97 this work was taken up by a new group of workers recruited from outside the ranks of 'revision souls' attached to the works, possibly from amongst those colonists who since Catherine's reforms of 1763 had been flooding into the Altai. Eschewing the constraints imposed upon the ascribed peasantry by the new labour legislation or the customary practices which limited the effectiveness of the master workers, a new, high-wage group was created whose proclivities for a high level of leisure preference were curbed by the imposition of rigid work norms. These were the task-workers or *urochniki*. They were predominantly found engaged in mine work or coaling. In the former activity some 1,274 were employed in 1795, divided between the Schlangenberg and Salairsk mines (966 and 308 workers respectively), whose labour not only sustained the increased mine output at these workings during the 1790s but also encompassed that work in the washing places which had previously largely been the preserve of the ascribed peasantry.[224] In coaling they took over completely. Almost 1,000 – 957 – *urochniki* each burning three *kuchi* in 1796/7 were able to process all of the wood (48,000 9.5-chetvert cubic sazhens = 129,341 actual cubic sazhens) supplied them by the ascribed peasant woodcutters and convert it into the 134,400 korobi of charcoal required at the works at that time.[225] Elsewhere they were found only in small numbers, 104 cutting wood, 45 carrying it, and 36 engaged in ore-carrying. This latter group, in spite of its diminutive size, played a critical role in the provision of transport services, however, carrying some 18,000 puds (290 tons) a year from the newly opened Riddersk mines to the Aleisk or Loktevsk smelters.[226] The management had resolved its remaining labour problems by creating a high productivity group of workers but at a cost, for to attract recruits it had to offer more than twice the going day-rate. The charcoal-burners, though contracting to make three charcoal beds of *kuchi* a year, received 12 roubles for each or 36 roubles a year. Nor were such rewards exceptional. Woodcutters and carriers were paid 35–8 roubles a year and the diminutive group of ore carriers 40–50 roubles.[227]

Unlike in the Urals, the Kolyvan management had resolved the labour problems created by the imperial edicts of 1779–82 but only by a major reorganization of the labour force. In 1796, the ascribed peasantry was responsible for only 40 per cent of the work which before the reforms had been largely their preserve – woodcutting, coaling, and carrying – in spite of a 75 per cent increase in their numbers. The remaining 60 per cent was now undertaken by those members of the ascribed peasantry who had become 'free' workers and the new peasant recruits, from the more distant areas of Tomsk province, who enrolled as task-workers or *urochniki*. Similarly, in relation to the smelting and mining activities of the master workers, as the numbers of this group of workers grew slowly and the proportion involved in mining activity declined, the 20 per cent shortfall in the number of mine operatives was made good by the recruitment of *urochniki*. The enterprise's manpower problems had thus been resolved but only because of its capacity to absorb a four-fold increase in its wage bill – an option not open to the entrepreneurs of the Urals.

The Kolyvan labour force: an overview   In lands which had only recently been a largely uninhabited wilderness and in spite of ill-judged, if morally laudable, government intervention, successive managements at Kolyvan–Voskressensk had, during the years 1727–97, created a massive and complex labour force commensurate with the enterprise's position as the supreme Old World silver-production centre. During the pioneering phase in the enterprise's history, which may perhaps be dated to the years 1727–38, however, it would have been a far-sighted man who could have anticipated the momentous changes to come. After the arrival of the original party of fifty workers despatched from Ekaterinburg to construct the works, numbers employed there rapidly increased, rising to a level of 400–500 workers during the 1730s and creating a settlement of some 900 'souls'. Yet in 1738 this was merely one of a number of fortified townships created since the 1720s as a flood of migrants had poured into the Altai, pushing up the total population of Kuznetsk district from 2,878 'souls' in 1722–4 to 6,374 in 1738. The denizens of the Kolyvan works, like other inhabitants of Altai, moreover, enjoyed very little security in their position, maintaining only the most tenuous of holds over the lands they had settled. In 1732 they had been totally dispersed when the works had been fired and in 1738 virtually the entire working population and their families were forced to reside within the palisaded walls of the works and ventured forth only in the summer when they combined mining and ore-carrying with the necessity of fighting a running battle with Kirghiz and Dzunghir raiders, aided only by the hundred or so cossacks stationed at the works. Life, to say the least, was precarious at Kolyvan as elsewhere in the Altai during the years 1727–38 and the small labour force, comprising no more than 2.5 per cent of the Altai population (Tomsk and Kuznetsk districts), sustained only a low level of copper production.

Even as Gmelin visited the beleaguered mining community in 1738, however, he observed the first indications that changes were under way as peasants began to establish small settlements beyond the walls of the Kolyvan fortress on the banks of the Barnaul, Charish, and Loktevsk rivers, heralding the beginnings of a new age of expanding production and a rapidly growing labour force. From 1739 to 1779 production increased dramatically in two phases, during 1739–51, as silver production displaced copper, and then, after a short-term crisis in 1751–8, again from 1758–71, as the mighty Schlangenberg rose to its position of Old World supremacy, before falling yields undermined its position in the period 1772–9. The impact of this mining boom on the populations of the Altai was profound. Steadily, yet relentlessly, they were drawn into the labour force. Over the years 1738–58 the proportion of the Kuznetsk district population attached to the works steadily increased before the entire population, together with that of Tomsk district, were engrossed into a remodelled jurisdiction of some 40,000 'souls' in 1759/62. Thereafter, the labour force of the works grew in line with the total Altai population, rising to 67,900 'souls' in 1782/5. The years 1738–79 thus witnessed a complete attachment to the works of the local Altai population whose numbers steadily increased from c. 36,000 'souls' in 1738 to 67,900 in 1782/5, seemingly unaffected by those Tartar incursions, which had wrought such havoc in their ranks during the pioneering epoch.

This security, which formed the background to the rapid population growth of the years 1738–1782/5, was not, however, the result of any diminution in the intensity of nomadic incursions. Rather it rested on the Russian response to these incursions, symptomatic of which was the cossacks' demilitarization and assimilation into the ranks of the ascribed peasantry, and their displacement by a regular military force first formed in 1747 and subsequently (in 1761) reorganized into the Kolyvan Mining Battalion – a force of three companies (400 men) of foot soldiers and a squadron (100 men) of dragoons.[228] A new order was being imposed on the population of the Altai, this regularization of the military presence presaging a similar restructuring of civilian life. By the height of the mining boom in 1771 frontier life in the Altai was rapidly becoming a thing of the past. A large group (4,775) of master workers and officials and their families now dominated mining and smelting activity and constituted a major element in four communities – Barnaul, Pavlovsk, Susunsk, and the Schlangenberg mining township – which, with populations of 2–5,000 persons, were large by any contemporary Russian standard.[229] Nor were the inhabitants of these communities, who also included some 8,000 peasants attracted there since the mid-century, entirely lacking in the amenities of life in spite of the remoteness of their situation. The local administration provided schools for the education of their children, a hospital for their health care and at Barnaul even an embryonic fire service. It also ensured them a level of provisioning commensurate with their high earnings. 'Civilization' had come to the principal communities of the Altai and as the local provisioning system extended throughout the region its 'benefits' were extended to the ascribed peasantry as during the years 1739–79, a complex agrarian regime emerged in the mining jurisdiction.[230] In the central region of the Altai, embracing the Kolyvan–Korbolicha mountains and the plain extending north-west therefrom (districts 4–6, map 3.5) which contained all the major communities and formed the 'ancient' core of the enterprise, in terms of this agrarian regime distinctions between 'town' and 'village', ascribed peasant and master worker, had by the height of the mining boom in 1771 ceased to exist. Each individual, wherever situated and of whatever status, maintained his house with its enclosure which was cropped in the summer as a cabbage garden, cut burning and building wood, part of which was brought to the market, and kept cattle which were stalled in the winter. The population's main food was thus milk, butter, and garden produce which could be supplemented, as incomes allowed, by the purchase of provisions at the enterprise's warehouses which were well stocked with the produce of the other agricultural regions. The most important of these agricultural supply regions lay to the south-east amidst the Ulba–Aleisk and Brobovskii mountains (districts 1–2, map 3.5) which formed the high ground dividing the lands of the Kirghiz from the Kolyvan government. The peasantry here were particularly well endowed: their gardens yielded a wide variety of produce including gherkins and melons; their stock, which as further north was in-wintered, extended from the ubiquitous milch cow to horses and sheep; and their fields produced a variety of grains of which 114,277 puds (1,836 tons) were despatched along the great highway to the central region in 1779 and sold to some 31,846 persons, each buying on average 3.5 puds at prices which compared favourably with other places in west Siberia.[231] Along that same highway also passed large herds of cattle drawn from a third agricultural region, encompassing the Solotarsk range and extending into the great steppe lands below the Aleisk and Shul'binsk mountains (district 3, map 3.5), where garden fruit was plentiful, cattle herds magnificent, and the miners and landsmen lived in the greatest freedom. The whole of the Altai had thus been subjected to the 'civilizing' influence of the principal mining and smelting centres. A stable agricultural regime had been established throughout the region and frontier life was confined to the south-west, where the landsmen practised agriculture, maintained small kitchen-gardens, and engaged in cattle-rearing in a manner indistinguishable from the Kirghiz, and the miners, living in cabins of poles covered with turf in which they dug out a room and stove under the earth, enjoyed a reputation as good huntsmen, excellent riders, and (when necessary) excellent soldiers.

The corollary of this 'civilization' of the countryside and improvement in the material position of the ascribed peasantry during the years 1739–79, was, however, a loss in their freedom. As they were steadily engrossed within the jurisdiction of the mine authorities they were, in a situation of acute labour shortages, forced to commit twice the amount of time as their Urals counterparts to industrial work, and in the years after 1761/2 were subject to a massive increase in their workloads which resulted in a major transformation of the Altai labour force. As mine output rapidly increased during the years 1758–79 and their 'real' earnings fell whilst those of the master workers increased, the ascribed peasantry flocked in growing numbers to join the latter group, which increased some six-fold, forming that prosperous element in the larger Altai communities that has been described above. Those remaining in the ranks of the ascribed peasantry, a group which increased only three-fold, who sustained equally high incomes from the new agrarian regime described above, also provided a six-fold increase in their industrial work by an intensification of their labour effected by changes in their task assignments. They thereby permitted a balanced growth within the industrial labour force and a parallel six-fold increase in mine output. The imposition of an increasingly authoritarian regime upon the more settled and secure populations of the Altai during the years 1739–79 thus sustained not only a high level of industrial output but also a high level of agricultural production, so that even if the population lost its liberty as the frontier closed it did so in the process of becoming increasingly wealthy.

Then during the closing years of the century, as the government increasingly intervened in the affairs of the enterprise, the whole situation changed. Even before 1797, when it emancipated a number of the ascribed peasantry in Tomsk district from industrial work, the state began a process which, when reinforced by the non-attachment of new settlers to the region, would ultimately break the enterprise's monopoly over local labour supplies and reduce the peasants attached to the works to the position of a minority group in the population of the Altai. Long before this protracted process had undermined the position of the ascribed peasantry, however, their significance in the industrial labour force had been destroyed. In 1779, in an attempt to regulate their condition in 'a manner just and conforming to humanity', Catherine II had statutorily reduced their previously high levels of productivity, and in the process transformed the whole structure of the labour force. By diminishing their levels of labour intensity to a quarter their former size, and making it easier for them to render their taxes, she had completely stifled their migration into the ranks of the master workers. This had doomed that group to a slow phase of self-sustained growth, and had created acute labour shortages in those spheres of work which had previously been exclusively their preserve. Free workers and *urochniki* now displaced the attached peasantry, by 1796 undertaking almost two-thirds of the work which had been previously their exclusive preserve. Thus as the mining enterprise moved into a mature developmental stage in which its work force was slowly submerged in a sea of peasant migrants, the mainstay of that labour force, the ascribed peasantry, were rendered unimportant by government intervention in 1779 creating an acute short-term labour shortage. By the recruitment of new elements into the labour force the management resolved this problem created by Catherine II's reforms, but only by transforming the labour force at a cost of a four-fold increase in its wage bill.

The Kolyvan–Voskressensk enterprise: expenditure and receipts

For the first twenty years of its existence, from 1727 to 1747, whilst the enterprise remained under the control of the Demidovs, the primary operational objective was copper production. The main works at Kolyvan, for a total outlay of 34,930 roubles (including 8,000 roubles in capitation tax collected from the 6,318 'souls' inhabiting the district), in 1733–4 produced 94 tons of the metal which in a refined state was sold for 40,887 roubles and yielded a return on initial outlay of some 47 per cent, a figure which would have evoked envy amongst contemporary Urals producers confined as they were to the exploitation of low-grade gravels. These high, although notably unstable, returns from the Demidovs' enterprise in the period 1727–45, however, paled into insignificance next to the returns made by the Crown when in 1746/7 it took over the works and, on the basis of the newly discovered Schlangenberg ores, began silver production (table 3.27).[232]

Table 3.27 The Kolyvan–Voskressensk enterprise: expenditure and receipts 1733–99

	Receipts (roubles)	Expenditure (roubles)
1733–4[a]	20,443	13,465
1746	204,560	60,000
1747–65[b]	373,753	65,790
1766–76[c]	1,719,330	260,387
1783	847,602	407,495
1799	1,250,400	652,736

Notes: a Annual average, expenditures less capitation tax. b–c Annual average.

From 1746/9 ore production expanded rapidly, interrupted only during the years 1751–8 and 1779–84, until at the end of the century output attained almost 3 million puds (47,366 tons).[233] In the smelting of this ore the transition from copper to silver production, at the time of the state takeover, brought about a marked increase in returns and, at constant mineral yields, subsequent increases in ore production to 1771/2 resulted in a parallel increase in the output of the unseparated precious metal (blicksilver) until it reached 20,910 kg a year at the latter date (figures 3.6 and 3.7) yielding 19,460 kg of silver and 831 kg of gold on refining (figure 3.7) worth 1,628,551 roubles. In the short space of a quarter of a century, 1746/7–71/2, largely on the basis of the output of the Schlangenberg mine, revenues derived from the production of precious (silver and gold) and base (copper and zinc) metals increased 8.5-fold establishing the enterprise as the Old World's primary precious-metal-production centre. By a reorganization of the industrial production base with an associated concentration of mineral output in the one great mine and an intensification of the ascribed peasantry's labour, moreover, unit costs were halved, and the increase in total costs was limited to a rise of 4.3 times so that during the boom years returns over prime costs were of the order of 500–600 per cent and the enterprise contributed almost 1.5 million roubles a year to the Treasury – a not inconsiderable element (6–7 per cent) of the state's budgetary income.[234]

Subsequent years, however, saw the collapse of this mighty industrial empire. From 1771/2–83 mineral yields steadily declined so that when, after a short-term production crisis during the years 1779–83, ore production was restored to the levels of the early 1770s, silver production

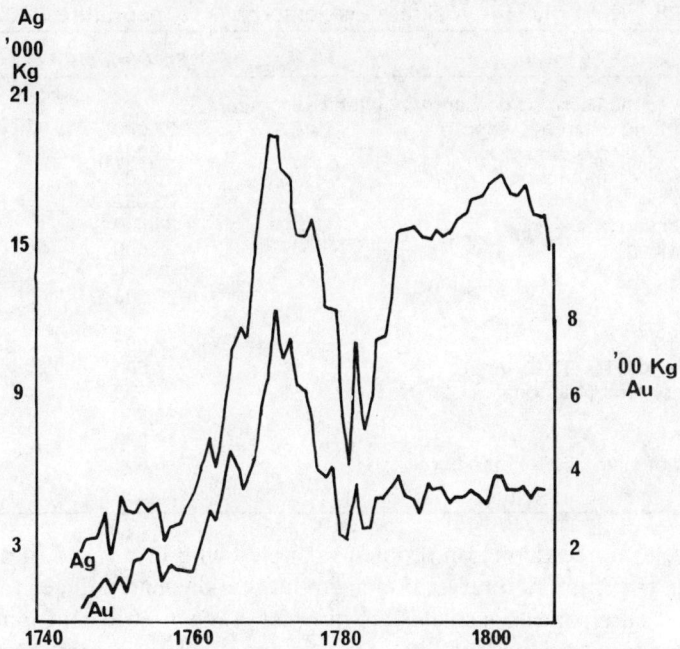

Figure 3.7   Kolyvan refined silver and gold production

and receipts underwent a disastrous (50 per cent) collapse, thereby doubling unit costs at constant input productivities and prices. Yet input productivities and prices did not remain constant, as these years saw the implementation of Catherine II's labour legislation of 1779–83 which, in reducing the labour intensity of the ascribed peasantry and raising wage rates, enhanced the wage bill for their work – woodcutting, coaling, and carrying – about four-fold, increasing their share of total costs from a fifth to a half, and raising costs per ton of ore produced and processed by some 60 per cent (table 3.28).[235]

By 1783 production was but a fraction of its former peak level and unit costs were more than three times higher than at that time, a significant element of the enhanced costs arising from a dispersal of funds to the peasantry which otherwise would have gone to the Treasury. The enterprise had not only lost its position amongst international specie producers but had also ceased to make a significant contribution to either the state's mineral revenues or its total budgetary income.[236] Nor did this situation change during the remaining years of the century. Locked within a continuing high-cost wage structure created by the labour legislation of 1779–83, as ore production doubled and yields once again declined there was a less than commensurate rise in revenues and a further enhancement in unit costs. By 1800 the golden days of the early 1770s were a distant

Table 3.28   The Kolyvan–Voskressensk enterprise: expenditure, 1783

Item	Cost (roubles)	
Payments to the ascribed peasantry and their successors		
Woodcutting, charcoal wood	27,000	
roasting wood	12,000	
mine wood	6,000	
Coaling	23,895	
Charcoal carrying	34,000	
Ore carrying	98,600	
Sub-total	201,495	49%
Salaries	125,000	31%
Materials		
Lighting	5,500	
Blasting powder (17.68 tons)	7,500	
Nerchinsk lead (402 tons)	32,500	
Various	30,500	
Sub-total	76,000	19%
Transport of silver to St Petersburg	5,000	
Total	407,495	

memory; resource-depletion problems coupled with ill-advised, if morally laudable, government intervention in labour markets had reduced the once important enterprise to a sorry state in which it made little contribution to either international production or government revenue, and prevailing high levels of output served merely to enrich an already prosperous peasantry.

Gold production in the eighteenth century

The history of independent gold production must perforce take the form of a postscript to this story of Russian precious-metal production in the eighteenth century, since most of the yellow metal produced at that time derived from the refining of auriferous silver produced at Nerchinsk and Kolyvan. Independent production, derived either from auriferous coppers (Voits and Shilovo–Issetsk/Chusovsk mines) or gold-bearing quartz (Beresov, Pishminsk, and Uktussk works) were of little significance in the eighteenth century (figure 3.8).[237] Indeed in the context of the subsequent rapid development of these fields in the nineteenth century the major question confronting the historian must inevitably be why there was so little development of these deposits in the eighteenth century.[238]

Auriferous copper production

Voits, near Olonets

Discovered in 1737 the mine was first exploited in 1742 for its violet pyritical coppers which were found from 1744 to contain auriferous quartz,

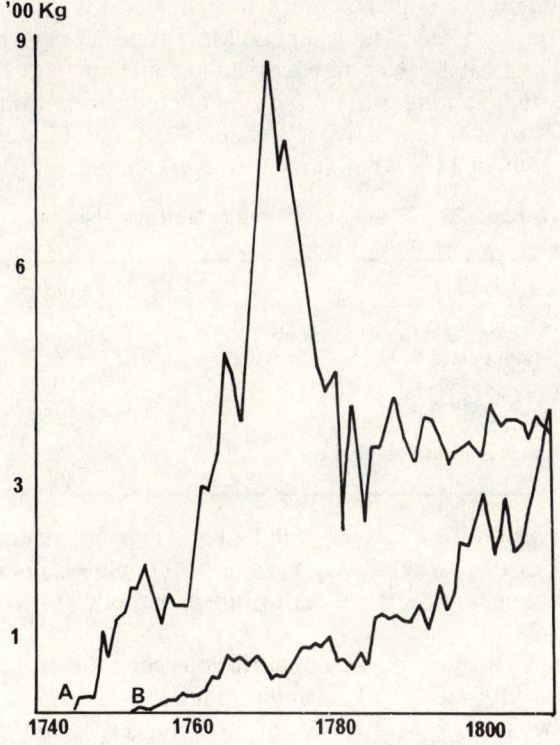

Figure 3.8 Russian gold production

A Kolyvan and Nerchinsk
B Ekaterinburg

thereby ushering in the beginnings of gold production there. This was an intermittent process during the years 1744–68, 1772–83, and, after the installation of a steam engine to drain the workings, finally in 1794 and yields were such that the enterprise consistently ran at a loss, expenses during the years 1744–76 exceeding receipts by some 80,800 roubles.[239]

The Shilovo–Issetsk mine and the Kriliatkov mine near Chusovsk, Perm Province

The situation was not significantly different in the case of the Issetsk and Kriliatkov mines exploited during the years 1745–56 and 1803–10 respectively. In the first instance production began after one Zakharii Shtor, a foreman at the neighbouring Sisertskii works, discovered the existence of gold in the copper ores in November 1745. Initial working revealed seemingly rich deposits, ores which yielded 1.4 zolotniks per ton; but these were soon eclipsed when in May–June 1746 a rich nest of nuggets, weighing 50 lb, was discovered. Production, accordingly, was placed on a regular footing, ore output steadily increasing from 134 tons a year in 1745–7 to 208 tons a year between 1748 and 1756 with a peak output of 386 tons in 1751 when forty men were employed (table 3.29).

Table 3.29 The labour force at Shilovo–Issetsk, 1751

Category	Number
Foreman and under-foreman	2
Miners class 1	5
class 2	19
class 3	7
Trainees	5
Carpenter and stable boy	2
Total	40

Yet initial expectations were not fulfilled and average yields were poor, the 2,156 tons of ore raised during the life of the mines (1745–56), apart from the occasional nugget, rendering up a pathetic 0.2 zolotniks per ton.[240]

Similarly the Chusovsk deposits of auriferous copper discovered in 1803 about 26 miles south-east of Ekaterinburg initially yielded 1–1.25 funts of gold per ton, because of the presence of small nuggets, again giving rise to high expectations, which, with the establishment of regular production amounting to 3,600 tons a year between 1803 and 1810, were not fulfilled. Average yields during these years amounted to no more than 0.1 zolotniks per ton.[241]

In the case of the auriferous copper ores, therefore, whether worked at Voits, in the region between Lake Onega and the White Sea, and Shilovo–Issetsk, in the Urals province of Perm, in the eighteenth century, or the Kriliatkov mine near Chusovsk in the nineteenth century, the reasons for the pattern of erratic and small scale production are rather obvious. Whilst such deposits might on occasion yield up occasional rich nuggets, average yields, if the mines were put into regular production, were pathetic, averaging only 0.1–0.2 zolotniks of gold per ton which with equally meagre copper yields rendered the workings uneconomic.

Auriferous quartz production

Beresov, Perm province (map 3.9)

Very different to the auriferous copper deposits were the gold-bearing quartz-deposits of the Urals which encompassed the Beresov mining

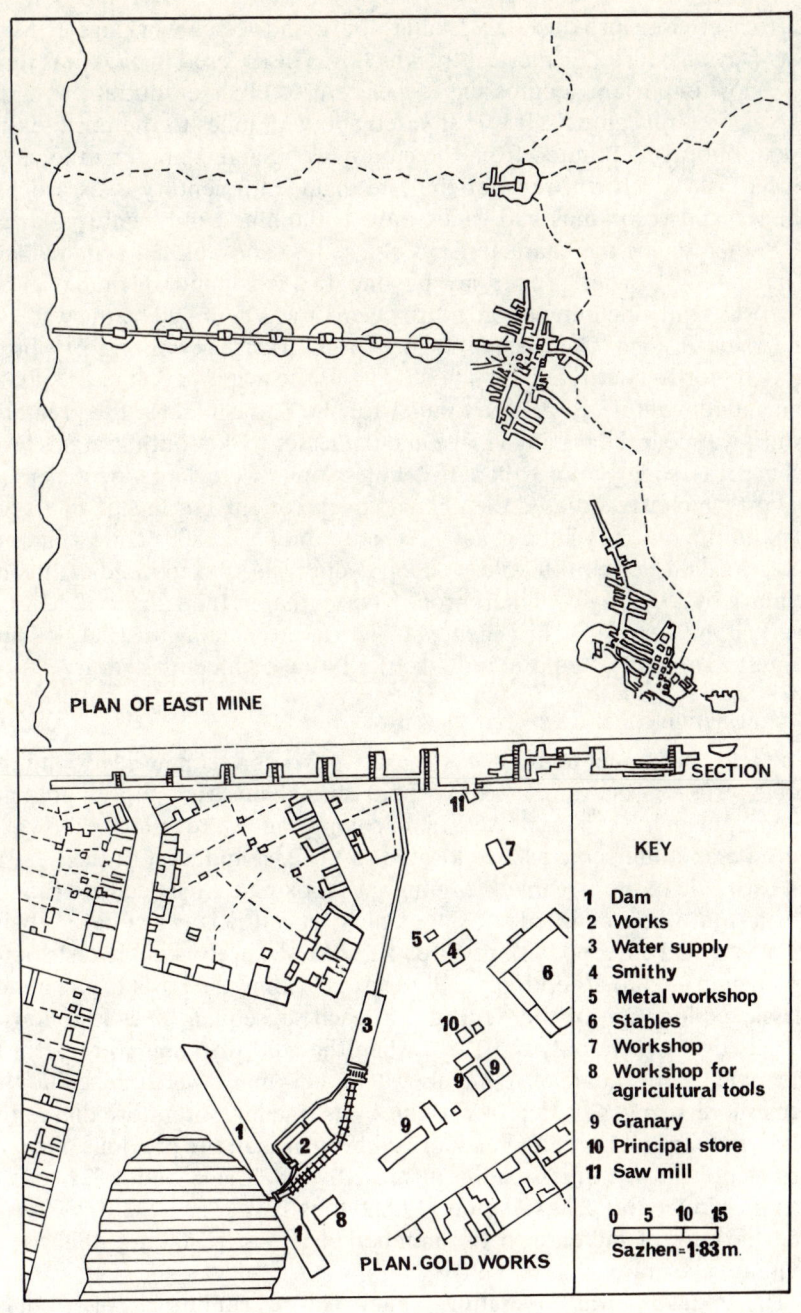

PLAN OF EAST MINE

SECTION

KEY

1 Dam
2 Works
3 Water supply
4 Smithy
5 Metal workshop
6 Stables
7 Workshop
8 Workshop for
agricultural tools
9 Granary
10 Principal store
11 Saw mill

0 5 10 15
Sazhen = 1·83 m.

PLAN. GOLD WORKS

Plan 3.9 Beresov mine

district in close proximity to Ekaterinburg and lesser workings at Miask and Neviansk. The former centre which was discovered in 1752 was by far the most important, engrossing 85 per cent of Urals output at the end of the eighteenth century. It was situated about 10 miles to the north-east of Ekaterinburg, separated from the city by woodland containing the village of Sartash which, throughout the late eighteenth century, was a den of thieves and vagabonds and which only as the nineteenth century dawned began to assume the characteristics of a settled and well-cultivated village. The mining district proper lay beyond the woodland, forming a level district slightly inclining to the north where it was bounded by the waters of the river Pishma. Across the span of this district a rather sorry brook flowed north–south fed by marshy pools, which was canalized in the nineteenth century to form an outlet for the Sartash lakes, the greater of which was near the mining village and the lesser to the south. On this level, which was slighly more than five miles long, were large peat deposits amidst which was situated the mining village, on the borders of an artifical tank in the middle of the canal, where up until the 1800s amidst the new houses could be seen the old wooden cabins raised in the mid-eighteenth century by a colony of miners from Klausenthal in the Harz who had first worked the deposits. This, then, was the Beresov mining district, Russia's largest autonomous gold mining district in the eighteenth century.[242]

Mining activity at Beresov

The basis of mining activity was a gold-bearing ore commonly found in a cubic form in a quartz matrix in a bold quartz vein, rich in both gold and copper, which was located in a decomposing white gneiss known as beresite and intersected slates located to the east and west of the beresite on each side of the stream.[243] Mining thus took place in a declivity running north–south between beds of slate which ranged northward before being displaced and covered with granite. The ore was pretty evenly distributed and penetrated to a depth of *c*. 100 feet, providing the basis for an almost classic exploitation of the deposits. In each subsequent decade following the discovery, save only the 1780s when the gold workings were beset by the same labour problems found elsewhere, new mines were opened up, but each successive generation of workings was smaller, with more dimunitive reserves of poorer ore, and of shorter life span than the previous ones. As production of ore thus steadily increased from 32,446 puds (522 tons) a year in 1754–6 to 1,231,480 puds (19,800 tons) a year in 1800–3, yields fell and the rate of increase in the number of mines, sustaining this rise in output, accelerated (table 3.30).[244]

The industry was operating under classic conditions of resource depletion but instead of production following a regular course of slowly diminishing returns the early nineteenth century saw a marked output discontinuity. Of total exploitable reserves amounting to 658,633 tons only

Table 3.30 Mining operations at Beresov, 1740–1800: discovery of new deposits

	Number of mines discovered	Average life of new mines (years)	Average annual output and yield of ore during operational life	
			Yield (oz/ton)	Output (tons)
1740–9	1	100	0.47	262
1750–9	8	91	0.50	510
1760–9	3	74	0.50	425
1770–9	3	78	0.28	62
1780–9	0	—	—	—
1790–9	15	34	0.35	215

36 per cent (241,157 tons) was raised in the period 1748–1800, a far larger amount (417,476 tons) being excavated in the terminal phase of operations, the period 1801–60.[245] In the eighteenth century production had been extensive in character but also superficial, leaving large reserves for subsequent excavation. New mines had been opened up all over the open plain creating an infinity of conical heaps of mining rubbish overspreading its entire extent in a manner suggestive of the working of a horizontal mineral bed immediately below the surface (like the copper schists of Mansfeld or Harz or the copper sands of the western Urals) but in fact indicative only of a shallow working of deep deposits. The main shaft of the typical mine was sunk the full 100 feet or so depth of the deposits but the main gallery within which people worked was driven off from the main shaft only half-way down, leaving the main elements of the deposit inaccessible. The reasons for this seeming neglect of potentially rich reserves resided in the peculiar formation of the site geology. The surrounding slates served to channel water to the main workings where it ran through the gneiss rapidly and in such quantities as to prevent steady working, thereby creating problems which only efficient pumping equipment could resolve.[246] The solution to these problems had proved elusive in the eighteenth century and it was only in the years after 1799 that they were resolved, initially with the construction of three horse-whims with a capacity to raise 0.18–6.1 million litres of water in twenty-four hours and subsequently with the installation of two steam engines, each with a capacity to raise 3.6 million litres of water in 24 hours, bought from an Englishman Major and a local mechanic Uberfeld.[247] The impact of the installation of this new pumping equipment at the mines was instantaneous. Production of ore which had slowly risen from c. 3,400 tons of ore a year in the early 1760s to c. 6,600 tons a year in the early 1790s thereafter within the short space of decade increased 4.5 times to 29,770 tons in 1806.[248] Mining activity had been totally transformed with the result that more ore was raised in the decade 1800–9 than in the whole previous history of the mines.

The labour force at Beresov

As a result of this transformation of mining activity in the Beresov district in the opening decade of the nineteenth century a large labour force, comprising some 3,579 master workers, was assembled there during the years 1807–10.[249] In the eighteenth century, when the mines were plagued by endemic flooding problems, the situation was very different. Data available for the closing years of this early phase in the mine's history, 1778–92, would suggest that, as output was maintained at *c.* 6,300 tons a year, about 1,200 persons were employed at the works (table 3.31).[250]

Table 3.31   The labour force at Beresov, 1778–91

	Master workers	Ascribed peasants
1778	415	625
1779	442	459
1780	508	505
1781	468	252
1782	515	75
1783	810	157
1784	672	499
1785	747	390
1786	954	277
1787	954	277
1788	1,285	388
1789	1,925	333
1790	1,125	201
1791	1,125	128

Prior to the reforms of 1779–83 the ascribed peasantry and master workers had made almost exactly equal contributions to the enterprise's labour supply but with the former's enforced reduction in labour intensity an acute shortage of labour emerged during the reform years 1779–83 which was reflected in contemporary production levels. Only as more and more master workers were recruited, displacing the now impotent peasantry, did production recover and a pattern of recruitment become established which would establish the supremacy of the master worker during the boom years of the first decade of the nineteenth century.

Metal production at Beresov

The ore once raised and collected was carried to the washing place which was located in close proximity to the mine workings, the central group of mines situated to the south-west of the mining village being served at a distance of 0.3–2.7 miles by the Beresov washery and the northern Pishminsk–Kluchevsk mines, some 4.5 miles distant from the mining village, delivering up their ore to be carried a similar distance to the washeries on the Pishma. At the washery the collected ore was poured into long water-filled troughs where it was beaten with iron stampers, the water which flowed continually through the trough carrying off the fine powder over the washing tables which were laid out like slightly inclined terraces below the trough; the heavier particles remained in the sump of the stamping trough or on the upper tables.[251] By such methods the auriferous quartz was graded at the three washeries which processed the output of the mines until *c.* 1797 (table 3.32).[252]

Table 3.32 Productive capacity at Beresov to 1797

Washery	Stamping works		Washing tables	
	Works	Stamps	Number	Groups
Beresov (1753)	2	18	469	3
Pishminsk (1764)	5	45	450	3
Uktussk (1759)	2	18	106	1

Constructed during the initial production boom of 1752–66, when ore output rose from 442 tons a year in 1753–4 to 2,095 in 1758–60 and 3,224 tons in 1764, this capacity proved quite sufficient to accommodate the subsequent doubling of output in the years to 1797 but was found totally wanting in the subsequent production boom which led to a more than four-fold increase in the quantities of minerals excavated. To accommodate this vast increase in output, resulting from the de-watering of the mines, processing capacity had to be both extended and reorganized, a process which gradually gathered momentum over the years 1800–10.[253] The Pishminsk washery, originally built to process the ores of the Stanovskii mine in 1764, was completely rebuilt in 1800 and further extended by the building of a new water channel and summer washery in 1804. At the Beresov works a similar reconstruction took place in 1803, three new dams being built on the newly canalized stream each with its own washery (the Beresov, Alexandrovsk, and Kluchevsk) collectively containing five crushers with forty-eight stamps. Within the existing works, capacity had been enhanced twofold and with the opening up of new works at the Elizabetinsk iron works, where gold was discovered in 1802, and at Nizhne–Issetsk, replacing the old works there in 1804, as well as at Ekaterinburg, where the new plant constructed in 1798 was further extended after 1807 by the building of a new washery, overall capacity was increased three-fold (table 3.33).[254]

In the 1800s the industry had been transformed. Ore production, from the newly de-watered mines, rose on occasion to some 29,770 tons or some 4.5 times more than in the early 1790s, and consistently during the years 1800–9 maintained an average annual output of 22,200 tons or slightly more than 3 times higher. To process this output new capacity had been brought into existence, the primary hydraulic 'straining' of the auriferous quartz taking place in washeries which were now constructed in such a way as to permit both summer and winter operations.[255] Similar innovations may also be discerned in the secondary process of separating the gold from the quartz. During the initial phase of activity (1748–97) this had been effected by heating the fine grains of gold/quartz with lead until the gold was taken up by the lead which was then ready for separation by cupelation, processes which took place in a series of hearths (*Stoffherden*) built at Beresov in 1753 and subsequently reconstructed in the opening years of the nineteenth century. From 1807, however, only part of the crushed and washed ore was so processed, auriferous-iron gravels passing instead to a newly constructed amalgamation works at Ekaterinburg where 'the first introduction of this art into Siberia' took place.[256] The refineries of Beresov and Ekaterinburg thus now processed large quantities of crushed and washed ore producing, by a variety of methods, some 20 puds of blick, or legatur, gold, which was then despatched to the imperial laboratory at St Petersburg for the separation of its constituent elements – gold, silver, and iron. That subordination of the Beresov product to that of Nerchinsk–Kolyvan, which had characterized the eighteenth-century gold industry, was at an end. Against a background of declining gold production at the latter centres, output at the former steadily increased during the years 1797–1805/7 until at the latter date, with a combined output of refined gold from the Beresov mines of Perm province and the Miask works of Orenberg of 20 puds (354.5 kg), the Urals industry stood on a par with its Siberian counterpart and in 1808/9 even exceeded it.[257] A new age was dawning when Russian gold, not silver, would dominate national precious metal production.

Table 3.33 Productive capacity at Beresov and neighbouring works in 1810

| Works | Stamping works | |
	Works	Stamps
Beresov factory		
Beresov washery (1753/1804)	2	18
Alexandrovsk washery (1803)	2	20
Kluchevsk washery (1803)	1	10
Pishminsk factory		
Pishminsk washery (1764/1800)	8	75
Pervopavlovsk washery (1804)	2	20
Uktussk factory (1759/1800)	2	18
Ekaterinburg factory		
First washery (1798)	2	18
Second washery (1807)	4	40
Elizabetinsk iron works (1803)	2	32
Nizhne–Issetsk factory (1804)	4	32

Precious-metal production in the eighteenth century

Initially discovered during the first great production cycle of the European precious metal industry, the mines of Nerchinsk in the Amur valley of south-eastern Siberia (opened in 1704) and Kolyvan–Voskressensk in the Altai region of that province (opened in 1729), had at first been of but little importance. From about 1745/6, however, they went from strength to strength. Formed, at that time, into a massive, interlocking and inter-dependent, mining and metallurgical complex, thereafter, with the Uralian gold mines of Beresov (opened in 1752), they contributed an ever-growing production of precious metals to the second great production cycle of the European industry until, at the height of the production boom in the early 1770s (with a combined output worth 2.2 million roubles, equivalent to 40,000 kg of silver a year), the mines, located in the vast wilderness of Russian east Siberia, produced and delivered to St Petersburg more of the precious metals than were produced in all of the rest of Europe put together (figure 3.9). Thus began the first great period of Russian precious-metal production, Russia's 'Age of Silver', when, for a quarter of a century from c. 1745/6–71, a flood of precious metals poured forth from the Siberian mines to the St Petersburg mint for refining and minting into coin, establishing the nation as Europe's primary producer of precious metals.

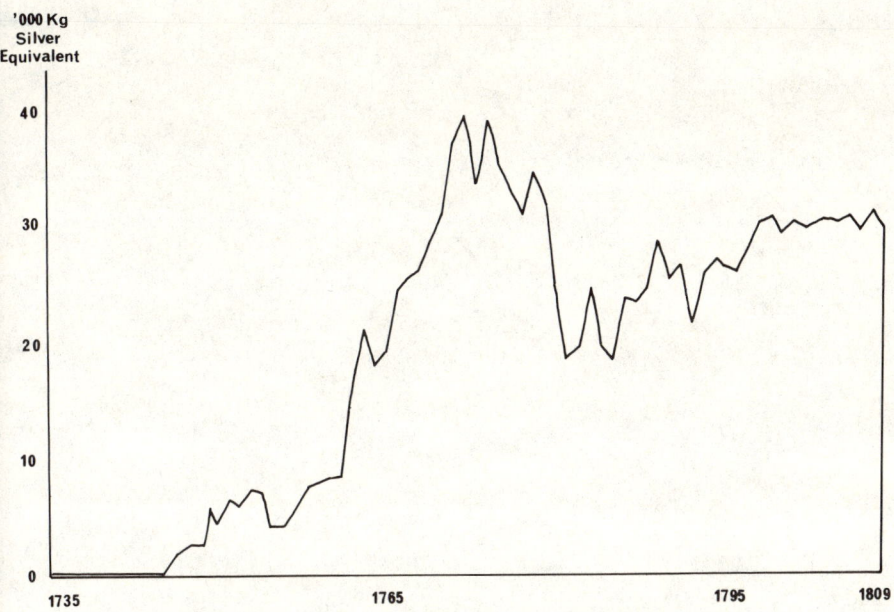

Figure 3.9 Russian precious-metal production

Then, against a background of recovery in the Americas, which at first stabilized international specie prices and then as the century drew to its close caused them to fall, the Russian boom came to an end. Acute depletion problems in the resource base of the main Siberian mines, coupled with severe labour problems occasioned by ill-advised, if morally laudable, government intervention in the labour market, all served to undermine the position of the main production centres, precipitating a major crisis during the years 1771–85, from which recovery was slow and only truly made effective by a structural displacement of activity in the years after 1797 to the Uralian gold fields. Amidst crisis conditions in a silver industry undergoing a difficult transition to a new resource base in conditions of high labour costs, the first decade of the nineteenth century thus witnessed the embryonic beginnings of that gold boom, which on the basis of autonomous production would once again in the nineteenth century place Russia on the world stage.

Chapter Four

Russia. Money supply

For some fifty years (*c*. 1745–95) increasing quantities of precious metals, derived from indigenous mining and metallurgical enterprises, passed to Russian mints contributing to a major transformation in that nation's position in international specie markets. Imports of monetary metals were displaced. On the basis of indigenous supplies of such metals, mint output and monetary supplies increased enormously.

The foreign trade sector[1]

During the years 1670–1700 Russian trade continued to conform to the patterns first established in the 1570s when commercial exchanges between east and west were first dominated by the effects of the enormous inflows of American silver which, creating massive imbalances in silver stocks throughout Europe, caused the appearance of major divergences in the purchasing power of that metal between regions. The routes along which merchants travelled and the ports they visited might alter as first, between *c*. 1580 and 1630, Archangel rose to prominence and then, from the 1630s, the Baltic ports servicing the Russian heartland reasserted their dominance, but whether the trade passed by way of the northern or eastern passages the nature of the exchanges was always the same.[2]

Throughout the late sixteenth or seventeenth centuries any given quantity of silver brought to Russia from the west would purchase four to six times more goods than in the merchant's home country. In such circumstances the attraction of shipping silver eastwards was overwhelming, and in all of the ports used in the Russia trade one finds merchants exchanging specie against goods, the balance of trade being highly favourable to the host nation. In its heyday the commerce of Archangel displayed precisely these characteristics. Foreign, predominantly Dutch, merchants brought gold and silver, as well as small quantities of price-inelastic luxury wares, which they exchanged for Russian commodities, especially furs, leather, and skins.[3] Nor did this situation change when the traffic through Archangel changed, decline setting in and the trade through the Baltic havens increasing to a high point in the 1690s. In each of the Baltic ports – Riga, Reval, Narva, and Nyen – the predominant form of commercial exchange was of specie against goods. Thus at Riga during the years 1650–2 foreigners brought riksdalers by the sackload, importing 2,000–2,700 sacks of coins (each containing 400–1,000 pieces) each year, to buy flax and hemp, and thirty years later the situation had not changed. In 1683 again one may discern foreign merchants importing about 800,000 riksdalers and taking away in exchange flax and hemp.[4] Nor was the situation different at Narva or at Reval. At the latter port during the first half of 1683 some 50,000 riksdalers had to be imported to balance the trade.[5] In each of the ports on the southern shores of the Gulf of Finland, which dominated the Russia trade during the closing years of the seventeenth century, the main form of commerce involved the exchange of specie for commodities.

This pattern was replicated, moreover, in those transactions which took place between the citizens of these Swedish-controlled ports and the Russian merchants who visited them. At Riga in 1679–80, for instance, the value of Russian imports into the town exceeded the value of the goods they received by at least two to one, causing much of the specie which had been brought to Riga from abroad to be trans-shipped to Russia. Throughout the late seventeenth century those Russian towns (like Smolensk and Novgorod) which enjoyed active trading links with the eastern Baltic ports received a steady inflow of precious metals.[6] Until c. 1699, therefore, trade continued in much the same way as it had for the last 130 years, foreign merchants bringing precious metals and taking away goods.

As far as these foreigners were concerned success in the trade depended almost entirely upon their ability to acquire specie, and in this respect the years 1670–1700 witnessed dramatic changes in the fortunes of the various groups involved in the Russian trade. As the effects of the Central and South American silver 'crisis' began to make themselves felt on western European specie markets during the years 1670–1710, causing acute shortages of that metal, the position of the Dutch and English, who were dependent upon these markets for supplies of silver was undermined.[7] As the trade through the Baltic to Russia thus grew the Anglo-Dutch component in this commerce declined in both absolute and relative terms, the 1690s witnessing the most dramatic decrease in shipping numbers as the western European 'crisis' reached its height. A new age was dawning and the Hanseatic towns from Hamburg to Lübeck, as well as a host of lesser centres in Schleswig and southern Holstein, were the main beneficiaries. As the Oberharz and Saxon silver mines poured forth increasing quantities of the white precious metals, as well as abundant amounts of copper, a flourishing trade developed along the Elbe bringing these wares to the towns at the base of the Danish peninsula from whence they were distributed either south towards the Low Countries or east, the latter trade flow allowing Lübeck merchants in particular to run an adverse balance in their expanding commerce to Russia.[8] Intra-Baltic shipping during the 1680s and 1690s thus filled the place vacated by the Dutch and English, the numbers of their ships involved in the Russia trade rapidly growing and the Lübeckers in particular increasing their share of this rapidly expanding trade.

On the eve of the Great Northern War the Swedish-controlled, eastern Baltic ports had established their dominance in Russia's overseas trade. In these havens, moreover, ships from intra-Baltic ports now outnumbered their Dutch and English rivals, the Lübeckers in particular having made especially spectacular advances over the previous decades. The German-speaking merchants' commerce was, however, in no way different from that of their western European competitors. Trade continued to be characterized, as it had been for the last 130 years, by an exchange of specie for Russian commodities. The only difference was that the silver shipped was now derived from European sources – from the mines of Oberharz and Saxony. As the English and Dutch found it increasingly difficult to obtain Central and South American silver as a result of the developing 'crisis' in the New World, and experienced a corresponding decline in their commerce to Russia, the Germans, by tapping central European sources of silver supply, went from strength to strength, establishing their hegemony in the rapidly growing Russia trade.

In this long-established pattern of commerce the Great Northern War (1700–21) merely afforded a brief hiatus. The number of ships travelling by way of the Baltic to arrive at ports servicing the commerce of the Russian heartland was dramatically reduced and although there was some compensatory growth of shipping at Archangel overall trade declined. The impact of these changes upon Russian monetary systems was disastrous. As the flow of silver into the realm dwindled so also did monetary stocks of that metal, an endemic debasement of the coinages masking the declining command of the Russian populace over silver. During the opening decade of the war *per capita* silver availability in the form of coin was reduced to the metallic content of 1.12 (1764) roubles or approximately one-third of pre-war levels.[9] Nominal prices thus declined slowly but silver-commodity prices collapsed in line with the decline of silver stocks. The severity of the crisis was such, however, that amidst the turmoil caused by the war the seeds of monetary recovery were already being sown. For those who could transport silver to Russia the pickings were rich. The amount of Russian produce they could acquire was enormous – but with Swedish raiders prowling the Baltic the risks were high. In most years, accordingly, few would venture from their home ports and there was little demand to take up specie on the markets of western Europe or the Hanse towns, but when, as during the years 1704/5 or 1708/9, the prospects of trade brightened there was a rush to acquire the precious metals, pushing up premiums to as much as 60–70 per cent above the level of the official exchange.[10] Such activity was, however, erratic as long as the war continued and it was only as hostilities drew to their close in the late 1710s that normality returned. Especially during the years 1718–20 the prospect of obtaining abundant quantities of 'cheap' Russian wares led to a frenzied activity on western and northern European specie markets, again pushing up premiums towards the peak levels of the 1700s and creating a massive trade in specie to Russia which resulted in a rapid replenishment of monetary stocks and an enhancement of silver-commodity prices there. Subsequently, premiums fell and the inflow of silver slackened, but as late as 1726 a million roubles' worth of bullion entered the recently annexed Baltic provinces through Riga, and an equivalent amount entered the Russian heartland through Archangel and St Petersburg.[11] All seemed set for a restoration of trade in the patterns of the late seventeenth century with the Dutch and English regaining some of the ground that they had lost at that time as the recovery of Central and South American silver production from the low levels of 1690s allowed them to more successfully promote their commerce.[12]

The years immediately following Peter's death in 1725, however, saw this whole system of commerce shaken to its very foundations. The collapse of Oberharz silver production, by reducing silver-commodity prices in the Hanseatic towns and closing price differentials within their trading systems, reduced dramatically the outflow of precious metals and precipitated crisis conditions in the recipient nations.[13] Silver stocks which had been growing rapidly during the early 1720s thus fell and stabilized at a lower level, inducing deflationary pressures in their economies. In the newly-annexed Russian Baltic provinces, where the German-speaking merchants had re-established their commercial hegemony during the years before 1725, a new trading equilibrium was now, with difficulty, achieved. The domestic silver-commodity price within these provinces remained at 40–50 per cent below pre-war levels but now as the purchasing power of specie in the merchants' homeland increased, interregional price differentials (measured in silver) closed and the recovery in trade to the Russian Baltic provinces, which had proceeded rapidly in the years before 1725, aborted.[14] Silver continued to flow to these provinces from the Hanseatic towns but the number of vessels carrying the white precious metal there from intra-Baltic ports now stabilized at a level 50–60 per cent below pre-war levels. Initially, western European merchants were in no position to provide compensatory supplies of specie. The crisis, occasioned by the downswing and collapse of mercury production at Almaden which crippled Central and South American silver production in the mid-1720s, resulted in a withdrawal of Dutch and English merchants from Riga and Reval during these critical years. Only as the first production cycle (c. 1725–55) of the new Central and South American long cycle ran its course, increasing the availability of silver on western European specie markets, could the Dutch and English intervene to stabilize conditions in the Baltic provinces.[15] At no time, however, before 1760 does the increasing number of western European ships seem to have been sufficient to compensate for the decline in intra-Baltic shipping. Vessels carrying silver from the west might increasingly displace those originating from within the Baltic but overall the numbers arriving at Riga and Reval remained some 20 per cent below pre-war levels during the years 1725–1760. An acute crisis in the central European silver industry, which assumed particularly severe dimensions in the years 1725–1729 and 1735–1738, coupled with a protracted stagnation in American silver supplies after the crisis of 1725–9 thus doomed the Baltic provinces to a very slow recovery of monetary stocks and prices which before 1760 never exceeded two-thirds of the levels attained before the Great Northern War.

Only as the second production cycle of the new long cycle pushed up production levels within the South American silver industry to new heights during the years 1755–85 did this situation change as a greater availability of silver on western European specie markets encouraged the Dutch and English forcefully to promote their trade to the Russian Baltic provinces. For the first time in the late 1750s their silver imports through Riga grew at a rapid rate, pushing aggregate trade in specie through the ports beyond the level of 1726.

Table 4.1 Specie imports at Riga, 1762–96[16] (millions of roubles per annum)

1762–8	1.72
1775–7	1.80
1783–4	3.10
1785–7	2.23
1788–90	2.23
1790–3	1.71
1794	3.05
1795	4.72
1796	4.95

Year after year the amounts of silver entering Riga increased to 1785. The inter-cyclical crisis which beset the Central and South American industry in the mid-1780s then left its imprint on the silver flows to Riga, but as the upswing of the third Latin American production cycle (c. 1785–1815) began the now long-established pattern of silver imports into the Russian Baltic provinces was resumed, the quantities of specie entering Riga exceeding previous peak levels after 1794 when in successive years 4.7 and 4.9 million roubles' worth of precious metals was entered therein.

In the Baltic provinces, incorporated into the Russian Empire at the end of the Great Northern War, which remained in terms of their administrative, monetary, and economic systems separate and discrete entities, the pattern of trade which had been established first in the 1570s continued unaltered throughout the eighteenth century. Merchants visiting Riga and Reval still brought silver which they exchanged for commodities which had been obtained from within a trading network which extended far to the south.[17] Initially, during the years 1670–1725 as 'crisis' conditions prevailed on western European specie markets, German-speaking merchants, through their access to central European silver supplies, established their dominance in the trade. With the decline of Oberharz production after 1725 their influence waned and that of the Dutch and English increased once more. It was not until the 1760s, however, that this latter group established their supremacy, their imports of Spanish American silver into Riga thereafter steadily growing until in the 1790s they reached almost 5 million roubles' worth of specie a year.

By the opening years of the nineteenth century western European coins once more dominated the circulating media of the Baltic provinces, Oddy noting in 1805 that '. . . accounts are here kept in rixdollars and ferdings, 80 of the latter to a rixdollar. The silver money here circulating, is mostly Dutch, Spanish and some of the Ecclesiastical princes of Germany'.[18]

These coins, moreover, continued to provide the basis for the trade to the ex-Polish and Lithuanian provinces to the south, the coinage going at a premium of up to 7 per cent when the caravans were being formed to undertake this commerce. In all respects, save one, commerce was conducted at the beginning of the nineteenth century in much the same way as it had been for the last 230 years. The exception was the provincial trade with Russia, for at this time, in 'normal' circumstances, this was conducted without the necessity of specie transfers. Only when a particular type of coin was required by the Russians, valued for its design rather than its intrinsic value, did specie flow from the Baltic to Moscow, provisioning of the Chinese or Persian caravans with Dutch ducats, for instance, causing those specific coins to go at a premium on the Riga market.[19] Yet such transfers were exceptional, for in the late eighteenth century the Muscovite heartland of the Empire no longer required imports of precious metals from western Europe, whether obtained directly or via Riga, to maintain its monetary system.

During the years following the death of Peter in 1725 a minor revolution had taken place which had transformed the position of the Muscovite heartland within the network of international specie supplies, rendering it independent of imports of foreign silver and gold. Until that date, as has been shown, Muscovy continued to receive through Archangel, St Petersburg, and Narva considerable quantities of silver, replicating a pattern of trade whose history could be traced back a century and a half. In 1726 almost a million roubles' worth of bullion, or half the total imports of the Russian Empire, entered the ports servicing the trade of the Russian heartland.[20] During the years 1766–7 about 1.72 million roubles' worth of specie entered the lands of the Russian Empire but now *all* passed to the Baltic provinces through Riga. The Muscovite lands had ceased to import silver, establishing a situation which would continue to prevail until the century drew to its close.

Few, if any, at the time of Peter's death in 1725 could have anticipated, however, these dramatic changes which would totally transform Russia's position within international monetary systems. In that year the trade recovery which affected all the ports of the Russian Empire was in full swing. Foreign merchants continued to flock to the 'Russian' ports (Archangel, St Petersburg, and Narva), seeking to take advantage of the contemporaneously high purchasing power of silver in the Russian lands, and amongst their number the Dutch and English particularly gained ground due to the greater availability of silver on western European specie markets in the early 1720s. Each group, moreover, brought with them silver with which they could buy eight times more goods than in their homelands. As exports of commodities from Russia thus increased rapidly, silver was imported, augmenting monetary stocks, which more than doubled the level of 1701–10 causing an enhancement of silver-commodity prices, which by 1725 had reattained a level which was some 70 per cent of that prevailing pre-war (figure 4.1).[21] As in the Baltic provinces, so also in the Russian heartland during the years before 1725 a major trade boom paved the way for a restoration of pre-existing commercial systems.

Before the *status quo ante bellum* was re-established, however, the trade recovery aborted. The combined effects of the Central/South American and central European mining 'crises' of 1725/6 wrought havoc in Russian commercial and monetary systems.[22] As Hanseatic, Dutch, and English merchants found it increasingly difficult to acquire the necessary supplies of silver to pursue their trade they suspended operations. Shipping numbers at the 'Russian' ports, as at those of the Baltic provinces, fell until in 1729 the numbers of vessels visiting Archangel, St Petersburg, and Narva was barely 60 per cent of the number in 1726. Silver flows to the Russian lands thus dwindled and an acute shortage of the metal began to appear. Free-market silver prices, accordingly, rocketed upwards and, as mint prices failed to keep pace, mint activity ground to a halt, hoarding increased, and monetary stocks and silver-commodity prices declined (figure 4.2).[23] The prevailing low commodity prices (measured in terms of silver) once more in 1730–1 provided an environment for commercial recovery, and as western European merchants once more flocked to Russia, channelling their trade through Narva, premiums again emerged on transactions involving the acquisition of riksdalers for the Russia trade. Dutch and English merchants were again avidly buying up the silver pieces to transport eastward and as this flow of specie entered the Russian heartland monetary stocks were rebuilt. As in the Baltic provinces,

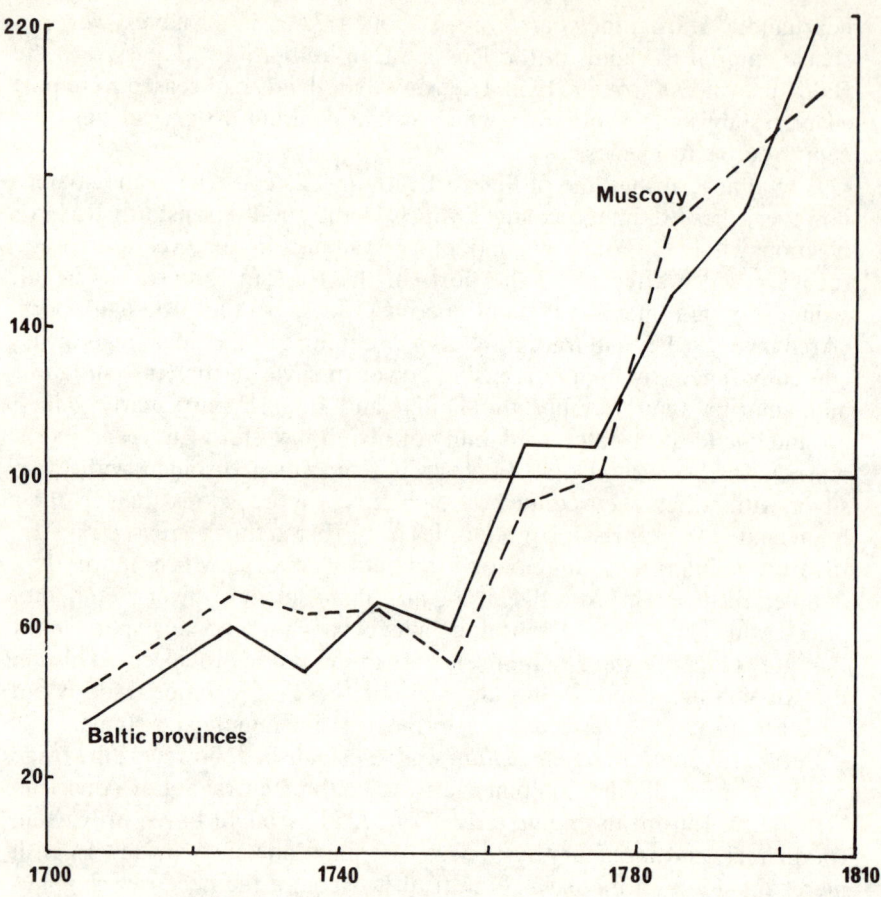

Figure 4.1 Russian silver-commodity prices in 1764 roubles (1690/
9 = 100)

however, at no time during the 1730s was this inflow sufficient to raise
Muscovite silver-commodity prices to the levels of the early 1720s.
Following the crisis of 1725–9 the equilibriating mechanisms of a
monetary–commercial system whose history stretched back more than 150
years were once more operating, but now, as western European monetary
stocks remained in a depleted state and silver-commodity price differentials
between regions closed, the flow of silver from west to east was reduced. In
1730/1, in both the Muscovite heartland and the recently annexed Baltic
provinces, the old mechanisms continued to operate but silver imports
were reduced to a very low level.

Yet in a number of respects that decade witnessed significant changes
within the Muscovite sector of the prevailing monetary–commercial system

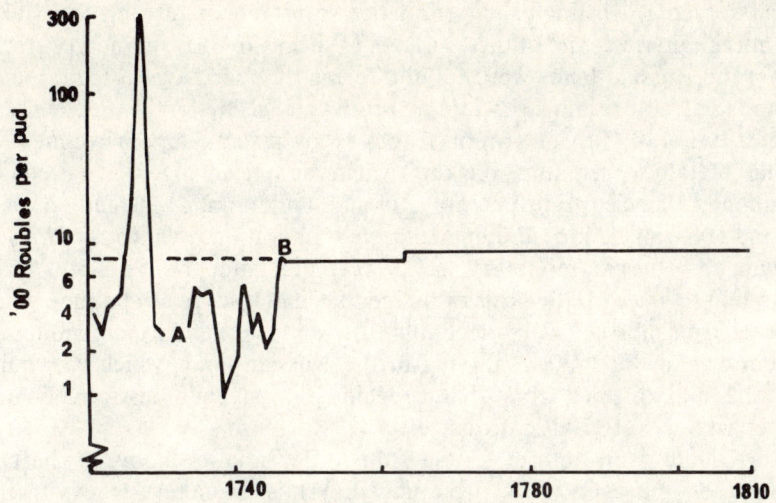

Figure 4.2A Russian silver prices

A Market Price (from 1747 equals mint purchase price)
B Mint purchase price

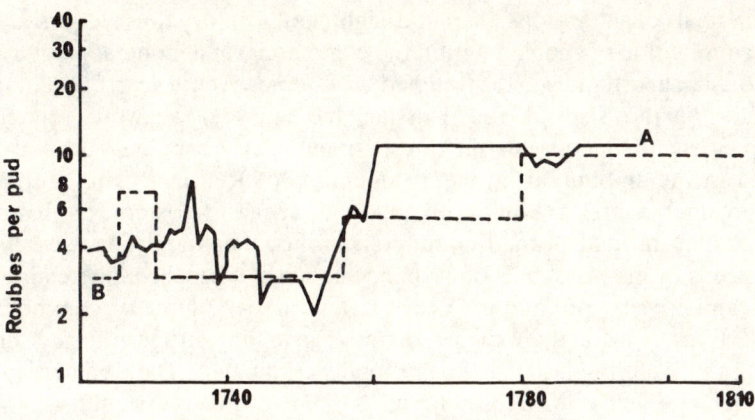

Figure 4.2B Russian copper prices

A Market price
B Mint purchase price

which served to distinguish it from the sector encompassing the Baltic provinces. First, in spite of the low level of specie imports, monetary stocks within the Russian lands were rebuilt during that decade to a higher level than in the Baltic, almost reattaining pre-war levels by 1740.[24] Second, this reconstitution of Russian (rouble) stocks took place in an environment of falling metallic premiums on the Amsterdam exchange as merchants abandoned specie transfers to Russia, undertaking instead normal commercial transactions and financing their operations by means of bills of exchange.[25] Russian monetary stocks were thus after the crisis of 1725–9 beginning to grow on the basis of indigenous metal supplies at a time when imports from abroad were declining. A new commercial and monetary system was in the 1730s emerging in the Russian lands which was quite separate and distinct from that prevailing in the administratively and monetarily discrete Baltic provinces.

Subsequent years, moreover, saw this differentiation between the two regions become even more pronounced. Whilst monetary stocks in the Baltic provinces grew slowly, barely exceeding pre-war levels by 1760, in the Muscovite lands pre-war stock levels had been reattained in the 1740s and twenty years later were some 75 per cent above late-seventeenth-century levels. In the Baltic provinces, moreover, stock augmentation was still based upon imported supplies of silver whereas in Russia such supplies had been entirely dispensed with. The trend of declining specie transfers to the Muscovite lands and associated fall in premiums on the Amsterdam bourse, observable in the 1730s, continued apace until, with the final cessation of silver imports to the Russian lands in the 1740s, movements of interest rates on the Amsterdam bourse were conditioned solely by commercial considerations. The mid-eighteenth century thus witnessed the final culmination of changes within the commercial and monetary system of the Russian heartland which stretched back some seventy years. Gradually, changes in the size of the Russian monetary stock were rendered independent of fluctuations in international specie supplies and became linked to the amounts of monetary metals produced in Russia. Equally, foreign merchants gradually abandoned specie transfers to Russia, undertaking instead normal commercial transactions and financing their operations by means of bills of exchange issued at normal commercial rates unencumbered by premiums. Year after year the volume of commercial transactions grew at the expense of those involving specie transfers until during the 1740s the latter had been totally eliminated. The trend was not, however, a linear one. Beginning in the late seventeenth century, it was interrupted during the war years and their immediate aftermath and it is only in the 1730s that these trends may again be clearly discerned, the pace of change accelerating thereafter until in the 1740s a completely new commercial and monetary system had been created in the Russian lands.

Nor was this new system seriously disturbed during most of the remaining years of the eighteenth century. The monetary stock of the Russian lands continued to fluctuate in accordance with variations in the domestic production of monetary metals, although with the issuing of paper notes – assignats – in 1769 intersectoral imbalances within the stock were reduced. Similarly, within the commercial sector of the economy of the Russian lands commodity exchanges remained the norm. Merchants, having ceased to import silver, conducted their business on the basis of trading goods against goods, necessary finance being raised on bills of exchange negotiated at 'normal' commercial rates. Within the Russian lands the commercial and monetary system created in the 1730s continued until at least the 1790s.

As monetary stocks within the Baltic provinces were rapidly augmented in the years after 1760 as a result of massive imports of specie from western Europe, moreover, the two previously separate sectors of the imperial economy began to draw closer together until during the late 1780s and early 1790s, a state of silver-commodity price equilibrium was created between them. In such circumstances a new unitary market structure was born within the Tsar's domain within which there was a single system of commercial finance characterized by a single market rate.[26]

Yet even as this unitary structure was created, relations between it and the external world were changing. The acute crisis within those industries producing monetary metals at the close of the eighteenth and in the early nineteenth centuries undermined the internal production base of the Russian monetary system.[27] The increasing issues of paper assignats, whilst meeting transactional demands within the territorially expansive economy, accordingly went at a discount in relation to silver until, during the years 1802 and 1803, large-scale imports of silver from western Europe led to a brief stabilization of the assignat rouble. As total imports of specie into the Russian Empire rose from about 5 million roubles a year (entirely through Riga) in 1795–6 to 10.7 million in 1802 and 11.1 million in 1803 (through both Riga and St Petersburg) precious-metal stocks once more grew and at a faster rate than the stock of assignats, causing the discount rate on the latter to fall.[28] In spite of the developing crisis in those Russian industries producing precious and base monetary metals, the metallic base of the monetary system was again being augmented but now, after an interval of sixty years, once again on the basis of imported specie supplies.

Subsequent inflows of bullion between 1803 and 1809 led to continued growth of metallic monetary stocks, but as a flood of assignats was issued destabilization began to occur.[29] This reached its height in 1812 as a result of the events which ultimately culminated in the Napoleonic invasion of that year. As the French crossed the Nieman silver and gold roubles disappeared from circulation, either into hoards or abroad, leaving the population dependent upon copper and paper assignats to meet their monetary requirements.[30] The currency had become separated from its metallic base, and whilst the subsequent defeat of Napoleon resulted in dishoarding of coins and the reinstitution of mint activity metallic monetary stocks remained acutely depleted; this led to the emergence of a dual monetary system involving the operation of both a metallic standard (the silver rouble) and a paper one (the assignat rouble).[31] For some thirty years the two operated side by side until the monetary reforms of 1839/43 paved the way for the creation of a convertible specie-backed paper currency (the credit rouble).[32] The year 1812 thus brought to an end an epoch in Russian monetary history. Henceforth, at any given level of specie stock, the management of the paper currency became the primary objective of monetary policy.

In the eighteenth century the situation had been very different. The levels of monetary stocks had been directly related to changes in the supplies of monetary metals, obtained, initially at least, by means of foreign trade. In the early eighteenth century, as for the previous 130 years, monetary stocks in both the Russian heartland and the Baltic provinces (annexed from Sweden at the end of the Great Northern War) changed in accord with variations in the supplies of internationally-traded precious metals. From the 1730s, however, this situation had changed dramatically. Whilst in the Baltic provinces, which remained in terms of their administrative and monetary systems separate and discrete entities, the old ways continued, their monetary stocks changing with variations in international specie supplies, in the Russian lands the size of the monetary stock was rendered independent of such fluctuations. Imports of specie diminished until finally they were eliminated in the 1740s, foreign merchants undertaking instead normal commercial transactions and financing their operations by means of bills of exchange. Monetary stocks grew but now on the basis of indigenous production of monetary metals. Within the Russian heartland a new commercial and monetary system had been created which continued to operate for the next sixty years (*c*. 1735–95). Only when domestic Russian production of monetary metals declined was this system displaced and the Russian lands, now incorporating both the Baltic provinces and the Muscovite heartland in a single system, once again became recipients of foreign silver (*c*. 1796–1809).

The domestic economy

For some sixty years (*c*. 1735–95) monetary stocks of the Muscovite heartland of the Russian Empire were rendered independent of international specie supplies. Imports of precious metals rapidly dwindled and by the 1740s had completely been eliminated. Russian monetary stocks grew on the basis of increasing supplies of indigenously-produced monetary metals.

The circumstances surrounding this new departure were not, however, highly propitious ones. Following the acute monetary crisis of 1725–9 normal market mechanisms had brought, during the years 1730–4, a countervailing inflow of specie to the realm allowing a reconstruction of monetary stocks and causing a sharp fall in silver prices.[33] Indigenous silver production, which had made a very minor contribution to mint activity during the crisis of 1725–9, accordingly came to a complete halt during the next quinquennium. Nor was its subsequent recovery after 1734 of any great significance.[34] The indigenous Russian silver industry, then confined to production at Nerchinsk, was certainly in no position to assume market leadership, and the country was seemingly remarkably ill-equipped to provide an indigenous substitute for foreign suppliers of silver. The increased mint activity after 1734/5, which was to render Russia independent of international supplies of monetary metal, did not depend, however, upon increased indigenous production of precious metals (gold and silver) but upon the expansion in the supply of a base metal (copper) to the mints, thereby developing upon, and accentuating, a trend first established more than a decade before. In the wake of the collapse in Oberharz silver production, which had engendered successive specie crises in Russia in 1725–9 and 1735–7, the mints had turned out more and more copper pieces, thereby laying the foundations not only of an independent monetary system but also of an industry which was to become the largest in the world.[35]

The Petrine dream. Towards an independent copper coinage, c. 1725–45

The origins of the new developments of the 1730s may be traced back to the opening years of the previous decade when the 'hot-house' Urals industry, created by Peter to displace Swedish copper imports and satisfy his martial needs, was, in the aftermath of the Great Northern War, subject to an acute depression. With the fall-off in state orders its high priced wares, produced from low-grade Urals gravels, found few customers and were despatched almost entirely to the Moscow mint where they were sold at prices which did not even cover production costs. In such circumstances activity within the industry was both intermittent and diminutive.[36] In response to these difficulties and to the decline in the rate of silver imports Peter, accordingly, set in motion a grandiose plan to create a coinage from indigenous sources of metallic supply. Central to this scheme, formulated during the years 1723–5, was a reorganization of mint activity, creating about the Moscow mint other workshops at St Petersburg, to strike coins from both domestically produced and imported gold, silver, and platinum and, at Ekaterinburg, coining copper supplied by the Urals works.[37] To supply the latter, new plants were also created (at Polevskii, Yagoshikhinskii, Pisorskii, and Lyalinskii) which together with the copper-smelting capacity created at Ekaterinburg encompassed some thirty-two hearths capable of producing about 225 tons of copper a year. Few other than the Tsar, however, had much enthusiasm for the scheme. Hennin, who was in charge of the Ekaterinburg works, dragged his feet with regard to the building of the mint which, although authorized in 1725, did not come into operation until ten years later.[38] Nor were private entrepreneurs any more willing than they had been during the war to exploit their copper leases.[39] Events, moreover, proved that their judgement was sound. To 1725 the state plants never operated at more than a third of their capacity and then only by selling an increasing quantity of their wares to the Moscow mint at prices which did not even cover costs.[40]

Only as the silver crisis, occasioned by the shortfall in Oberharz silver production, began to take effect, causing a curtailment in Russian silver imports, did the situation change. Silver prices rocketed upwards closely followed by copper prices.[41] Both metals became the object of intense speculative activity. From 1725–9 activity at the state plants was transformed, production being sustained at a level approaching full capacity, turning out supplies which could be sold at the Moscow mint at a price which usually was more than 80 per cent over production costs. Nor were private entrepreneurs now unwilling to join the fray.[42] The Demidovs, who had secured rights to exploit copper deposits in the Viso–Kaya and Kungur regions in 1702/9 and 1720 respectively, now set about bringing them into production, obtaining a confirmation of their former rights in 1726/7 at the same time as authorization to work the newly discovered Kolyvan–Voskressensk deposits. They brought the new plant on line in 1727 (Viiskii) and 1729 (Suksunskii and Kolyvan–Voskressensk), and were also joined by others.[43] The Strogonovs opened their Tamanskii works in 1726 and the Osokins, salt producers and merchants, their Irginskii plant in 1728/9. Before the crisis drew to its close, moreover, plans were being laid for yet another extension of capacity, the Osokins beginning construction of two new plants (Yugovskii and Biziarskii) and a newcomer Tuchaninov staking his claim to participate in the boom by laying the foundations of two more (Talitskii and Sisertskii).[44] Thanks to the higher copper prices offered at the Moscow mint during the years 1725–9 a massively increased supply of that metal had been produced by an industry whose capacity had rapidly increased from about 225 tons in 1725 to 325 tons in 1726/7 and about 500 tons in 1729 when a further 335 tons of smelting capacity was in construction.

Even by then, however, the writing was on the wall. The monetary crisis, having peaked in 1727/8, thereafter subsided as a countervailing inflow of specie caused silver prices to fall during 1729–31. Stocks finally stabilized at below the level of 1725 to take account of the increased quantity of copper in circulation, and the new bi-metallic (silver–copper) ratio was encapsulated in the monetary reform of 1730/1.[45] For a moment prices of the two metals stabilized, but as the new capacity came on stream prices plummeted until in 1732 copper was being sold at the mint at about 10 per cent below production costs.[46] The folly of the investment decisions of 1729/30 was now only too clear. In spite of the cessation of production at Kolyvan–Voskressensk, due to the fire which gutted the works in that year, capacity was far in excess of demand. From 1730 to 1734 most plants were working well below capacity, production in 1732–3 amounting to only about 313 tons a year of which 195 tons were derived from state works and 118 from private ones. Judging by the state-owned sector of the industry, moreover, most of this metal not either utilized or stock-piled at the works was despatched to the mint. In an industry working at barely one-third capacity scarcely 10 per cent of output (20 tons) could be sold, either cast into bells or manufactured into bars, to private customers.[47] The dependence of the industry upon the mint was only too clear and, called into existence to make good the deficiencies in specie supply during the years 1725–9, with the stabilization of monetary stocks between 1730 and 1734 the mint could absorb but little of the metal. This pushed down prices below production costs, causing an acute depression in the industry, and bringing about a reduction in capacity which by 1735 fell to 750 tons.[48]

When in 1734/5 silver prices once more increased and copper prices again rose in sympathy, therefore, the industry was far better placed than ten years before to take advantage of the new situation. As silver imports thus slowly recovered, the copper industry began to operate at full capacity turning out some 750 tons of the metal in 1735.[49] This supreme effort revealed all the weaknesses of the prevailing industrial structure, however, and caused the initiation of a major reorganization during the years 1736–8 when, as production was disrupted, silver imports contributed to the stabilization of the currency.[50] As foreigners once again resolved Russia's monetary problems, native producers laid the foundations for the realization of Peter's dream. In the state sector of the industry, under the Empress Anna, the Ekaterinburg mint was finally commissioned and the original foundry, of three copper smelting hearths completed in 1725, was augmented by the construction of another six hearths. Elsewhere capacity was extended against a background of rationalization. Outlying plants, like the Polevskii works, operational since 1725, were closed down and converted to iron production. New plants were opened (Nizhne–Yugovskii and Visimskii) and with existing works in the Kama-Iren' region (Yagoshikhinskii, Pisorskii, and Lyalinskii) were grouped about a new copper-refining and fabricating plant (Motovilikhinskii). By 1738/9 a formidable state copper complex had been created (map 4.1), rendering superfluous the retention of the Kolyvan–Voskressensk works which had first been seized from the Demidovs in 1735 and which had contributed some seventy tons of copper to state output during the years 1735–7.[51]

A similar process whereby production organization was rationalized may also be discerned in the private sector within the enterprises of the Demidovs and Osokins. In 1734 Akinfi Demidov had been in possession of

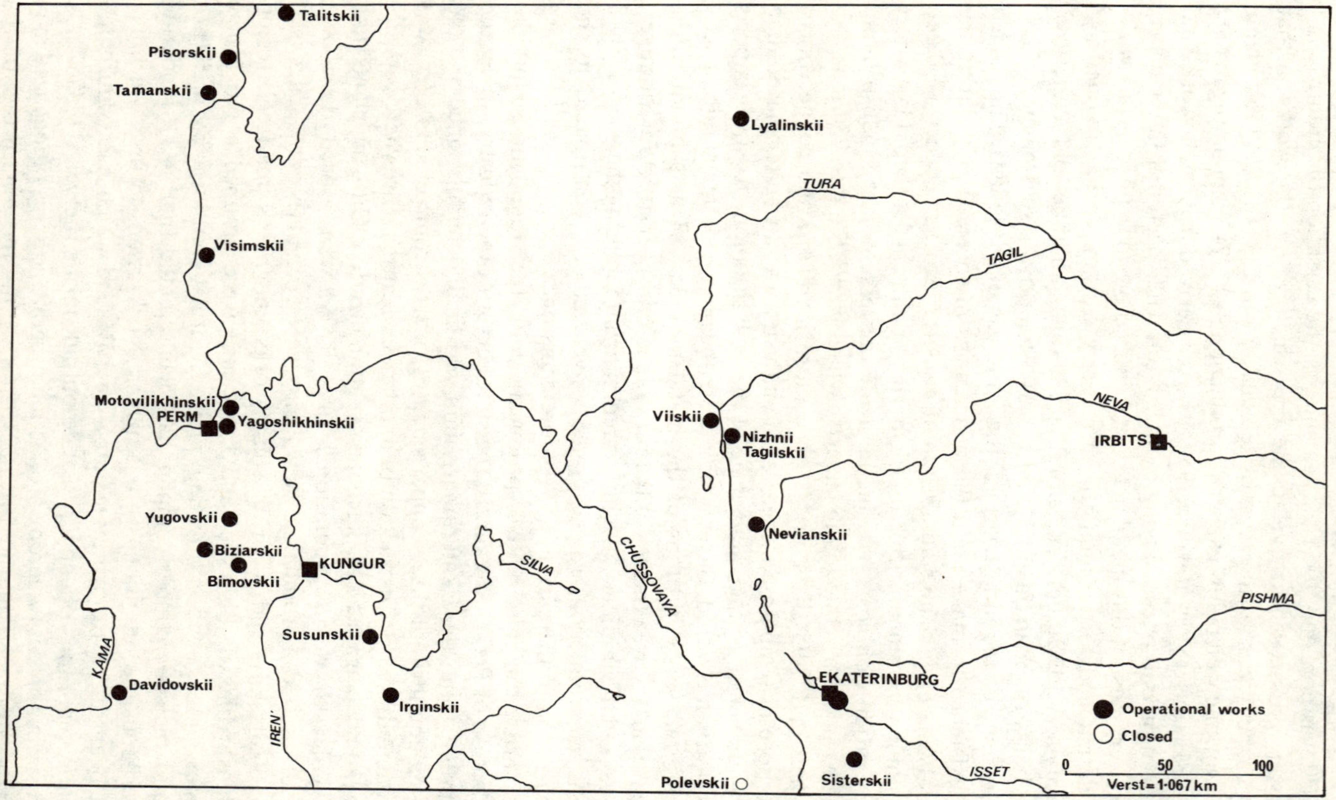

Map 4.1 Central Urals copper industry, 1738/9

three copper-smelting and -refining plants in the Urals (Suksunsk, Viisk, and Nizhne-Tagil) processing both local ores and black copper from the Kolyvan works.[52] The state's seizure of Kolyvan in 1735 dealt a serious blow to this integrated structure, precipitating an acute crisis in the eastern sector of the Urals enterprise and causing a reorganization of capacity in the west where the Suksunskii plant was provided with alternative supplies of black copper from a new plant (Bimovskii), exploiting deposits discovered and licensed in 1705 but only now brought into operation (the licence being renewed in 1733 and construction of the works being undertaken in 1734–6).[53] By 1738, with the reacquisition of the Kolyvan works, therefore, Akinfi possessed a copper empire which potentially rivalled that of the state.

Nor were the Osokins in any way different in their implementation of a similar scheme to create a major complex about their Irginskii works. This, like the Demidovs' western Urals complex, would have a complementary character, black copper from the Biziarskii works opened in 1732/3 being refined at Irginsk. Three large enterprises thus dominated the Urals copper industry at the close of the 1730s about which were a number of lesser enterprises. The Strogonovs and Tuchaninov each had a single works in operation, whilst a new plant (Kushvinskii), having been constructed by the state during the years 1735–8, was brought into operation in 1739 by Schoenberg, president of the newly created general directory of mines, the works having been granted him on 5 April of that year.[54]

A new, restructured industrial complex had been created, capable of producing some 884 tons of copper yearly for the recently constructed Ekaterinburg mint. Peter's dream had been realized. Both in terms of present capabilities and future potential the possibility existed for the creation of an indigenously produced copper coinage. Yet few circumstances could have been more unpropitious for the launching of such a venture. In the period of industrial disruption caused by the restructuring of copper production from 1736 to 1738 foreigners had seized the opportunity to import large quantities of silver, enhancing premiums on bullion dealings at Amsterdam and causing Russian silver prices to fall until in 1738/9 the market for monetary metals showed all the signs of being saturated.[55] The new copper plants thus came on stream just as demand for their products collapsed and prices fell to an all-time low, dooming the new works to a period of acute depression, alleviated only by an increase in sales to private customers who, having bought about 20 tons a year from 1732 to 1733, bought 54 tons annually in the period 1736–8 and in 1739, as prices dropped below 3 roubles per pud, 180 tons.[56]

Declining Oberharz silver production between 1725 and 1740 had thus heralded a series of acute silver crises in Russia as, during the years 1725–9 and 1735–7, declining imports of precious metals created acute shortages and enhanced the price of gold, silver, *and* copper. On each occasion the production of silver at Nerchinsk had increased but with little effect on mint supply. High copper prices, however, engendered a major response from an industry which only a few years before, in the aftermath of the Great Northern War, had lain in ruins. Capacity within the industry had been extended and production increased, rising to about 325 tons in 1727 and 750 tons in 1735. During each successive silver crisis, accordingly, the native copper industry had made a greater and greater contribution towards resolving the nation's monetary problems and imports of specie had correspondingly declined. As monetary stabilization was achieved, in 1730–4 and 1738–9, however, the new capacity brought into existence during the preceding crisis was found to be far in excess of current requirements either for coin or other non-monetary uses. Prices fell and plants closed or were forced to operate at far below their potential capabilities. During both periods (1732–3 and 1738) the industry operated with 66 per cent excess capacity and at the latter date the figure was not higher only because more copper could be sold to private customers.

In order to resolve this problem and perhaps to implement the policy of monetary self-sufficiency which could be traced back to the original Petrine scheme of 1723–5, the Crown took an extra-ordinary step in 1739, undertaking, by an ukaz issued on 3 March, to take two-thirds of the output of the private sector of the industry at current prices for the mint.[57] In one simple step the problem of excess capacity was shifted from the producer to the consumer and the nation had guaranteed itself an indigenously produced coinage – but not, in the short-run at least, without incurring considerable cost. By undertaking to acquire two-thirds of the private sector's output it provided entrepreneurs with a licence to expand production as long as the residual proportion of output could be sold at prices in excess of costs. Nor during the 1740s was this a major problem as producers found that at a constant price of 3 roubles per pud, they could sell 180–190 tons of copper.[58] A new age was dawning in which the state, by guaranteeing a market for the produce of the private sector, created the conditions for a new bonanza.

In the vanguard of the advance were the Demidovs who, building upon the foundations laid during the reorganization of 1736–8, now created possibly the most formidable industrial complex within the central Urals copper industry. Almost as the new law was promulgated work was begun on building a new plant (Shakvinskii) to supply black copper to the Suksunskii works, the licence for its operation only being sought the following August and obtained in 1743, three years after production had started. In 1741 the construction of yet another works – the Ashapskii – was begun and made operational some three years later. Before the quinquennium, 1740–5, drew to a close, therefore, a major industrial complex had been created in the Kungur region comprising three works (Bimovskii, Shakvinskii, and Ashapskii) each supplying black copper over a distance of some 40–7 miles to a central refining and fabricating plant (Suksunskii) which with Nikita Demidov's Davidovskii plant had a capacity of about 320 tons of copper a year.[59] A similar reorganization may also be observed in the other private enterprises. Piotr Osokin similarly built a new works (Kurashimskii) in 1740 which, like the neighbouring Biziarskii plant opened in 1732/3, produced black copper for his Irginskii works and which with his brother Gavriil's Yugovskii plant created yet another complex in the Kungur region.[60] Within the private sector of the industry the previously widely scattered distribution of works was giving way to a high degree of concentration to which the Strogonovs' building of their Yugo–Kamskii works merely added a final element. By the mid-1740s the Kungur region reigned supreme and within its bounds most plants formed part of highly efficient, integrated industrial complexes (map 4.2).

The reasons behind this drive towards greater concentration and efficiency during the years 1739–45 lay in changes taking place far to the east where the Demidovs, parallel to their reorganization of the Urals works, had restructured the Kolyvan–Voskressensk works in the Altai which had been restored to them in 1737. Production had not only increased, rising to 52 tons in 1740, but plans first laid in 1727 to build new capacity at Barnaul and Shul'binsk were finally realized. By 1744 a major new industrial complex had been built in the Altai capable of producing 182 tons of copper a year. Whilst the works were being constructed (in 1739–44) much of the black copper produced was refined in the Demidovs' eastern Urals complex, but whether processed in the Urals or the Altai this metal, produced from high-grade white-copper ores, posed a major threat to the western Urals industry.[61]

As this copper from the Kolyvan–Voskressensk works was placed on the market prices fell and the state works were driven entirely out of the private market for copper, henceforth only maintaining production levels because of the state's willingness to buy their wares and cover losses on current account.[62] From 1740 consumers satisfied their demands for copper entirely from works in the private sector of the industry whose sales increased both as a result of an overall extension of the market and a displacement of state supply. From selling 11–12 tons of copper a year during the period 1736–8 private entrepreneurs suddenly found they could sell 100 tons a year on average in 1739–40 and 160–190 tons thereafter. With the State being willing from 1739 to match each private sale with two public ones they were afforded the opportunity to expand production rapidly from 163 tons in 1738 to 308 tons a year between 1739 and 1740 and 490 in 1746 but, as has been suggested, initially in conditions of acute inter-regional competition.[63] As Altai copper flooded the market it was accordingly not only the state works which felt its impact but also those private plants of the western Urals which only by the most vigorous rationalization were able to maintain a place in a market within which they

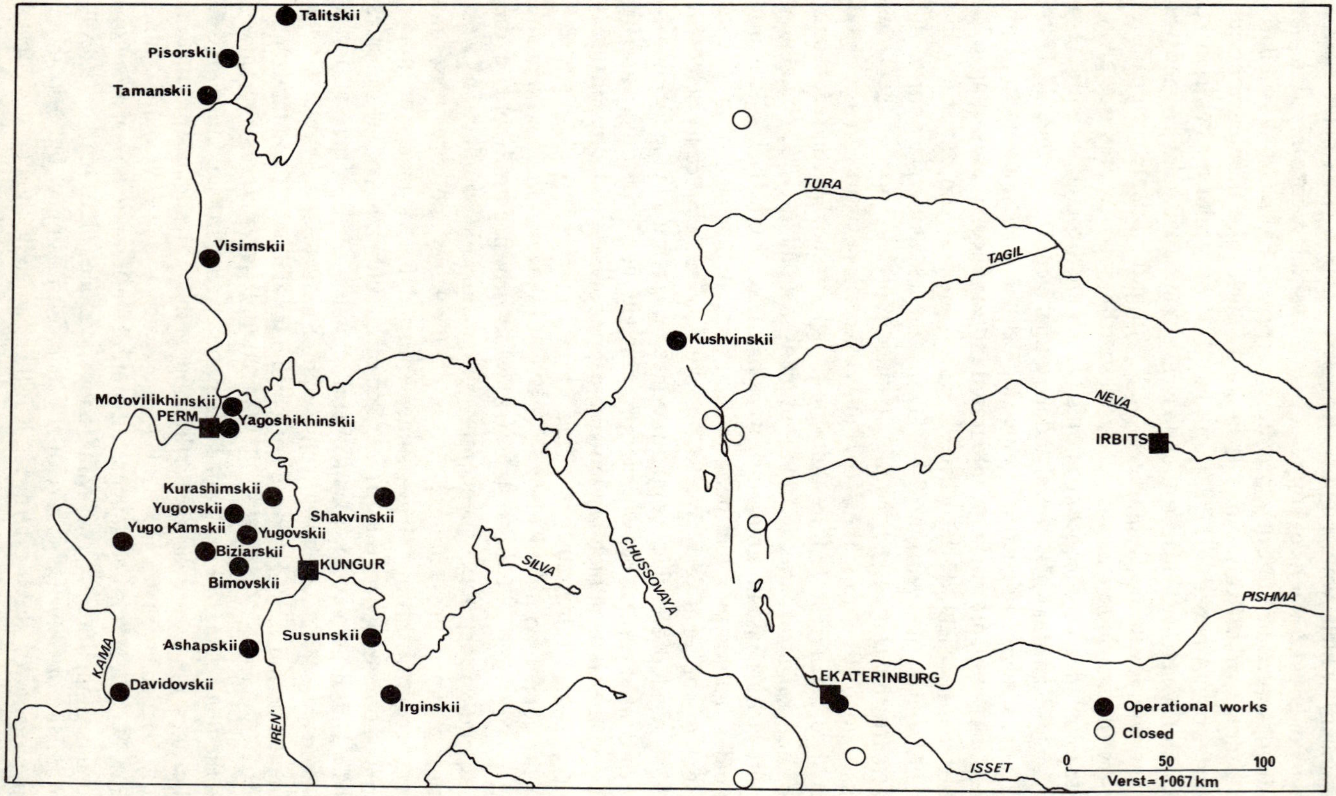

Map 4.2 Central Urals copper industry, 1745

were forced to cede a third of sales. During the quinquennium 1740–5, therefore, the Demidovs' Altai works dominated the expanding Russian copper industry, forcing private entrepreneurs to restructure their enterprises in the Urals completely and the state to subsidize its plants which ran at a continual loss.

Faced with this fiscal burden it is perhaps not surprising that, taking advantage of the contemporary investigation into the Demidovs' affairs which had been initiated in 1738, the state seized the Kolyvan works in 1744 and suppressed copper production therein.[64] Yet if this action was intended to raise prices and make the state 'factories' once more financially viable, then the Mining College officials who arranged it were to be sadly disappointed. After a brief rise in prices in 1743/4 they once more fell, dooming the state works to yet another phase of endemic loss-making.[65] The major beneficiaries of the removal of eastern competition were the private entrepreneurs who during the years 1744–5 carried through the final stages of their rationalization programmes.[66] Able to reduce costs of production and prices they came to dominate private copper markets, their plants by 1746 operating at close to full capacity (490 tons) with some 163 tons of copper being sold to private consumers and the residue to the state.

The early 1740s had thus witnessed a major transformation of the Russian copper industry. Underwritten by state guarantees, the private sector of the industry had expanded rapidly, production increasing from about 163 tons a year in 1738 when it was operating at barely one-fifth of total capacity, to 490 tons in 1746, when all plants were fully employed. In 1746 the works operating in this sector were, moreover, much more efficient than a decade before and were capable of undercutting the state sector of the industry which only maintained an annual output of 330 tons with the aid of heavy state subsidies. In comparison with the previous decade vastly greater quantities of copper were being produced and by undertaking to acquire the whole output of the state 'factories' and two-thirds of the copper produced in the private sector the state had at its disposal more than enough of the metal to produce a domestic coinage and finally dispense with foreign imports of monetary metals. From 1739/40, accordingly, domestic silver prices stabilized and the mint began to coin more and more copper pieces as mint activity steadily increased, rising from about 138,000 roubles a year in the period 1736–43 to 262,000 in 1744–5.[67] After some twenty years Peter's dream had finally been realized as, in 1744–5, an increasing mint output was manufactured almost exclusively from supplies of domestically-produced copper.

Copper minting during the first Russian age of silver, c. 1746–56

The boom which had begun in 1739–40 and which had continued to 1746, when output within the copper industry had risen to about 820 tons a year, thereafter showed no signs of faltering. New plants were opened up and within existing ones capacity was enhanced by the building of new hearths. In the central Urals the merchant Sekunov constructed the Uiskii works in the Sylva basin in 1749, but more significantly in 1745 the Simbirsk merchants Tverdichev and Miasnikov established their Voskresenskii works to exploit the rich southern Urals copper deposits at Kargalinsk.[68] Production, accordingly, again increased to c. 900 tons in 1751 and the productive potential of the rich south Urals deposits was for the first time realized. During the next quinquennium, moreover, the southern Urals fields really came into their own. Between 1750 and 1755 of twelve new plants constructed ten were in the southern Urals, only the Strogonovs continuing to extend industrial capacity further north with their building of the Domrianskii and Pozhevskii works to complete their industrial complex on the middle Kama. Activity about Kargalinsk was intense, the leading role in the development of the area being undertaken by the Simbirsk merchants Tverdichev and Miasnikov who, with the building of their Voskresenskii works in 1745, had first staked their claim to exploit the southern riches. Between 1750 and 1755 barely a year passed without their building a new works – the Preobrazhenskii (1750), Bogoyavlenskii and Blagovestchenskii (1751), and the diminutive Arkhangelskii (1753). Where the Simbirsk merchants led, others followed. The Tula merchants Ivan and Maxim Masalov and Gregori, Ivan, and Piotr Krasilnikov constructed their Kano–Nikolskii and Arkhangelskii–Saranskii/Isheryakovskii works in 1751 and 1754, Piotr Osokin his Troitskii in 1754, and Shuvalov and von Sivers their Pokrovskii and Vosnesenskii works in 1754 and 1755 respectively. By 1755 eleven new plants were operating on the southern ore fields at costs which could fall as low as one rouble per pud, and although, as prices fell, growth was associated with rationalization, the Demidovs' Davidovskii plant being closed in the face of southern competition, total output within the copper industry for the first time surpassed the 1,000-ton-a-year mark.[69]

This boom saw total output within the copper industry increase from 820 to more than 1,000 tons a year in 1746–56, but paradoxically, even as it began, wild rumours began to spread amongst the foreign merchant community about a possible withdrawal of the copper coinage. During 1744–6 the exchange for copper was suspended and those holding such coins frantically tried to get rid of them.[70] The merchants' fears were not entirely unfounded for, save during the years 1748/9–50, copper minting was steadily reduced until its final cessation in 1752–4, and new issues began to circulate only in Siberia.[71] The reason for this unexpected turn of events resided in the appearance of competitive supplies of metal to the Russian mints. Throughout the late 1740s and early 1750s silver and gold replaced copper as the primary monetary metal fabricated at the mints. This specie was not of foreign provenance but was derived from within the empire where, far to the east, production at Nerchinsk and Kolyvan began, for the first time, to increase rapidly.[72] In a pattern of booms (c. 1745–8 and 1751–4) and slumps (1749–50) the quantity of gold and silver derived from these eastern mines increased, engendering an inverse pattern of copper minting until, during the years 1751–4, with the realization of the vast riches of the Altai mines, which pushed annual Russian silver production up to about 400,000 roubles a year, copper minting at Ekaterinburg came to a halt. Within a monetary system which still retained that autonomy which had first been established in 1739/40, precious metals now reigned supreme and the copper industry originally created by the state to satisfy its requirement for monetary metals no longer had any raison d'etre: the rapidly accumulating stocks of unwanted copper left on the state's hands were merely a financial embarrassment for a state plentifully endowed with alternative supplies of monetary metals.

Russia, which only a short time before had been entirely dependent upon foreign supplies of monetary metals, had thus by the 1750s not only rendered itself independent of such foreign imports, but also had experienced a crisis, occasioned by a surfeit of such domestically produced metals for which a solution had to be sought. For the state this meant, in the short-term, abandoning its expensive commitment to buy copper for which it no longer had any use. Accordingly, even as silver roubles flooded from the mint, the mints for some three years (1752–4) ceased production of copper pieces and the state suspended purchases of that metal, only resuming its acquisitions in 1755 and 1756 at a reduced level when it undertook by proclamations issued on the 7/8 March 1755 and 9 October 1756 to purchase one-third of the private sector's output which it monetized, together with deliveries from the 'state factories', at reduced mint prices.[73] By such means supplies of copper from the private sector and state demands for a subsidiary copper coinage were brought into accord.

The problem of what to do with the copper produced in the state's own 'factories' remained, however, and, in the new circumstances prevailing in the early 1750s, the Treasury was much more amenable to listening to the advice of the commissioners appointed to inquire into the affairs of the state works than it had been before, undertaking during the years 1751–8 a rigorous programme of 'privatization'.[74] In 1757 the more profitable Yugovskii plants were disposed of to Count Ivan G. Chernishev. In 1758 the whole of the Kama river copper complex, comprising the Pisorskii, Yagoshikhinskii, Visimskii, and Motovilikhinskii works was sold to Count Mikhael I. Vorontsov. By 1760 only the Ekaterinburg complex remained in the state's hands. Within two brief years the state had virtually completely withdrawn from a direct involvement in the copper industry, transferring all of its problems to the entrepreneurs of the private sector.

That these entrepreneurs were willing to acquire these works which had been so burdensome to the state, however, suggests that during the 1750s market conditions were very different from twenty years before. As has already been noted, since that time non-state demand for copper had steadily increased, from 11–12 tons a year in 1736–8 to 100 tons in 1739–40, 163 tons in 1746, and 194 tons in 1751. Now, during the early 1750s, as prices fell and incomes amongst the populace rose, sales to private customers increased once more – to 700/750 tons a year.[75] Manufacturers for the first time found a ready market for their wares both at home and abroad. Exports which had averaged 2 tons a year from 1743–50 increased to 134 tons a year between 1751 and 1755 and peaked at 220 tons in 1753.[76] Domestic sales, which had approached 190 tons a year at the end of the 1740s averaged 600 tons during the next quinquennium, 1751–5. That industry, which less than twenty years before had been entirely dependent upon mint purchases of its wares for its survival, had come of age and was able to survive even the complete withdrawal of state support during the years 1752–4. As the mints turned out more and more gold and silver pieces during the years c. 1745–56, copper minting had steadily declined, before being finally extinguished during the years 1751–4, but private sales, of bells and copper bars for the manufacturers of Tula, had increased. A major boom, based upon the demands of private consumers, was under way and as entrepreneurs sought entry to the industry the possibility of 'privatization' of state plants became a reality.

The reforms of 1756/7 and the copper boom of 1756/7–62 In terms of the state's policy to rid itself of the burdensome Treasury copper plants and to establish a subsidiary copper currency on the basis of diminutive compulsory acquisitions of that metal levied upon the private sector of the copper industry conditions in the mid-1750s seemed highly propitious. Rapidly expanding copper production, ensured by burgeoning sales to private consumers, provided, on the basis of a one-third industrial levy, a regular supply of that metal to mint the new heavyweight copper coins (a kopeck coin weighed almost three-quarters of an ounce!) first issued in 1755. Similarly, boom conditions, in enhancing the value of industrial assets, seemingly proffered opportunities to dispose of state plants. Yet to some degree the boom was more precarious than it seemed, for growth was also associated with a process of market-induced rationalization. As prices fell from the 3-roubles-per-pud level of the 1740s to an average of 1.7 roubles in the mid-1750s the least efficient plants, with production costs of 2.25 roubles per pud, were rendered sub-marginal and only the new southern Urals works, operating at costs as low as 1 rouble per pud, enjoyed supra-normal profits.[77] The industry as a whole might have become independent of the state but as the southern Urals producers assumed market leadership those producers who continued to operate in the central Urals experienced considerable difficulties and the Treasury's plants, which were at best highly marginal units, were particularly exposed, rendering them far from attractive propositions for would-be entrants to the industry. To ensure the success of its 'privatization' programme and to facilitate the sale of its rather dubious assets the state, accordingly, was forced once more in 1756 to intervene.

For a brief moment in 1756 monetary and industrial policy became inextricably intertwined. Gambling upon that real growth within the economy which was creating strong deflationary pressures, new enhanced targets for mint output were set. Instead of the 350,000–400,000 roubles' worth of coin which had been produced each year solely from precious metals in the period 1752–4, the mint, having cleared existing stocks of copper during 1755–6, was henceforth to manufacture about 1 million roubles' worth of coin.[78] This expansionist monetary policy, moreover, was backed up by a profligate credit operation organized through the intermediary of the Copper Bank and Bank of the Artillery and Engineering Corps created in 1758, which loaned some 3.2 million roubles against copper reserves derived from new production and old copper cannon discarded from the arsenals. To meet such mint targets and to provide adequate banking reserves, the coining of copper once more became inevitable, and when, during the years 1757–60, production at both Nerchinsk and Kolyvan declined it became an absolute necessity.[79] Monetary considerations were thus probably paramount in the formulation of the reforms of 1756/7, as the state attempted to offset deflationary pressures in the economy whilst at the same time lowering interest rates, but their implementation could not but affect the success or failure of the Treasury's 'privatization' schemes. To ensure an adequate supply of copper to meet its new monetary and credit targets, in conditions of declining specie production, the state not only had to revive its former pre-emptive rights, once more acquiring two-thirds of the industry's output, but it also had to raise mint prices, striking 16 roubles' worth of coins from each pud of copper instead of the previous 8. As a result of these measures supplies of copper to the private market contracted sharply, and as prices more than tripled the state was forced to adjust its own acquisition price upwards to 5.5 roubles per pud.[80]

Market conditions were transformed with an immediate effect on the industry. Previously unrealistic schemes to sell state plants were now rendered viable. Even virtually worthless establishments, like the Nizhne– and Verkhne–Yugovskii works, were now worthy of acquisition, even if only for the stocks of ore which had been accumulating at their depots for some years past.[81] By its enhancement of prices the state had rendered even the most marginal of plants viable and and paved the way for the sale of its somewhat dubious industrial assets. It had also afforded intra-marginal producers considerable profits and, as a rising demand for copper amongst the populace caused prices to go on rising until they reached

11–12 roubles per pud on the private market in the 1760s, the conditions for a major boom were created.[82] As prices rocketed upwards the impetus to increase production was overwhelming. More and more new works were built and output rose, reaching c. 3,500 tons a year in 1762 when some 1,100 tons of copper were sold to private customers.[83]

Once again, in the new market conditions of the late 1750s, there was intense activity in the southern Urals. Tverdichev and Miasnikov now completed their large complex on the Belaya by building their Verkhotorskii works in 1759, and branched out to develop the deposits in the vicinity of Bakalskoe by constructing the Simskii works in the same year. The Osokins by building their Nizhne Troitskii and Usen' Ivanovskii works in 1760, similarly completed a major complex on the diminutive Belebei deposits where they were joined in 1762 by one Yaguzhinskii with his Kurganskii works. These major industrialists were now joined by others, moreover, one Luganin building the Zlatoustovskii works in 1760 and the merchant Tevkeleev his Varzino–Alexeevskii works in the same year. By 1762 the southern Urals had become heavily industrialized, the copper industry (with some eighteen 'factories') dominating metallurgy in the area.[84]

Trends already apparent in the early 1750s towards establishing the industrial ascendency of the southern Urals were thus brought to a final conclusion under state tutelage during the years 1757–62. The associated process of rationalization within the industry which had threatened to eliminate much of northern capacity at the earlier date, however, did not materialize. By enhancing guaranteed prices for deliveries to the mint, the state also underwrote the fortunes of northern industrialists who, from 1757 to 1762, enjoyed an 'Indian summer'. At existing price levels even the counts Chernishev and Vorontsov, who had taken over the bankrupt state 'factories', could make profits and the former even found it worthwhile to construct a new plant (Anninskii, in 1759/60) to smelt the ores derived from mines servicing his Yugovskii works.[85] In such conditions almost anyone could make profits and, accordingly, unlike during the earlier part of the decade, during 1757–62 the central Urals sector of the copper industry also experienced a boom. The Strogonovs added yet another plant (the Khokhlovskii) in 1756 to their central Kama complex and were joined for the first time by the Golitsyns, who built their Nytvinskii works in the same year to the south-west on the Nytva, a tributary of the Kama. One Pokhodiachin, a merchant-distiller of Verkhoturie, constructed an extensive works (Petropavlovskii) of seven copper hearths in 1758 on newly discovered deposits to the north of his home town and, finally, in 1760 the petty merchant Glebov added the Sermeitskii works.[86]

Thanks to state intervention in 1756/7 growth had been achieved in the industry without structural change, and as production increased from about 1,500 tons in 1755 to 3,500 tons in 1762 both the central and southern

Urals experienced boom conditions. Thanks also to that intervention and to a commitment to mint the copper it acquired, the state became involved in an expansionist monetary policy. Having set mint production targets at a million roubles a year in 1757 it soon found these norms exceeded. During the 'silver crisis' of 1757–60 copper minting not only made good the shortfall in the minting of precious metals but raised total output to just short of 2 million roubles a year. Nor did this commitment to copper minting waver when silver and gold production recovered to its former level in 1761–2 and total mint output again increased – to 3 million roubles a year.[87] Not only had the continuance of the industry's growth thus been assured but Russia had also been provided with an autonomous coinage which was sufficient, for the first time, to restore prices to the levels of the late seventeenth century.

Russian copper production and minting, 1762

As a result of some thirty years of uneven development Russia possessed in 1762 a major copper industry, capable of producing about 3,500 tons of metal a year, which ranked second to none in the contemporary world.[88]

Within this industry the enterprises of the central Urals, which were grouped into a series of large industrial conglomerates, still maintained their dominant position in terms of the number of operational hearths, old and new intermingled (map 4.3). Here were the great state 'factories', now held by the Chernishevs (Nizhne– and Verkhne–Yugovskii and the complementary Anninskii works built in 1759/60) and Vorontsovs (Yagoshikhinskii, Pisorskii, Visimskii, and Motovilikhinskii) and those of the Demidovs (Viiskii, Suksunskii, Shakvinskii, Bimovskii, and Ashapskii), created at the birth of the industry and restructured during the decade 1736–45 which, with the works of the Osokin brothers (Yugovskii, Biaziarskii, and Kurashimskii) dominated the Kungur region and stretched far to the north along the middle reaches of the Kama. Only in this northern extension of their industrial empires was the position of these founding fathers challenged – by the Strogonovs who from small beginnings in 1726 (the Tamanskii works) built up, particularly during the years 1746–56, a formidable industrial complex (Domrianskii, Yugo–Kamskii, Kuse–Alexandrovskii, Pozhevskii, Khoklovskii, and Nytvenskii[89]) on the middle Kama which, when divided between Gregorii Dimitrievich's three sons, endowed each with businesses as important as those of the Demidov heirs. In 1762 the great aristocratic families had established a hegemony over the central-Urals sector of the copper industry which still encompassed two-thirds of the total number of operational hearths. Amongst the merchant-industrialists only the Osokins proffered a significant challenge to their position, and even they were busy diversifying their activities by building new plant in the south. Others had either to

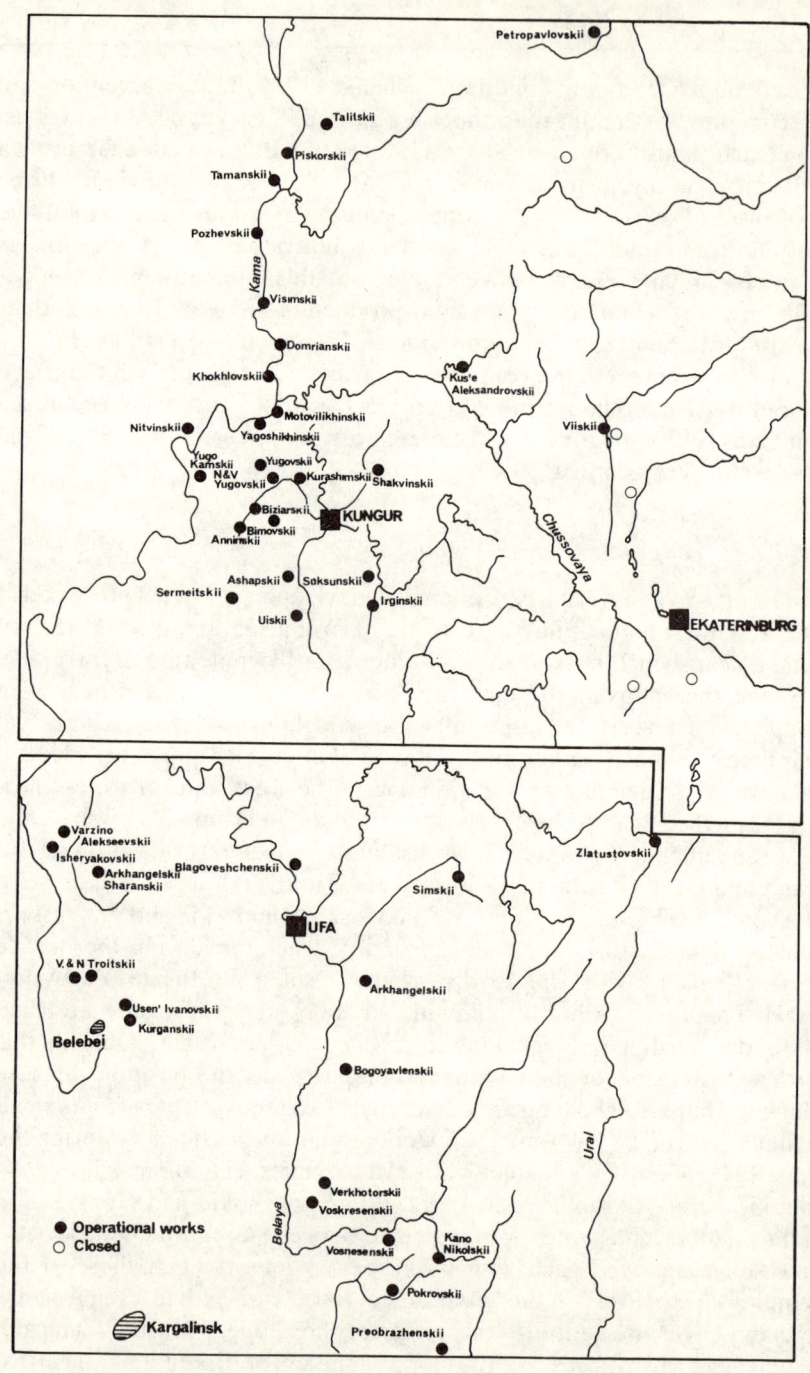

Map 4.3   Central Urals copper industry, 1762
Map 4.4   South Urals copper industry, 1762

content themselves with the crumbs from the aristocrats' tables, locating works on the periphery of the main production areas, or had to venture into remote areas to exploit new deposits. In the first category were Sekunov and Glebov, who built their Uiiskii and Shermyanskii (Sermeiskii) works to the south of the main Demidov complex between the Iren' and Babka, and Tuchaninov, who continued to operate his Talitskii (Troitskii) works in the upper reaches of the Kama to the north of the Strogonov complex. In the second category was Pokhodiachin who built his massive Petropavlovskii works in the under-populated forest lands between the upper Lyalya and Sasva, to the north of the ancient road from Solikamsk to Verkhoturie. Together, however, their contribution to total regional production was small, their works never encompassing more than one-third of the number of operational hearths in the central Urals.

In part the diminutive role played by merchant-industrialists in the central Urals was a result of their abandonment of this region in the 1750s for the rich deposits of the southern Urals, leaving the fortunes of their aristocratic counterparts to the vagaries of court intrigue. The initial pioneering venture in this area had taken place in 1745 when the local Simbirsk merchants, Ivan Tverdichev and his brother-in-law Ivan Miasnikov, laid the foundations of their future industrial empire by building their Voskresenskii works in the region of the upper Belaya. Thereafter, year after year, they built new works along the course of this south Uralian river to exploit the rich Kargalinsk copper deposits, creating what became by 1762 the greatest industrial complex producing copper in the Urals, comprising some six works exploiting the Kargalinsk ores (Arkhangelskii, Blagoveschenskii, Bogoyavlenskii, Verkhotorskii, Voskresenskii, and Preobrazhenskii) and one (Simskii) on the lesser Bakalskoe deposits, with a total annual output of c. 480 tons (map 4.4). Where local merchants showed the way others, moreover, soon followed. As early as 1754 the Osokins had begun their penetration of the south, building over the next six years a complex on the Belebeya deposits (Verkhne- and Nizhne Troitskii and Usen' Ivanovskii) where they were joined by one Yaguzhinskii who constructed his Kurganskii works in 1760. If the major ore fields had thus fallen to the already well-established merchant families, the newcomers, many of them from the old Tula industry intent on securing regular supplies of raw materials for their work to ensure their place in the contemporary consumer boom, were not slow to acquire the lesser deposits. In the north-west of the region on the tributaries feeding into the Belaya and lower Kama, the Krasilnikovs built two plants (Ishteryakovskii and Arkhangelskii–Sharanskii) and Tevkeleev one (Varzino–Alexeevskii). Far to the east, on the Aya, Luganin constructed his Zlatoustovskii works and to the south Masalov, another Tula industrialist, built his Kano–Nikolskii works. Everywhere in the southern Urals merchant-industrialists dominated the industrial scene, the aristocracy only being represented by two 'factories': Count Shuvalov's Pokrovskii works and von Siver's Voznesenskii.

By 1762 the focus of industrial activity was moving southward and existing counts of the number of operational hearths may well seriously underestimate the importance that the southern Urals had already assumed by that date, for whilst central Urals works like the Strogonovs' Pozhevskii plant or Tuchaninov's Talitskii works had already been reduced to working 1–1.25-per-cent ores, which reduced annual output per hearth to 7–8.75 tons, the new 'factories' of the Simbirsk merchants utilizing 3–4-per-cent ores from Kargalinsk produced 21–8 tons per hearth per year. This suggests, if such figures have any general significance, that, with total output running at *c*. 3,500 tons a year, the central Urals had already been eclipsed, contributing no more than 40 per cent (1,366 tons) in comparison with the southern Urals' 60 per cent (2,134 tons) to total production.[90] Yet whether the copper was derived from the somewhat tenuously placed northerly works or from the new, highly productive southern ones, with a combined output of *c*. 3,500 tons a year, the industry ranked second to none in the contemporary world and was capable of supplying not only a buoyant home market for copper, with some 1,100 tons of copper a year in 1762, but also of supplying the mint with enough metal to ensure a restoration of prices to the levels of the late seventeenth century.

Table 4.2 Estimated Russian monetary stocks, 1699–1762 (millions of 1764 roubles)

| | Baltic Provinces | | Stock | Import of silver | Muscovy | | Russian Empire |
	Stock	Increment Silver imports			Increment Dom. silver and gold	Prod. copper	Stock
1699			12	13.4	NIL	NIL	
1718/22			25.4	9.9	0.01	3.7	
1740			39	NIL	7.14	12.5	
1762	30.5		58.6				89.1

In matters of money supply, Russia was a very different country on the eve of the accession of Catherine II from the one which, during the years following the death of Peter I, had suffered from acute deflationary pressures as a result of the cessation of silver imports caused by the Oberharz mining crisis. Yet even at that time, during the years 1725–40, there were indications of the shape of things to come, as, in the face of contracting specie supplies, the native copper industry turned out enough metal to mint some 7 million roubles' worth of coin. Such sums were certainly insufficient to prevent a fall in Muscovite prices during the 1730s but they did at least temper the rate of decline, serving to differentiate Muscovite price movements from those prevailing in the recently annexed

Baltic provinces and elsewhere. Already by 1740 the Muscovite lands were beginning to experience an autonomous series of price movements. Subsequent years, moreover, saw this tendency become even more pronounced. As in the Baltic provinces and elsewhere prices began slowly to rise in Russia they continued to fall in the mid-1750s when, thanks to a change in state minting policy, the trend was reversed, prices finally stabilizing at a level 30 per cent above that of the early 1720s.[91] During the 1740s conditions had been dominated by the law of 1739 which ensured, during the first half of that decade, a level of copper minting sufficient to displace imports of specie entirely. As the decade drew to its close, however, the rapid growth of indigenous gold and silver production led to a displacement of the base metal in mint activity and raised hopes in government circles during the years 1745–55 for the restoration of a monetary standard based on gold and silver. By its commitment to a coinage based on precious metals, however, the state precipitated an acute deflation. Prices which had steadily risen to 1744 thereafter fell until 1756/7 when they stood at a level lower than that prevailing in 1737/8.[92] Only with the edicts of 1756/7 did this situation change when, as the state became committed to higher and higher mint output norms, rising to 3 million roubles a year in 1762, prices recovered, for the first time in forty years exceeding the levels of the early 1720s.

Copper minting during the second Russian Age of Silver c. *1762–80*

The commitment to high minting levels, which had characterized state policy during the years 1757/62, remained, in spite of marked fluctuations in the production of monetary metals, central to the economic policies of Catherine II during the early years of her reign (1762–80). In the 1760s and 1770s mint output rose from an annual production of 2.5 million roubles to 4 million, the rate of stock increase rising from 5.0 to 5.5 per cent per annum.[93] On occasion, in order to maintain short-term minting levels or to effect adjustments in the relative values of the metals comprising the monetary stock, the government had to alter mint prices; temporarily raising the price of copper in 1762 to make good a one-year shortfall in the supply of that metal or enhancing silver prices in 1763 to accommodate the declining importance of that metal relative to copper in the monetary stock. These were, however, merely managerial devices contributing little to the growth of the monetary stock. Similarly, the issue in 1769 of assignats – which could be obtained by the deposit of a corresponding amount of coin – whilst affecting the flexibility of the monetary system by allowing an adjustment of inter-metallic (copper-specie) stock balances, contributed nothing to the growth of the monetary stock.[94] The key, as before, to the massive augmentation of monetary stocks resided in changes in the production of monetary metals – and to the state's willingness to mint available supplies. In large measure, therefore, monetary expansion rested upon the increase in domestic precious-metal production which, as has already been noted, rose from about 400,000 roubles' worth of gold and silver in the late 1750s to one million in the 1760s and 2 million in the early 1770s.[95] Yet almost equally significant was the stability of copper supplies to the mints, in a period of major structural change within the industry which permitted total mint output to exceed 5 milllion roubles a year in the early 1770s and to average 4 million roubles throughout the whole of that decade.

Bearing in mind the troubled conditions prevailing in the copper industry during the 1760s and 1770s, the state's ability to maintain supplies of that metal to the mints was no small achievement. The industry was in chaos. Already in 1762 the central Urals industry of the Kungur region and the middle and upper Kama had experienced acute difficulties as a result of depletion problems in its resource base. As the metallic content of ores declined costs rose in the enterprises of the central Urals until they were operating at a level some two and a half times higher than their competitors in the south.[96] In conditions of rising prices between 1756/7–62, however, this had been of little import. When prices stabilized during the years 1762–80, it was a very different matter, particularly when the whole industry experienced an enhancement in costs as a result of changes in its labour market.[97] Wage rates which had remained stable to the 1750s thereafter began to rise. Workmen engaged in mining, woodcutting, charcoal burning, and carting, whose remuneration had remained unchanged for thirty years, began in the 1750s to demand, and receive, higher payments for their work. In the 1760s what had been sporadic demands became more general, and when in 1769 and 1778 the state intervened to enhance wages and regulate working conditions industrialists faced a massive increase in unit labour costs. No one was immune from the effects of the developing crisis.

Even in the southern Urals, where in the aftermath of Pugachev's rebellion (1773–4) there was an extensive rebuilding of copper plants which had been put to the torch, the more marginal works had been allowed to remain in decay (map 4.5). Shuvalov's Pokrovskii works and von Sivers' Vosnasenskii plant, poorly located in relation to the rich Kargalinsk ore deposits, which had been surrendered into the state's hands during the 1760s, and Yaguzhinskii's Kurgansk smelter, which had been intruded into the Osokin-dominated Belebei ore field in 1761 and had ceased production because of a shortage of raw materials in 1766/7, were all allowed to remain blackened shells after 1773, reducing the number of operational plants in the south Urals from eighteen to fifteen.[98]

Further north, in the central Urals mining region, the situation was far worse. Here the rise in labour costs seriously aggravated problems caused by a deterioration of the industrial resource base, and during the 1770s plant after plant closed down, normally after a protracted phase of declining production. Thus at Tuchaninov's Talitskii (or Troitskii) works, located on the upper reaches of the Kama, where in the face of southern competition output had been reduced to 11.25 tons a year between 1748 and 1752, production thereafter steadily fell to 9.64 tons in 1766/7 and 8 tons in 1770 before it was closed down.[99] Similarly the Demidov works, divided between Nikita Akinfievich and Alexandr Grigorivich, were already showing signs of depletion problems in 1766/7 when production at Nikita's Viiskii works was reduced to 5.4 tons a year before closures took place in the 1770s.[100] Most spectacular, however, was the wholesale closure of the Strogonov 'factories' on the middle Kama. First to fall was the Pozhevskii works, where in 1770 it was claimed production costs exceeded prices by as much as one rouble a pud, but others soon followed, both the older Tamanskii works and the newer Khoklovskii closing down in 1777.[101] Altogether with the closure of these plants and the Sermeitskii works, which had been sold by the merchant Glebov to Yakovlev in the 1760s, seven works or a quarter of the total number of central Uralian plants ceased production in the 1770s. The mighty industry which had dominated world production in 1762 was less than twenty years later in total disarray. Production was declining and the process of geographical relocation, suspended by government action between 1756/7 and 1762, was again under way.

Yet even as the southern Uralian works displaced those of the central Urals industrial region during the years 1762–80 both were subject to intense competition from new enterprises established on the eastern slopes of the mountain range. From about 1772 a whole series of new smelting furnaces were built in an arc about Ekaterinburg, normally on sites already occupied by iron-smelting furnaces. Thus Saava Yakovlevich Yakovlev, having bought the inheritance of Prokope Demidov during the years 1767–9, established two new copper smelting hearths at Neviansk and three new ones at the Byngovskii works before 1777.[102] Similarly Nikita A. Demidov, having acquired the Kaslinskii iron works, built by the Tula merchant Korobkov in 1752, installed two copper smelting works there in 1772. The Strogonovs added yet another two hearths at their Bilimbaevskii iron works and the state one each at its Kamenskii and Turinskii 'factories'. In all eleven new furnaces were built in these variously owned enterprises which, drawing on the same resource base as Tuchaninov's Polevskii and Sisertskii works, may have produced about 300 tons of copper a year, bringing the total output of this new eastern Urals copper industry to about 720 tons a year.[103] As the old central–southern Urals copper industry thus declined a new and important metallurgical complex was emerging on the eastern side of the range exploiting ores capable of yielding more than 4 per cent refined metal. The once-solitary Tuchaninov smelters had by 1777 become part of a major industrial complex.

Yet important as this new complex was, it was to be eclipsed as a result of changes taking place further north. The discovery of ores, capable of yielding 10–15 per cent refined metal in the vicinity of the old Lyalinskii works, allowed the Verkhoturie merchant Pokhodiachin to establish a plant – the Bogaslovskii – which would rank second to none in the industry and with his Petropavlovskii works produce a quarter of its total output.[104]

Once again, therefore, during the years 1762–80 the Urals copper industry had undergone a process of major structural change. In conditions of stable prices and rapidly rising costs the process of relocation towards the south, briefly interrupted during the years 1756/7–62, was resumed. As production within the central Urals mining region declined, both as a result of plant closures and declining output from the remaining operational plants, the output of the southern Urals enterprises was maintained. In spite of the closure of a number of marginal works and the resiting of one or two others, most remaining south Urals plants were able by 1780 to increase output sufficiently to compensate.[105] By that date, however, both the central and southern fields had been eclipsed by new industrial complexes in the eastern part of the range which now produced nearly half the industry's output. After two eventful decades production had thus once more been restored to the levels of the early 1760s but only after a major crisis which had seen output fall, at the height of the Pugachevshchina, to little more than 1,000 tons a year.

Whilst in the long term the state's supplies of copper had thus been secured, mint masters experienced a series of acute short-term copper supply crises during the years c. 1762–77, which not only created problems of maintaining mint-output levels but also caused difficulties of ensuring an appropriate inter-metallic (copper/gold–silver) balance within monetary stocks. To resolve these problems the state was forced to intervene. Initially, as copper production within the Urals slowly declined from its high point in 1762, this involved tapping alternative supplies from the Kolyvan mines where before 1764 'the greatest part' of the copper produced 'had hitherto been of no use'.[106] To facilitate the conversion of this metal into coins the construction of a new mint was authorized in 1763, and with its opening, at the Susunsk works of the Kolyvan metallurgical complex in May 1764, a new age dawned in the monetary history of Siberia.[107] Henceforth it enjoyed its own autonomous source of money supply which produced not only small quantities of gold and silver coin but also some 2,000–7,000 puds weight of copper pieces struck each year at the enhanced rate of 25 roubles per pud, and by 1781 had created a local, indigenously-produced currency of some 2.5 million roubles. Relieved of the responsibility for provisioning the Siberian lands with coin the Ekaterinburg mint from 1764 utilized available Urals copper supplies only to provision European Russia, its pieces being 'transported by water to Moscow, St. Petersburg and other parts' and becoming 'current through

Russia'.[108] In the short term the problems occasioned by the shortfall in Urals copper production had been resolved but with the deepening of the crisis as a result of the edict of 1769, regulating the working conditions of the ascribed serfs, more drastic measures were required. In that year, in order to maintain copper supplies at the Ekaterinburg mint, therefore, the state extended its pre-emptive rights over the industry, increasing the amount of its compulsory purchases from the private sector from two-thirds to three-quarters of total output.[109] As a result of these measures and an enhancement in the price paid for residual output delivered to the mint, copper minting at Ekaterinburg, which had fallen to about 1.47 million roubles' worth of coins in 1768, recovered during the years 1769–73 to the levels of the early 1760s. In the last year 2 million roubles' worth of coins were struck from copper supplies derived from state works, compulsory deliveries, and purchases from the private sector. The barrel was being scraped dry and in that year supplies to the private market were drastically reduced to 285 tons.[110] When, therefore, the industry felt the full impact of Pugachev's rebellion the next year the state had no room to manoeuvre and from 1774 to 1777 minting at Ekaterinburg fell far short of requirements, only recovering to the output levels of *c.* 2 million roubles a year thereafter as a result of expansion in the new eastern Urals industrial complex.

For two decades the state had struggled, not always successfully, to maintain copper minting at or above 2 million roubles a year. The main thrust of monetary expansion had, however, come from elsewhere as the metallurgical complexes at Nerchinsk, Kolyvan, and Beresov poured forth an increasing supply of gold and silver. Yet in spite of this flood of precious metals entering circulation, supply fell far short of contemporary demand. The demand for small denomination copper coins, for the purposes of petty transactions, which had grown steadily during the years *c.* 1730–62, when such coins comprised 32 per cent of the monetary stock, was now waning and as *per capita* requirements for such pieces fell so did their importance in money supply, amounting to barely 10 per cent of stocks in 1786. Contemporary demand for gold and silver pieces of large denomination, on the other hand, rapidly increased during the years 1762–88. To redress the imbalance between the currency demands of society and the existing structure of monetary stocks, and to alleviate the resulting shortages of gold and silver coins, the state, accordingly, in 1769 began to issue assignats – paper money obtainable on the deposit of an equivalent sum in coin. The measure was an instant success. Surplus copper coins were exchanged for assignats. Thus as total monetary stocks in the 'Russian lands' increased during the years 1762–86 from 59 million to 136 million roubles the amount of copper in circulation actually fell from 18 million roubles to 15 millions. The residual output minted at Ekaterinburg during these years was exchanged for assignats, augmenting the 80 million roubles' worth of gold and silver coins in circulation in 1786 with a further 46 million roubles' worth of paper notes.[111] Whilst the introduction of the assignat in 1769 thus altered the balance of monetary stocks, favouring larger denominational issue at the expense of smaller, it in no way before 1786 affected the rate of growth of monetary stocks which, having risen from 5 to 5.5 per cent per annum during the 1760s and 1770s, now, in the 1780s fell to 2.5 per cent per annum.

Deflationary policies thwarted: copper minting c. 1780–95

Even as the rate of growth of 'Russian' monetary stocks began to decline during the 1780s changes in the economy ensured that the impact of monetary changes on domestic price levels would be exaggerated. Indeed, after about forty years (c. 1735–75) of relative price stability in conditions of rapid stock augmentation, in the 1780s, even as the rate of stock growth began to decline, Russia experienced the beginnings of major inflation which was of sufficient magnitude to cause alarm amongst both the populace and its rulers.[112]

In this new situation the government acted with resolution, determining to reduce minting levels. In this objective it was aided by the developing crisis in the precious-metal industries where production, having peaked in the early 1770s, declined until c. 1785.[113] Yet the problem of copper minting, which had been sustained with so much effort at about 2 million roubles a year during the 1760s and 1770s, remained. On 28 June 1780, therefore, an ukaz was promulgated, reducing the state's pre-emptive purchases to 50 per cent of the production of the private sector of the copper industry, raising the contract price for its acquisitions from 5.5 roubles per pud to 10, and permitting the free sale of the residual output including its export on the payment of a small duty.[114] In one measure it thus attempted to encompass two policy objectives. At prevailing production levels within the industry, which during the years 1780–2 amounted to c. 190,000–200,000 puds (c. 3,050–3,251 tons) a year, it reduced the amount of copper passing to the Ekaterinburg mint from 160,000 puds (c. 2,560 tons) in 1780 to 116,800 puds (c. 1,877 tons) in 1782.[115] By raising the price it paid for its acquisitions it also went some way to alleviate the costs imposed on the industry by its labour reforms of 1779. At prices prevailing at the time of that reform producers had received about 7 roubles per pud for their copper. Now it was anticipated they would receive 10.5–11 roubles per pud, delivering half their output to the mint at a guaranteed price of 10 roubles per pud and disposing of the rest on private markets where prices had remained at 11–12 roubles per pud for twenty years or so. In such a manner the government proposed to reduce copper minting whilst at the same time protecting the copper industry, isolating it from the effects of the 1779 labour reform, and effecting its reorientation towards private markets.

In formulating its plans, however, the government had not taken into account the changes which were already under way at that time in the private market for copper. Already in the 1770s when, by increasing its pre-emptive purchases, it had reduced supplies to private markets, prices had not risen. Now in the 1780s as the copper placed on the market increased from *c*. 500 tons in 1780 to *c*. 1,500 tons in 1782 (or about one-third more than in 1762), prices fell – to 9 roubles per pud.[116]

The effects of this price fall, which left average sales prices only one-third above the level of the previous decade, were unexpected, encouraging a continuance of the rationalization process observable in the 1770s. In the central Urals the industry was decimated. The Chernishevs and Vorontsovs, having observed production decline at their Anninskii, Yagoshikhinskii, and Visimskii works, surrendered them into the Treasury's hands in 1780 only to witness their continued decline and closure during the year 1785–8. Nor were the other great aristocratic and merchant producers immune from the effects of the developing crisis. The Strogonovs, having closed down three major works (Tamanskii, Pozhevskii, and Kholklovskii) in the 1770s, now ceased production at their Domrianskii and Nytvenskii plants in 1786 and 1788. Both the Demidovs and Osokins also closed down smelting capacity at their Suksunskii and Irginskii works, converting these enterprises to fabricating metal from their remaining Ashapskii–Bimovskii and Yugovskii–Biziarskii works.[117] By 1790 the once-extensive central Urals industrial complex had become restricted to a diminutive number of plants located in the immediate vicinity of Kungur (map 4.5). As central Urals production declined and plant after plant fell into the state's hands, however, there was a countervailing expansion of output in the southern and eastern part of the range. In the Tverdichev–Miasnikov industrial empire on the Belaya, for instance, most works were operating after 1783 at levels well above that prevailing in the years 1769/73.[118] Aggregate production levels were, accordingly, sustained, continuing at about 3,500 tons a year through the 1780s and only falling slowly to *c*. 2,500 tons in 1795 as the southern Urals industry began to suffer from depletion problems.[119]

Yet whilst production was sustained and the state's purchases from the private sector were reduced, leaving each plant manager able to sell more to private customers, such sales did not materialize. As the balance between the state and private sectors of the industry was disturbed, due to private 'factories' falling into state hands, the production from publicly-owned works steadily increased from 223 tons in 1773 to 750 tons in 1785 and *c*. 1,100 tons in 1795 after the Assignat Bank had acquired the great Pokhodiachin industrial complex.[120] Private production correspondingly declined, and with it private sales from *c*. 1,500 tons in 1782 to 1,400 tons in 1785, 1,250 tons in 1790, and 720 tons in 1795. Each manager might sell

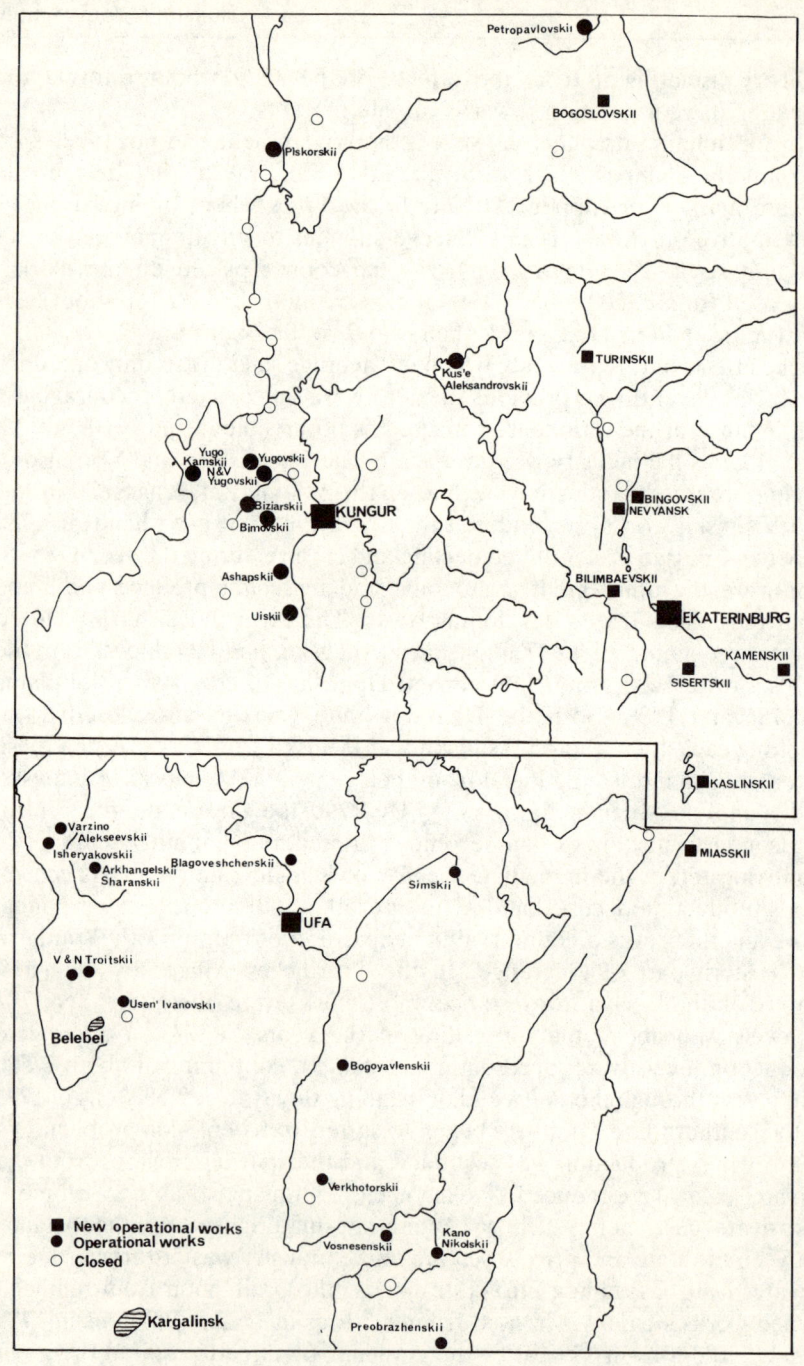

Map 4.5 Central Urals copper industry, 1795
Map 4.6 South Urals copper industry, 1795

more of his plant's output but there were fewer plants. Declining mint activity as a result of the reduction in pre-emptive purchases was being counterbalanced by an increased minting of copper from state works. Far from being dramatically reduced copper minting was largely maintained, continuing at about 2 million roubles' worth of coin during the 1780s and only falling to 1.8 million in the early 1790s.

The state's plans to reduce minting levels, so carefully laid in 1780, had thus by 1795 been brought to naught. Mint output had been sustained and metallic monetary stocks within the 'Russian lands' continued to grow, attaining a level of *c.* 141 million roubles in 1788 and 167 million in 1795. Copper coins surplus to the requirements of the population continued to be converted into assignats, at least until 1786, the amount of copper in circulation falling to *c.* 15 million roubles and the stock of secured assignats rising to 46 million. The fateful decision to issue 54 million roubles' worth of unsecured notes in 1787, however, totally altered this situation and presaged the course of events to come. Henceforth, as the issue of assignats continued, unrestrained by the necessity of a metallic backing until they reached a level of 150 million roubles in 1795, they increasingly surpassed the quantity of coins in circulation; and the notes, which until 1786 had passed at or very near to parity with 'silver', fell heavily in value to nearly 30 per cent below par in 1795.[121] In such circumstances no one would deposit coins against notes, and a dual multi-metallic and paper monetary system emerged.

The high levels of mint activity, first established on the eve of Catherine II's reign, were thus sustained for the next thirty-five years from *c.* 1760 to 1795, output rising to a peak of 5 million roubles a year in the early 1770s before slowly declining to 4 million in 1780, 3.5 millions in 1790, and 3.3 millions in 1795. Monetary stocks steadily increased but the rate of increase which had continued to rise to *c.* 1770 thereafter rapidly declined to 2.4 per cent annually in 1790 and 2.1 per cent in 1795. Unlike the years *c.* 1735–75 when the rapid growth of monetary stocks had been associated with relatively stable prices which only in *c.* 1775 reattained late-seventeenth-century levels, during the decade 1775–85 slow growth in that stock caused major inflation. Consumer demand for money was rapidly changing. The quantities of money in circulation were far in excess of what was required to satisfy transactions demand in the economy at constant prices. The coins produced were, moreover, totally inappropriate to market usage, a feature only rectified during the years 1769–86 by the issue of assignats which allowed the conversion of those copper coins which were surplus to the population's needs into high-denomination paper notes. Thus as copper minting levels were maintained and the copper coinage, which had enjoyed such a prominent position in the Russian currency in 1762, lost ground, the volume of assignats increased, augmenting specie stocks, which in 1786 amounted to 80 million roubles, with an additional 46

Table 4.3 Estimated Russian monetary stocks, 1762–95 (millions of 1784 roubles)

	Baltic provinces			Muscovy Increment			Domestic production			Russian Empire
	Stock	Increment Silver imports	Stock	Silver imports	Gold/ silver	Copper	Paper Secured	Unsecured		Stock
1762	30.5	30.0	58.6	nil	37.2	nil	46	54		89.1
1788	60.5	11.3	141.0	nil	6.8	18.3	nil	50		201.5
1795	71.8	71.8	166.9	35.3	51.2	34.4	46	104		238.7
Stock in 1795	71.8		166.9					104		238.7[a]
1795		71.8	158.3	86.5		34.4	80.4	150		

Sources: See appendix A2 (1), tables 3, 4–7.
Note: a unsecured notes are not included in monetary stocks which, as in tables 4.1–4.2 encompass metallic issues.

million roubles' worth of notes which circulated at or very near par with specie. Stock growth was, accordingly, associated with inter-metallic balance in the years to 1786. The decision in 1787 to issue a further 54 million roubles' worth of assignats without coin backing, however, introduced a new element into the system. Henceforth as the issue of assignats continued without the necessity of a metallic backing, until they reached a level of 150 million roubles in 1795, they surpassed the quantity of coins in circulation and went at a discount, falling to thirty per cent below par in 1795. A new age was dawning, and as silver-commodity prices began once more to stabilize Russia started for the first time to experience a paper inflation.

The end of an era: bi- and multi-metallic systems in disarray, 1795–1809

The years 1787–95 had seen the beginning of the end of the old monetary system. The government's subsequent prodigal use of the printing press in the period to 1809 and the resultant suspension of convertibility ensured its ultimate demise. Henceforth the old system was replaced by a new one as the burgeoning volume of unsecured paper was divorced from its metallic base. Spatially, the new system was no longer administratively confined to the Muscovite heartland but from 1802 encompassed also the previously separate Baltic provinces and the southern and western lands annexed in the late eighteenth century, welding their diverse monetary elements into a unitary 'rouble' area. Functionally it assumed a dual character, involving the operation of a metallic standard (the silver rouble) to which the other metals were related in what has been described as a 'parallel circulation' or a 'limping standard', and a paper standard (the assignat) which evolved independently of each other and were rendered exchangeable on the basis of constantly changing bourse quotations.

The transition between the two systems was a complex one although initially as far as the metallic system was concerned there were few problems. Domestically produced monetary metals continued to flow to the mints, if in smaller quantities than before. In all between 1795 and 1809 some 25 million roubles' worth of gold and silver coins, made from indigenously-produced metals, entered circulation, together with 16.5 million roubles' worth of copper pieces.[122] Supplies were now augmented from another source, moreover, as for the first time in sixty years foreign specie began to enter the Muscovite heartland of the empire. As has been shown, such imports had long been a feature of the trade of the Baltic provinces. Year after year from the late 1750s the quantities of silver entering these lands through Riga had increased, some 1.72 million roubles' worth of specie, in the form of coins and ingots, entering the port annually during the years 1758–68, rising to 1.8 million annually between 1775 and 1777 and 3.1 million a year in 1783–4. The inter-cyclical crisis which beset the Central and South American silver industry during the mid-1780s had then left its imprint on the trade, but by 1793 imports of specie had reattained the levels of the early 1780s and growth continued unabated to 1796. During the three successive years 1794, 1795, and 1796, 3.1, 4.7, and 4.9 million roubles' worth of precious metals entered the port for transmission to the monetarily autonomous Baltic lands and the recently annexed territories to the south.[123] At this point in time, however, traditional trade patterns were thrown into disarray. Imports of specie through Riga declined, and for the first time in sixty years significant quantities of foreign specie were shipped to St Petersburg.

During the period 1797–1801 imports of gold and silver at the Russian capital increased, rising to 3 million roubles' worth of bullion in 1801, before decline set in and Riga again established its hegemony over the trade. Thus during the years 1795–1809 an intermittent but not unimportant flow of gold and silver into Russia combined with the domestic production of these metals to augment the nation's monetary metal stock which, rapidly growing in the 1800s, attained a level of 272 million roubles in 1802 and 337 million in 1807.[125]

Table 4.4 Specie imports into Russia, 1796–1803[124] (millions of roubles)

	Riga	St Petersburg
1796	4.9	0.3
1797	0.2	1.2
1799		1.2
1800	n.a.	2.8
1801		3.0
1802	9.2	1.5
1803	10.1	1.0

This not unspectacular growth in the stock of monetary metals was, however, completely overshadowed by the increasing issue of unsecured paper notes which during the years 1795–1807 flooded the economy, leading to a collapse of the old monetary order and the emergence of a new one. Administratively until 1802 change took place within the old 'rouble' exchange area, encompassing the territories of the Muscovite heartland, but even with the incorporation of the Baltic coinages in a common imperial monetary stock the volume of assignats being issued was such that, after a brief appreciation of the paper notes in 1802–3, they thereafter continued their relentless fall in value. If the creation of the monetary union had not stayed the internal fall in the value of the assignat, however, it did go a long way towards preventing its external collapse. Until 1802 the foreign exchange value of the assignat, tied to the level of specie rather than total coin stocks, had always fallen more rapidly than its internal value until on the eve of the union the two diverged by as much as 30 per cent,

Table 4.5 Estimated Russian monetary stocks, 1795–1807 (millions of 1764 roubles)

| | Baltic provinces | | | | Muscovy | | | | | Russian Empire |
| | Stock | Increment specie imports | Stock | specie imports | Gold/ silver | Domestic production | | Increment Paper | | Stock |
						Copper	Secured	Unsecured		
1795	71.8	5.0	166.9	5.1	13.5	9.6	nil	80.5		239
1802	76.8	– 7.8	195.0	54.5[b]	11.7	6.8	nil	302.9		272
1807	69.0		268.8							337
Stock in 1807	69	69	268	95.1	76.4	50.8 / 96.8	46[a]	487.4 / 487.4		337
1807		240.5	240.5	171.5	50.8	50.8		533.4		

Sources: See appendix A2(1), tables 3.5–7.
Notes: a As in tables 4.1–4.4 monetary stocks include only metallic or metallic-backed issues. b Following the integration of the two monetary zones Muscovy imported specie not only directly (11.4 million roubles) but also via the Baltic provinces, receiving not only new supplies (35.3 million roubles) but also part of existing Baltic stocks (7.8 million) which were now reminted.

creating a basically unstable monetary situation. Now with the rapid augmentation of specie stocks as a result of the incorporation of the 'Baltic' coinage into the imperial monetary stock, the differential between the two closed, paving the way for the monetary reform of 1810. In that year the bi- and multi-metallic systems of earlier years were swept away, and the non-argentiferous coins (gold, platinium, and copper) were relegated to a 'parallel circulation' only loosely related, by a 'limping' standard, to the primary metal.[126] As a result of these measures the internal and external values of the currency converged but, with the reduction of the metallic monetary stock which was now based only on silver, they also fell, the value of the assignat rouble declining in the course of 1810 from 44, to 33, to 25 silver kopecks. A new order had been born and whilst the total monetary stock continued to encompass all silver, gold, platinum, and copper in circulation, whether domestic or foreign, only silver was recognized as providing the monetary standard, for fixing the external exchange value of the currency (which remained as it had been since 1763, in the case of Britain, 6.32 roubles per pound sterling or 38 pence per rouble) and evaluating the assignats in circulation (which exchanged in 1810 at four assignat roubles per silver one). Stabilization had been achieved. Now began the process of retrenchment under the new system, for the objectives of Sperenskii's reforms were not only stabilization but indeed the wholesale restoration of the assignat to its face value. To this end, he proposed a redemption operation financed from the proceeds of sales of state lands and from loans floated on domestic money markets. He was not entirely unsuccessful, being able in the course of 1811 to retire 5 million roubles' worth of notes from circulation.[127] The years 1810/1811 thus witnessed the last wistful acts of a conservative Ministry of Finance to restore the old order. As events were to prove, however, they were fruitless, and the same years in fact witnessed the beginning of a new one, based on a double silver and paper standard.

Postscript. Monetary policy under a double standard, 1810–39/43 During the years 1810–43 this new Russian monetary regime was structured about a double silver and paper standard, and although an examination of how it operated is beyond the scope of this study it is perhaps permissible to delineate some of its salient features in order to set the pre-existing system in context.

During the years 1810–43 this new Russian monetary regime was structured about a double silver and paper standard, and although an examination of how it operated is beyond the scope of this study it is perhaps permissible to delineate some of its salient features in order to set the pre-existing system in context.

Having started on the path of retrenchment in 1811, a year which saw the assignat rouble actually appreciate to 26 silver kopecks, Sperenskii saw his whole scheme fall apart under the brutal impact of Napoleon's invasion. As French troops crossed the Nieman silver and gold roubles totally disappeared from circulation, leaving the populace – if contemporary estimates of the magnitude of the flight bear any resemblance to reality – dependent upon diminutive supplies of copper coins and a rapidly growing volume of assignats to meet their monetary needs.[128] The monetary system was thrown into total turmoil and the government was forced to intervene, temporarily suspending the silver standard and in 1812 making the assignat legal tender on all transactions at the contemporary bourse rate of 4 assignat roubles to the silver rouble. For three years the assignat rouble became a *de facto* 25-silver-kopeck note, exchanging in practice not against silver but at a value conditioned by the volume of notes in circulation. Thus as between 1812 and 1815 the volume of assignats in circulation increased by 27 per cent so the value of the assignat rouble fell by a corresponding amount, five notes being exchanged where four were required at the official rate.[129] In 1815 the Russian monetary system was in total chaos – but Napoleon was defeated and with the restoration of public confidence the way lay open for a reconstruction of that system. During 1816–17 gold and silver coins were released from hoards and activity at the mint was intense, 43.6 million roubles' worth of gold and silver coins being struck, largely on the basis of foreign specie supplies.[130] Specie stocks were recovering rapidly and on reaching a level of *c.* 196 million roubles, commensurate with the excess paper issues of 1812–17, in 1817 the internal value of the assignat was restored to its conventional rate (four assignat roubles to a silver rouble) and assimilated with its external value.[131] The boom did not stop at this point. Indeed in 1818 metallic money was available in such quantities and circulated so widely that many people found it difficult to accumulate the assignats which, since 1812, were required for the payment of taxes, and the government was forced, on occasion, to accept silver, but at the discounted rate of 3.6 assignat roubles to a silver rouble, reflecting the contemporary level of silver stocks of *c.* 230 million roubles.[132]

By 1818 specie stocks had been restored to the pre-crisis levels of 1809 but the legacy of the events of 1812 remained as the denizens of the St Petersburg bourse continued to adhere to the conventional exchange rate which undervalued the assignat and in the new circumstances existing in 1818 reversed prevailing specie flows. Even as specie stocks adjusted to equilibrate the internal and external values of the currency, however, government intervention served to transform a short-term market-induced outflow of silver into an endemic flight of that metal from Russia. Debate over the state of the currency had been an intrinsic feature of discussion in court circles ever since Sperenskii's reform of 1810/11, but with the restoration of normality in 1817/18 the Ministry of Finance merely resumed the policies abandoned in 1811, Gurev proposing to withdraw assignats from circulation with money raised on domestic and foreign loans.[133] Thus as during 1818 the outflow of silver served to lower the foreign exchange value of the assignat and to bring it into conformity with the 'conventional' internal rate, the government's policy of reducing the number of assignats in circulation operated to enhance the exchange, maintaining differentials and encouraging a continuing outflow of specie. From 1818 to 1823 silver flowed out of the realm, until with the abandonment of the policy of assignat withdrawal the outflow was halted. Differentials once more closed at an internal–external rate of 3.78 assignat roubles per silver rouble, and monetary stocks stabilized at 595.8 million assignat roubles – and, 158 million silver roubles or two-thirds of the level of 1809, 1818. The level of specie stocks had finally stabilized at a level consistent with the perpetuance of the 'conventional' exchange.

It now only remained for Gurev's successor, Egor Frantsevich Kankrin, to hold the line, and remarkably successful he was at the task. Following the difficult period up to 1823, the monetary system involving both metallic (the silver rouble) and paper (the assignat) standards formally maintained a remarkable stability until the restoration of convertibility twenty years later. The stock of assignats, having been reduced to 595.8 million roubles as a result of Gurev's redemption operation, thereafter remained unchanged until the reforms of 1839/43. Stocks of silver also having been reduced, to two-thirds of the 1818 level in 1823, thereafter slowly increased to 168 million roubles on the eve of the reforms. Exchange rates between assignats and silver, accordingly, slowly rose until they attained a level of 3.52 assignat roubles to a silver rouble on the St Petersburg bourse in 1838/9, leading Kankrin to anticipate that the notes would move to par – in *180 years*.[134]

The deployment of the double standard currency in commercial rather than monetary transactions posed somewhat more difficult problems for not only was the course of the bourse quotations subject to short-term temporal fluctuations, but also there was no single ruling exchange rate, the market being highly fragmented. There thus emerged a series of 'popular exchange rates' which provided a three-dimensional embodiment of prevailing local market conditions, defining not only the exchange relationship between assignat and silver roubles but also that which existed between these currencies and commodities priced in 'conventional' roubles.[135] Goods, accordingly, in this system of accounting were evaluated in independent units which were rooted in the official exchange of 1812 when four assignat roubles had been said to be worth one silver rouble. To all intents and purposes, as far as customers were concerned, goods were priced in assignat roubles valued at 25 silver kopecks, temporal and spatial deviations from this rate being accommodated by the discounting of prices by a 'popular agio' which divided the difference between the conventional and actual exchange rate between payers in silver and paper in accordance with the relative scarcity of the two currencies. Thus in 1837 in the capital, where the bourse rate prevailed and the populace were adequately supplied with both metallic and paper currencies, customers buying commodities, priced in 'conventional' roubles, were surcharged if they paid in silver, having to render 26.67 silver kopecks to 25 'conventional' roubles, reflecting the appreciation of the assignat embodied in the bourse/local market exchange rate of 3.52 assignat roubles per silver rouble and dividing the 12-per-cent enhancement equally between payers in silver and paper. Elsewhere the situation was very different. In the western and south-western provinces, more highly monetized and particularly well supplied with assignats, those paying in notes only received a discount of 4.75–5 per cent as against those paying in silver who were surcharged the same 6 per cent paid in the capital, the abundance of paper money being reflected in the market rate of 3.6 assignat roubles per silver rouble and the small net discount (− 1.–1.25 per cent) reflecting the greater availability of money in general. In contrast, on the markets of the lower Volga, where money was in generally short supply (relative to goods), the silver surcharge was small, amounting to 2.5 per cent, indicating that the market had not benefited greatly from the increased supplies of that metal, and the discount on assignats was large – 10 per cent – reflecting their scarcity. Overall the high level of net discount (− 7.5 per cent) was indicative of an acute shortage of coins and notes but the market rate of 3.51 assignat roubles per silver rouble reflected a balance within the depleted monetary stock which was broadly similar to that prevailing in the capital. Through a double mechanism, in theory, both temporal and spatial variations in the values of the two currencies could thus be related to the official bourse rate which was based on stable but heavily depleted silver stocks.

In practice the pattern of long-term stability was more fragile than it appeared. Total monetary stocks, far from revealing that depletion and stability which characterized silver stocks, steadily recovered from the war-time crisis and its immediate aftermath, reattaining the levels of 1809 and 1818 during the 1830s, and by the 1850s exceeding them by some 40 per cent as a flood of non-argentiferous coins (predominantly gold) entered into 'parallel circulation' with silver.[136] In a normal bi-metallic situation the resultant imbalance in monetary stocks would have resulted in an enhancement in the price of silver (or a calling down of gold) to accommodate the declining importance of the white metal relative to gold. Because of its commitment to a silver standard, and because of fears of destabilizing the assignat, however, such a course of action was not open to the government. Yet where it would not take the lead, the market would. From c. 1830 Russian silver coins of good weight were gradually replaced in circulation by foreign coins of lower intrinsic value which exchanged at par. *De facto*, silver stocks were being depleted and the value of the metal was being enhanced as market forces sought to create a new, though unofficial, bi-metallic equilibrium. Nor in this situation could the assignat–silver exchange rate remain unaffected by these changes, and whilst in St Petersburg the official exchange rate prevailed, in Moscow and the market fairs, at the very centre of Russian commercial activity, where as early as 1833 the merchants complained that foreign coins had virtually replaced Russian ones in the market place, silver, which everywhere else was declining in value (i.e. payers were being surcharged at rates which varied between parity with the conventional valuation and 6 per cent), rose rapidly in price; payers in that metal received a discount of 5 per cent in 1837/8 on the conventional valuation. Full-weight Russian coins were clearly becoming difficult to obtain, and, as the debased foreign issues began to dominate the local circulating media, assignats also began to disappear, the notes going at a heavier and heavier discount until they stood at 83 kopecks per conventional rouble in Moscow and 82 in Yaroslavl and Nizhne–Novgorod in 1837, and 80 in Moscow in 1838. Russian money was disappearing from the markets of the commercial heartland of the empire causing net discount rates to increase rapidly to − 22/3 per cent in 1837 and − 25 per cent in 1838 whilst the particularly heavy fall in the supply of assignats dramatically enhanced the exchange value which rose to 2.5 per cent above the official bourse rate in 1833 before stabilizing in 1837 and rising to 4.75 per cent again in 1838. Markets were in chaos, and whilst the government attempted to bring some stability to the situation during 1833–5 by publicizing the true value of the foreign coins and declaring that it would accept them in payment of taxes at their intrinsic value it was running against market forces which required an upward revaluation of silver to create a new bi-metallic equilibrium.

In this difficult situation, where the official Russian silver and paper currencies were being driven out of circulation by foreign silver and domestic gold coins, reform was clearly necessary but the state's intervention in 1839 to peg the exchange at a rate far below the 'popular' one (initially proposed at 3.6 and subsequently raised to 3.5:1 in conformity with the St Petersburg rate), and to accept silver and notes for a new note issue (the deposit rouble) at this rate, proved disastrous. Holders of assignats made an immediate heavy loss, causing a recession at the fairs which was compounded by the effects of the appalling harvest of 1839. Money became tight and few would accept the losses involved in acquiring the new deposit notes issued in limited quantities with government backing in 1840, preferring to utilize the gold coins then entering circulation in increasing numbers. By its obdurate adherence to a silver standard and refusal to accept gold for its new issues reform was brought to naught. The circulating media remained dominated by the officially-unrecognized foreign silver and domestic gold coins. The commercial economy was thrown into an acute crisis. Only when, on the Tsar's insistence, both gold and silver were accepted in 1843 in exchange for yet another new issue (the credit rouble) could reform of the monetary system be accomplished. Initially the reform plan of 1843 was conceived in traditional terms. The new credit notes, issued in small denomination bills of 1–25 roubles as well as 50 roubles, would be available on deposit of an equivalent sum in coin. The assignats in circulation would gradually, over a five-year period, be withdrawn and destroyed, being replaced with credit notes with a backing initially of 4 million roubles transferred from the military reserve fund which would subsequently be augmented with a further 25.6 million as the deposit notes issued since 1840 were destroyed and their backing, largely provided from government funds, was transferred to the new reserve. In such a manner the existing assignats would be replaced by credit notes with a one-sixth metallic backing created largely at the state's expense. In the light of the experience of 1839/40 no reliance was made in 1843 on the populace's willingness to exchange silver and assignats against the new notes at the official rate of 3.5:1. In the event, however, the Tsar's insistence on the acceptability of both silver *and* gold for acquiring the new notes transformed the situation. At the official exchange and bi-metallic

Table 4.6 Estimated Russian monetary stocks, 1795–1860 (millions of roubles)

	Paper notes		Silver	Other metals	Total metallic money
	Amount	Bourse rate[a]			
1791–1800	150	1.42	77	90[d]	167[e]
1801–10	230	1.51	102	93	195
1811–20	712	4.00	178	—	—
1821–30	596	3.84	161	—	—
1831–40	596	3.67	164	184	348
1841–50	180	1.12[c] 1.02[b]	161	175	336
1851–60	304	1.64 1.05	185	290	475

Sources: See appendix A2(1), table 10.
Notes: a 1769–1840 assignat roubles per silver rouble by St Petersburg bourse rate calculated until 1810 in relation to total amount of coin and thereafter only in relation to silver. b credit roubles per gold rouble. c credit roubles per silver rouble on the foreign exchange. d copper coins in circulation 1795 34.4 million roubles, 1802 44 million, the residual forming the reserve of assignats issued before 1786. e Until 1802 the figures relate only to the Muscovite region and exclude stocks held in the Baltics. Thereafter the figures reflect conditions in the empire as a whole.

ratio, whilst few would exchange silver for the new issues many would exchange the contemporaneously cheap gold. In only two years 70 per cent of the old paper currencies (assignats and deposit notes) had been exchanged. By 1852 only 2 per cent of the original quantity of assignats remained in circulation. Because of the populace's willingness to exchange gold or assignats brought with gold at the official rates for the new notes, by 1853 reserves amounted to 47 per cent of the credit note issue, far in excess of the 16.67 per cent required by law. Thanks to the Russian gold boom of the second quarter of the nineteenth century the metallic base of the nation's currency had, after some fifty years, been restored. Thanks also to that boom the new currency system was able to survive the crisis induced by the government's prodigal use of the printing press during the Crimean War and its immediate aftermath. Between 1853 and 1858 the quantities of notes in circulation more than doubled and reserves fell to their legal minimum in 1859, leading to the suspension of convertibility, but confidence in the new notes was sutained as the burgeoning volume of gold in circulation ensured that they remained at an exchange rate with gold at or very close to par.[137] Internal currency stability had been achieved. External had not. On the foreign exchange, where the rouble remained tied to silver, the rouble ran at an endemic and increasing rate of discount. A new age was dawning in which the bi-metallic problems, first discernible in the 1830s, were unresolved but were externalized.

Russian monetary stocks

In contrast to the situation prevailing in the nineteenth century, when Russian monetary systems were plagued by excessive issues of inconvertible paper notes, in the eighteenth century the levels of monetary stocks had been directly related to changes in the supplies of monetary metals delivered to the mints. Initially these had been obtained through the medium of foreign trade. In the early eighteenth century, as for the past 130 years, monetary stocks in both the Russian heartland and the Baltic provinces (annexed from Sweden at the end of the Great Northern War) changed in accord with variations in the supplies of internationally traded precious metals (figure 4.3, A).[138] From the 1730s, however, this situation had changed dramatically. Whilst in the Baltic provinces, which remained in terms of their administrative, economic, and monetary systems separate and discrete entities, the old ways continued, their monetary stocks changing with variations in international specie supplies, in the Russian lands the size of monetary stocks was rendered independent of such fluctuations. Imports of precious metal steadily diminished until they were finally eliminated in the 1740s, foreign merchants abandoning specie transfers and undertaking instead normal commercial transactions which they financed by means of bills of exchange. Russian monetary stocks continued to grow but now solely on the basis of indigenously produced supplies of monetary metals.

For some sixty years (c. 1735–95) a burgeoning volume of indigenously produced monetary metals flooded to the Russian mints, laying the foundations for a massive expansion of Russian monetary stocks. As early as the late 1720s and 1730s there were indications of the course of future events when, in response to acute contractions in specie imports (in 1725–9 and 1735–7), the native copper industry had begun to turn out enough of

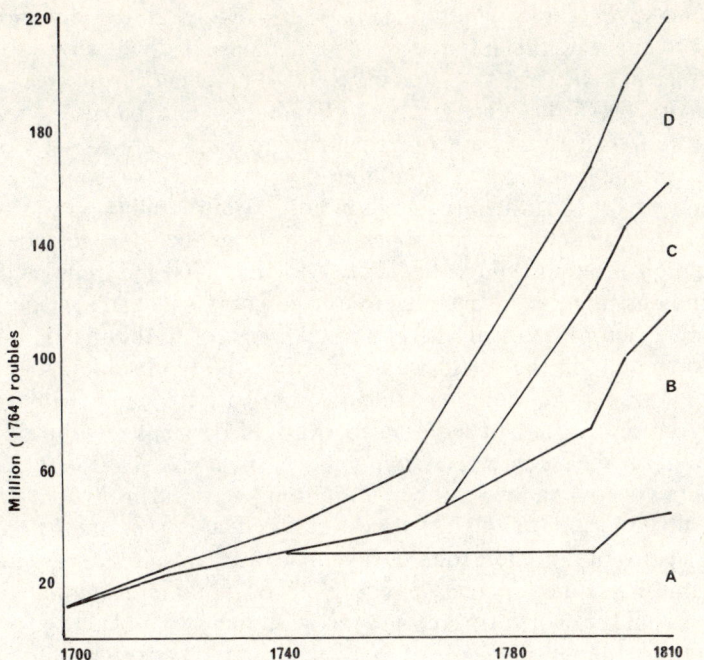

Figure 4.3 Russian monetary stocks

A Imported specie
B Domestically produced specie

C Copper-backed assignats
D Copper

the base metal to mint some 7 million roubles' worth of coins. Such sums were certainly insufficient to prevent a fall in Muscovite prices at that time but they did at least temper the rate of decline, serving to differentiate the Muscovite price movements from those prevailing in the Baltic provinces and elsewhere. Subsequent years, moreover, saw this tendency become even more pronounced. Whilst monetary stocks in the Baltic provinces seem to have remained below their late-seventeenth-century levels until at least 1760, in Muscovy that level had been reattained by 1740 and twenty years later it had been exceeded by some 75 per cent. The onus of providing supplies of metal to sustain this phase of rapid stock accretion had originally rested on the Russian copper industry which, particularly after 1739, provided a regular and rapidly growing supply of the base metal which was of sufficient magnitude to permit the mints to totally dispense with foreign supplies of specie (figure 4.3-D). As the 1740s drew to their close, however, the rapid growth of indigenous gold and silver production, which ushered in Russia's first age of silver (c. 1745–55), led to a displacement of the base metal in mint activity and raised hopes in government circles for the restoration of a monetary standard based on

213

gold and silver. By its commitment to a coinage based on gold and silver, however, the state precipitated an acute deflation and only with a complete volte-face in policy in 1756/7 was the trend of falling prices reversed. By the edicts of 1756/7 the state broke with the past and became once more committed to a rapid rate of growth in monetary stocks, setting mint targets initially in 1757 at a million roubles' worth of coin a year but subsequently raising them to 2 million in 1760 and 3 million in 1761–2. The effects of this change in policy were dramatic. As precious-metal production briefly declined between 1757 and 1760 the onus of meeting government targets fell once again on the copper industry which rapidly expanded output until, on the eve of the reign of Catherine II, it ranked second to none in the world, supplying the mint with some 2,400 tons of copper a year (= 2.4 million roubles) and private customers with a further 1,100 tons annually. Nor with the recovery of precious metal production, ushering in the second Russian age of silver (c. 1762–80), did this commitment to high levels of copper minting waver. Throughout the early years of Catherine II's reign every effort was made to sustain the output of copper coins at 2 million roubles a year with the result that, as silver and gold minting rapidly increased (figure 4.3-B), total mint output rose to 5 million roubles' worth of coin a year in the early 1770s and averaged 4 million throughout the whole of that decade. Mint activity was intense and even the government's attempts to reduce mint output, in the face of acute inflationary pressures during the 1780s, had little impact as monetary stocks continued to increase rapidly until c. 1795. For about forty years (1756–95) as Russian monetary stocks rapidly increased on the basis of indigenously produced supplies of monetary metals, the major problem confronting governments had thus not been one of ensuring an adequate metal supply but rather of maintaining an appropriate inter-metallic balance in accord with consumer demand which initially induced a rise in precious-metal prices in 1763/4 and in 1768/9 led to the introduction of the assignat which permitted the conversion of small-denomination copper coins, surplus to the population's requirements, into large denomination notes.

Only with the collapse of indigenous production of monetary metals after c. 1795 did the spectre of supply shortages once more appear. In the event, however, even as internal production declined, imports of precious metals once more, after an interval of almost fifty years, began to enter the Russian lands. Yet, as traditional patterns of monetary metal supply were re-established, the whole system which they supported disintegrated. The fateful decision made in 1787 to issue 54 million roubles' worth of unsecured paper assignats marked the beginning of the end for the old monetary order, and as the volume of inconvertible paper notes thereafter increased a new order was born based on a double metallic (silver rouble) and paper (assignat) standard.

Chapter Five

Russia. Economic growth

Against a background of declining Central and South American silver production, the years *c*. 1670–1760 witnessed a revival in the European production of that metal. Central European production, initially concentrated in the mines of Saxony and Oberharz (*c*. 1670–1740) and then in those of Hungary (*c*. 1740–80), steadily increased, attaining a level of 20,500 kg a year at the height of the first production cycle in *c*. 1725 and 36,700 kg a year at the height of the second in *c*. 1760.[1] Yet even as the central European production cycles ran their course, contributing between 4 and 7 per cent of total European silver supplies, events were taking place to the east which would cast a shadow over the whole European scene as new mines, opened up in Russia, rose in the ascendent. Initially discovered during the first production cycle of the European industry, the mines of Nerchinsk (opened in 1704) and Kolyvan–Voskressensk (opened in 1729), together with the Uralian gold mines of Beresov (opened in 1754) contributed an ever-growing production of precious metals to the second great production cycle until, at the height of the production boom in the early 1770s with a combined output worth 2.2 million roubles, equivalent to 40,000 kg of silver a year, the mines produced collectively more of the precious metals than all of the rest of Europe put together.[2] Contemporaneously, moreover, the Russian copper industry underwent a parallel process of expansion, until, with an output of *c*. 3,500 tons a year in the 1760s, it ranked second to none in the world.[3]

For some sixty years from *c*. 1735 to 1795, whilst specie continued to be imported into the recently annexed Baltic provinces, a flood of both base (copper) and precious metals (silver and gold) derived from indigenous mining and metallurgical enterprises passed to Russian mints, rendering the Muscovite heartland of the empire independent of international supplies of specie.[4] Imports of precious metals were displaced and Russian monetary stocks were rapidly augmented on the basis of domestic production, creating potentially strong inflationary pressures within the economy and threatening to transform the nation from being a net importer to an exporter of specie. Measured in terms of commodities the purchasing power of specie steadily diminished with dramatic effects. Domestic Russian industries, whose low production costs (measured in terms of silver) had provided 'protection' from foreign competition, were now seriously threatened. Imports from western Europe, such as cloth or colonial wares, which had always been four or five times more expensive than competitive Russian wares before *c*. 1730, thereafter, as a result of both 'real' and 'monetary' factors, became progressively cheaper until they closed towards parity after 1750 giving rise to a major import boom. Producers of export commodities were no less affected. The major export wares, flax, hemp, and rye, subject to internal inflationary pressures which threatened to push up silver commodity prices and close international price differentials, became potentially less competitive on international markets. The prospect of Russia's markedly active trade balance becoming adverse, leading to a reversal of specie flows and the development of a major export trade in precious metals, thus, after *c*. 1730, seemed a very imminent reality.

In the event, however, a 'revolution' in the Russian countryside and associated transformation of internal transport systems, prevented this occurring by reducing 'real' export prices.[5] The export boom continued and as imports rose Russia experienced a passive balance of trade during the years *c*. 1735–95.[6] The full impact of increasing monetary supplies during these years was felt, accordingly, within the domestic economy. For some sixty years, from 1735 to 1795, Russia acted as a sponge, absorbing the burgeoning flood of coins which poured from the mints, the vast quantities of money entering circulation being largely required to meet a rapidly rising transactions demand within the economy. Indeed, judging by estimates of changes in transactions demand, derived from an analysis of prices and monetary stocks which are confirmed by contemporary national income estimates, during the course of the eighteenth century Russia experienced a phase of very rapid economic growth.[7] Within contemporary boundaries national income increased more than five times, rising rapidly by some 3.4-fold between the closing years of the Great Northern War, 1718/22, and the accession of Catherine II in 1762 before slackening off, to a 57-per-cent growth, during her reign and that of her son, 1762–1801 (figure 5.1).[8]

Russian economic growth 1718–88

Annexation in the west

This initial phase of rapid economic growth from 1718 to 1762 can not be attributed totally to the incorporation of the rich Baltic and Ukrainian lands, whose inhabitants even as late as 1762 were almost half as rich again (+ 46 per cent) as their Muscovite counterparts, into the empire.

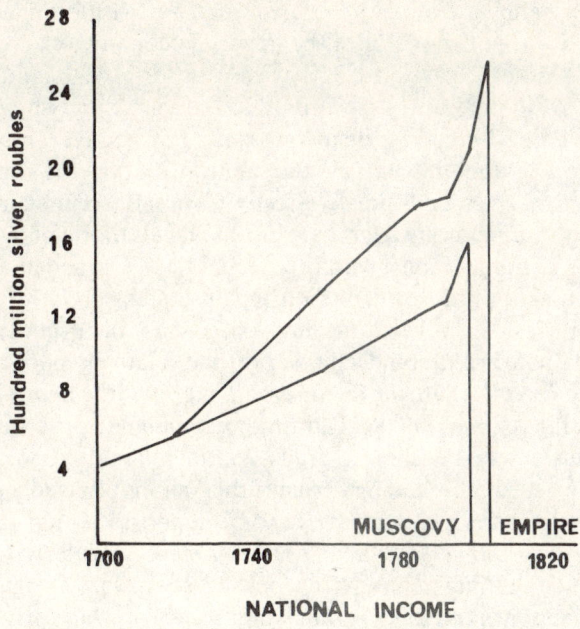

NATIONAL INCOME

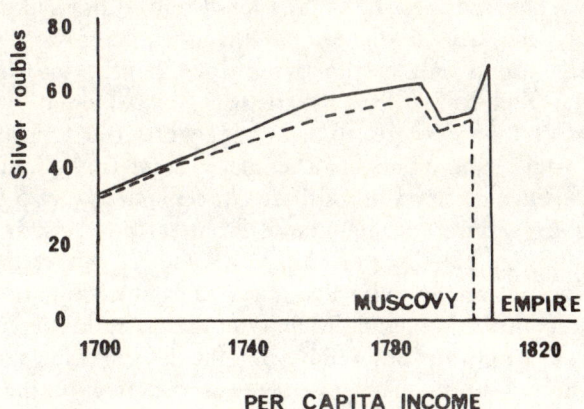

PER CAPITA INCOME

Figure 5.1 Estimated Russian national income, 1700–1807 (constant 1762 prices)

Important as the acquisition of these new territories was, it can only account for about half of the incremental growth in imperial Russian national income during these years and had the effect of raising average *per capita* income within the empire in 1762 by only 7 per cent. Similarly, during the reigns of Catherine II (1762–96) and Paul (1796–1801), whilst incomes within the Baltics and the Ukraine were maintained at relatively high levels, enhancing average *per capita* incomes within the empire, the lack of growth within this sector of the imperial economy, where aggregate income levels increased by only 22 per cent in forty years and *per capita* incomes fell by 26 per cent, served only to dampen the overall impact of growth elsewhere. Perceived in a long-term perspective embracing the whole of the eighteenth century the annexation of the rich western territories had a once-and-for-all effect on imperial Russian economic growth, initially augmenting aggregate income levels of the poor Muscovite state by some 78 per cent but, with their economies trapped at a high-level equilibrium, making little contribution to subsequent growth.

Contemporaries, however, did not enjoy the historian's ability to perceive the situation with hindsight and for the relatively poor denizens of the Muscovite realm, whatever the future might hold, the acquisition of the rich agricultural regimes of the Ukraine and Estland must have seemed very rich prizes indeed. Within the trans-Pontine plain there were very few such centres where high-yielding agricultural systems allowed a population of 16–25 persons per square kilometre to be supported (map 5.1).[9] In the territories of the Old Muscovite State they were singularly lacking. Its political boundaries might have been pushed southward in the course of the early eighteenth century – allowing some of the fortresses, first established by Alexis three-quarters of a century before, to be reoccupied – but life here was precarious, to say the least, and settlement assumed an almost totally military character. In the frontier towns themselves contemporaries found only martial activity. Of Penza, one wrote that it was 'a very large place in which the fortress is surrounded by a wooden wall' and the inhabitants were 'soldiers kept there to defend the country against the Tarters', and others made similar observations about Voronezh and Samara.[10] Outside the walls of these citadels there were a few settlers who, willing to endure the rigours of frontier life, raised population densities up to ten persons per square kilometre, but settlement was sparse and few could claim to be able to see the smoke from neighbouring villages. Settlers in the immediate vicinity of the forts did, however, enjoy some security. Those who ventured beyond the defensive lines to the south and east, where the Russian presence was confined to the fortresses protecting the Volga route to Astrakhan, did not. The Russian military presence here was not dissimilar to that found on the defensive lines but, as one contemporary made clear, imperial authority did not extend beyond the walls of the forts; he said that the Kuban Tartars

> make frequent incursions into the outparts of Russia, plunder and fire the villages and often carry off cattle and horses and people, by reason of which there is a great tract of land on the west side of the Volga all the way between the town of Saratoff and the Caspian Sea [which] lies wholly uninhabited, save only the islands of Atracan [Astrakhan] and the

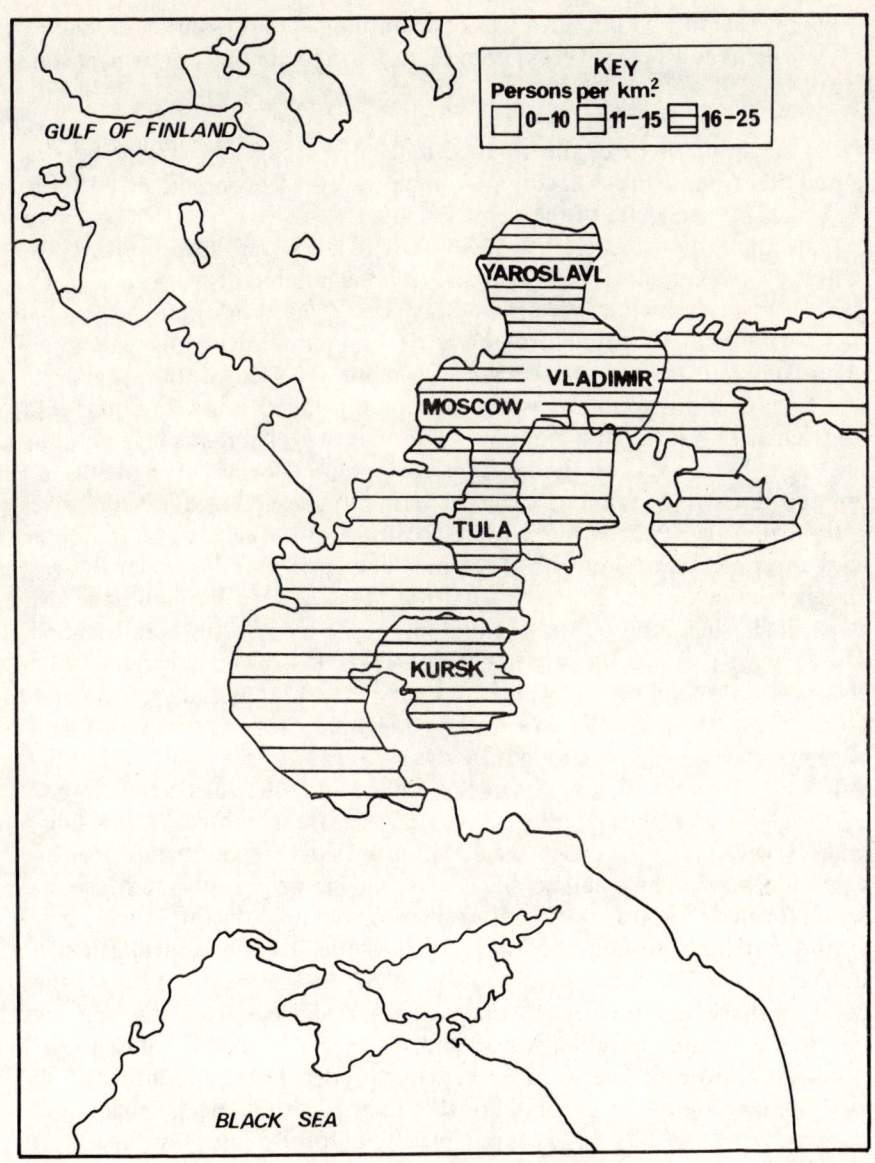

Map 5.1 Muscovy, 1725

people that live within the towns of Camishinka [Kamyshin], Czaritza [Tsaritsin] and Terki [Terek, in the Caucasian piedmont] where garrison are kept.[11]

Over the whole area of the wooded and open steppe the Tartar hordes posed a constant threat ensuring minimal levels of economic activity for 200 miles on each side of the defensive lines.

Only about the old capital of the Muscovite lands was there a sufficiently stable level of economic activity to provide the inhabitants with an average *per capita* income which was in any way comparable with that enjoyed by denizens of realms located further west. To contemporaries this region was almost ideally situated, its temperate climate and lack of hills, marshes, and forests ensuring that corn, fruit, and vegetables were all abundant, together with much cattle and honey, 'so that the inhabitants lack nothing but wine'.[12] Yet even here *relatively* intense forms of agriculture, embracing both field and garden crops, rapidly merged into less intensive forms of husbandry (map 5.2). Cereal cultivation, organized on the basis of a three-field system, with its appendant gardens and orchards, was largely confined to the provinces of Tula, Moscow, and Vladimir, the last in particular being noted for its situation 'in the most fertile country in all Muscovy'.[13] This was the heartland of the old Moscow aristocracy, 'which the Russes are fond of, and where they have their friends and acquaintance about them; their villages are near and their provisions come easy and cheap to them, which is brought by their slaves'.[14]

It was, however, an area narrowly confined and one which soon merged into regions of more extensive husbandry. To the south-west, in Kaluga and Orel provinces, the rich black soil afforded verdant pastures for an extensive animal husbandry, sheep grazing in the vicinity of Moscow, which formed the basis of a local woollen textile manufactury, giving way to the fattening grounds on the eastern slopes of the central Russian uplands where Ukrainian cattle bought at Bryansk were prepared for the Moscow market. Similarly, to the north-east and east cattle grazed the pastures extending between Kazan and Kostroma-Yaroslavl whilst to the north-west towards Tver a more heavily wooded landscape afforded an environment suited to all kinds of livestock – cattle, sheep, goats, and horses – and capable, in favoured places, of producing grain and such garden crops as vegetables, flax, and hemp; where soils were poor and a heavy forest cover remained, honey and beeswax were produced, and the inhabitants combined their agrarian pursuits with craft activity.[15]

In the vast reaches of the Muscovite lands, much of which remained only nominally under Greater Russian rule, therefore, only in the immediate vicinity of Moscow could there be said to be intensive economic activity, an agrarian-craft complex yielding the inhabitants of the area an income which was sufficient to raise average *per capita* incomes within the

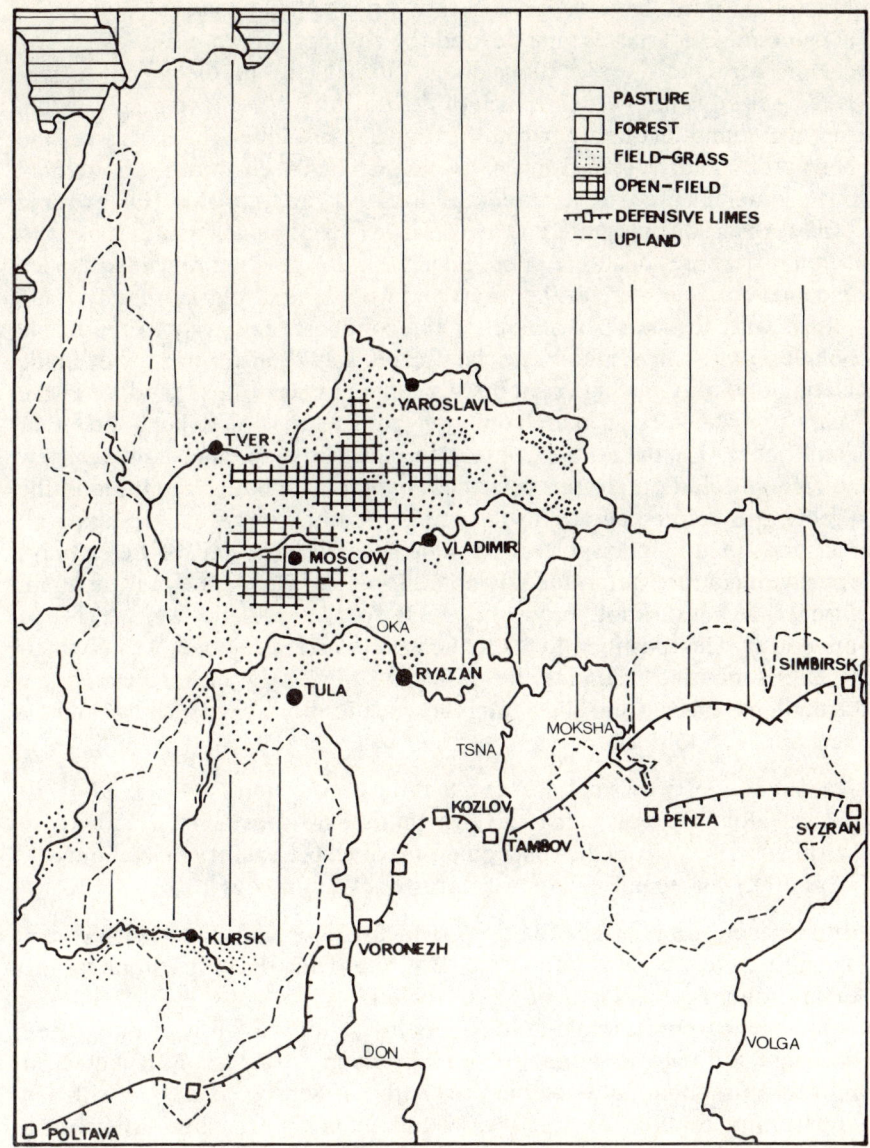

PASTURE
FOREST
FIELD–GRASS
OPEN–FIELD
DEFENSIVE LIMES
UPLAND

YAROSLAVL

TVER

MOSCOW VLADIMIR

OKA

SIMBIRSK

RYAZAN

TULA

MOKSHA

TSNA

KOZLOV

PENZA

SYZRAN

TAMBOV

KURSK

VORONEZH

DON

VOLGA

POLTAVA

Map 5.2   Muscovy: the central provinces, 1725

Muscovite lands to a level – 42 roubles a year in constant 1762 prices – which by any international standards was modest indeed.[16] Accordingly, from their perspective Peter's acquisitions, embracing as they did high-productivity agrarian complexes (very different in form from those of

Muscovy), must have seemed a rare prize, yielding assets capable of augmenting imperial income beyond their wildest dreams.

Nor were their expectations to go unrealized. In the north-western lands, seized from Sweden, much of the territory was encompassed within the woodland economy which extended through White Russia and Novgorod as far east as Tver and which, as has been shown, afforded an environment suited to all kinds of livestock and capable, in favoured places, of producing grain and such garden crops as vegetables, flax, and hemp, or, where soils were poor and a heavy forest cover remained, honey and beeswax. Yet here as the forest and woodland of Ingria, rapidly being settled with Russian peasants after the conquest, gave way to the fertile boulder clays of Estland the landscape was transformed, the Baltic German nobility having created a veritable granary whose produce in the years to 1718 'foreign merchants preferred to that of Poland and other countries'.[17] Following the consolidation of Russian rule, however, few foreigners could obtain this prized commodity, for whilst the trade in the products of the first agrarian zone – hides, leather, beeswax, and above all flax and hemp – transported through Riga, Narva, and St Petersburg, rapidly recovered after the disruptions of the war, the trade in grain through Reval did not. From c. 1717/18 to 1735 Estland rye, which had previously been destined for Stockholm or western Europe, was diverted to St Petersburg, becoming the principal source of grain for Peter's new capital on the Neva and alleviating those acute shortages which had caused Perry in 1716 to declare,

> all manner of provisions are usually three or four times as dear and
> forage for horses etc at least six to eight times as dear as it is at Moscow;
> which happens from the small quantity which the country thereabouts
> produces, being more than two-thirds woods and bogs.[18]

By his acquisition of the Baltic provinces Peter had obtained the rich agricultural regime of Estland which was to provide the cornerstone for the rapid economic development of the region.

By his incorporation of Little Russia in the formal fabric of the empire he similarly obtained yet another high productivity agrarian regime which provided the inhabitants of the area – the Dnieper Cossacks – with the opportunity to 'live in much ease and plenty'.[19] In this instance the prosperity of the region rested on the interdependent relationship existing between the pastoralists of the Polish and Russian provinces of Volynia, Podolia, Kiev, Chernigov, Poltava, and Kharkov and the more settled agriculturalists of Kursk. Ever since its rise to international prominence in the 1580s Little Russia had been a land of 'meadows, too extensive and luxuriant to measure, where the horned-cattle' organized into *Chereda* of 100–800 animals, 'scarcely appear to view such being the height of the grass', the herds, at least until c. 1650, yielding up young stock, 3–5 years old, to be driven west.[20] With the Zaporzhi's desertion to Russia in 1648–54, however, this situation had changed dramatically. The western cattle trade, henceforth, drawing on supplies from the Polish provinces alone was drastically reduced in importance, only *c.* 20,000 animals taking passage to Silesia in 1725.[21] The Russian herds, on the other hand, now passed to new markets. The young stock, culled each spring, were now brought to the Ukrainian cattle fairs at Kharkov and Kiev where they were sold on to those who would prepare them for their long drive north.[22] Thereafter over the summer they grazed the closes set about the neat little villages of white houses on the lower slopes of the Ukrainian uplands, manuring the land which ultimately yielded abundant crops of corn, flax, and hemp. Then in the autumn they were driven north, finally being sold at Bryansk to those who would finish them off for the Moscow market.[23] In such a manner an extensive pastoralism supported an intensive agricultural regime and allowed the cossacks to 'drive a great trade in hemp, potash, wax, corn, and cattle.'[24]

The Ukraine and Baltic provinces thus encompassed high productivity agrarian regimes and associated commercial systems capable of yielding the inhabitants of these regions high levels of income – 77 roubles a year at constant 1762 prices, or 83 per cent more than had been enjoyed by their Muscovite counterparts in 1718/22.[25] The effect of the annexation of these lands was, accordingly, enormous, enhancing the national income of the Russian Empire by some 78 per cent. Yet rich as they were the economies of the newly annexed lands lacked growth potential and, trapped at a high-level equilibrium, made little contribution to subsequent growth. From *c.* 1762, when reasonable reliable data first become available, output levels increased, enhancing aggregate incomes by almost 50 per cent in the years to 1792–6, but growth assumed an extensive character ensuring that a growing population could be maintained at high but stable levels of *per capita* income. This situation was, moreover, inherently unstable as events at the turn of the century were to reveal. The extension of low-yield cereal cultivation in response to rapid population growth, particularly in White Russia and Lithuania, undermined the pastoral/woodland – arable balance with the regional economies reducing manuring levels and adversely affecting yields and eliminating incomes derived from the exploitation of grasslands and woods. Prior to 1792/6 the impact of these changes had been slight, but the century ended in an acute crisis which, although concentrated in White Russia and Lithuania, seriously affected the average level of *per capita* income in the region. Viewed in the context of the period 1762–1802, the economies encompassed within the western provinces annexed in 1721, were, as a whole, stagnant, aggregate incomes rising by only 22 per cent and *average per capita* incomes falling by 26 per cent from the high levels of 1762. The incorporation of these territories into the Russian Empire thus had a once-and-for-all effect on national income growth, initially augmenting aggregate and *per capita* income levels at the time of their annexation but making little contribution to subsequent imperial growth.

Muscovite growth

The key to the rapid expansion of national income in the Russian Empire during the eighteenth century resided not in the annexation of new territories but in a transformation of the economy of the Muscovite heartland where, following the postwar recovery, aggregate income levels increased by 70 per cent between 1718/22 and 1762 and then by 70 per cent again between 1762 and 1802, expansion here accounting for almost two-thirds of the growth in imperial Russian national income during the years 1718/22–1802.[26] Year after year aggregate income levels increased within the Muscovite lands at roughly constant growth rates but within the period 1718/22–1802 two quite different and distinct phases of economic growth can be distinguished. From 1718/22 to 1788 the Muscovite lands experienced a phase of intensive economic growth, raising average *per capita* incomes by some 40 per cent during the years 1718/22–62 and then by a further 9 per cent between 1762 and 1788. Thereafter, from 1788 to 1802, intensive growth gave way to extensive as the economy expanded under conditions of diminishing returns.

Between 1718/22 and 1788 the Muscovite lands experienced an 'agricultural revolution' which set in motion a process of major structural change within the economy. The securing of the defensive lines between Poltava in the west and Samara–Simbirsk in the east, begun under Peter I and brought to a successful conclusion in the reign of the Empress Anna, brought that security to the northern wooded steppe which had previously been lacking. The building of a new transport system ensured that not only Moscow but also St Petersburg could gain access to the products of these lands (map 5.3).[27] Earlier links between the capital and the interior had been extremely tenuous 'by reason of the tediousness of the way, the being obliged to wait for Floods and Rains at several shallow Places, and the Vessals and Floats being often dash'd and staved to pieces against the Rocks and Falls that are by the Way, and the Goods often lost and spoiled', but the building of a canal system, linking St Petersburg to the Volga waterway network, transformed this situation. In this new system there were two artificial canals: the Vishnii Volochek, linking the upper Volga system with that of the Volkhov, and the Ladoga Canal, which skirted the southern shores of the lake joining the Neva at Schlusselburg. The former, completed in the closing years of Peter's reign, ran through the middle of the small town of Vishnii Volochek, as did the main St Petersburg–Moscow road, making it a place of great activity. The traffic along the waterway was one-way and throughout the navigable season barges, each having been drawn up the Tvertsa to Vishnii Volochek by ten horses at a rate of 10–12 miles a day, passed from lock to lock before beginning the descent of the Msta and Borovichi rapids where 'the stream is so violent that the boats not infrequently shoot along this space within the hour'.[28] Thereafter descending the Msta to lake Ilmen they went down

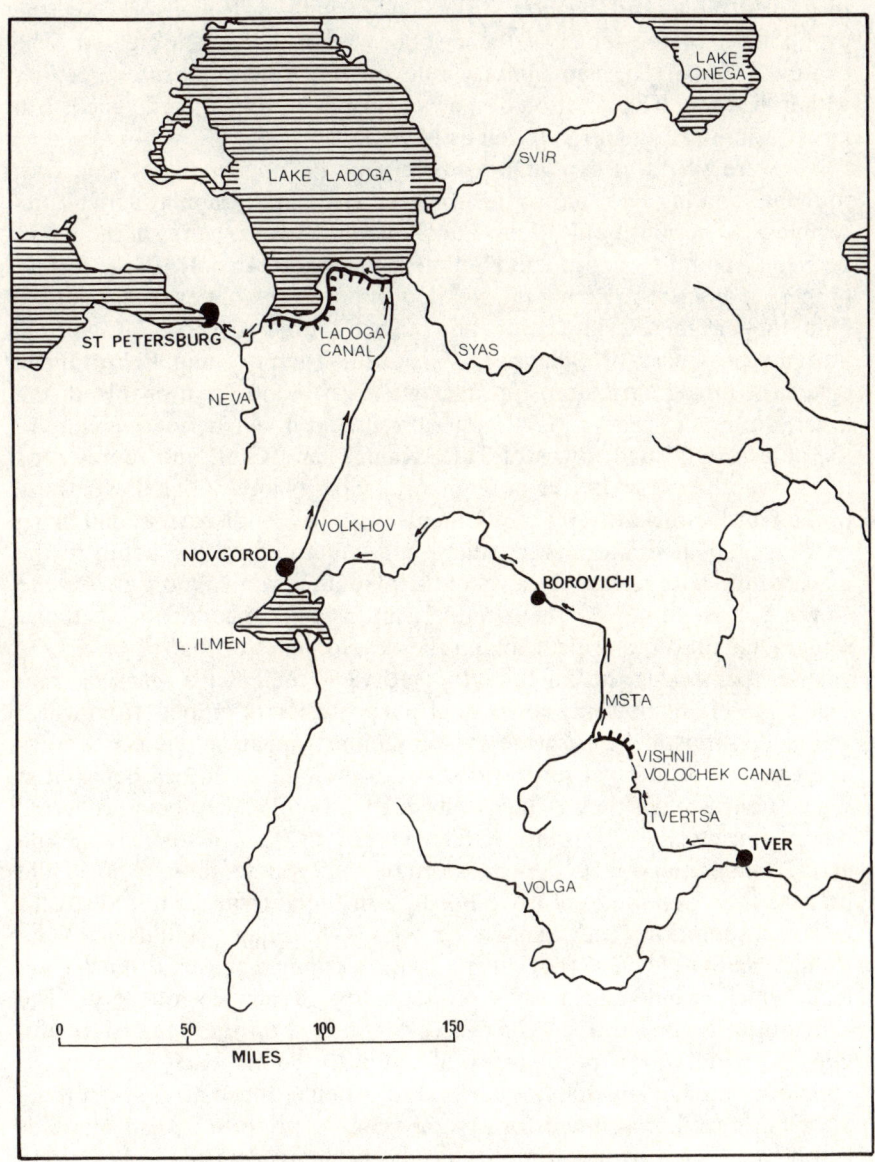

Map 5.3   Volga–Neva canal system

the Volkhov to lake Ladoga. Here they entered the Ladoga canal, completed in 1732, passing some 67.5 miles along its course before arriving at Schlusselburg and the Neva. By the mid-1730s, therefore, St Petersburg enjoyed direct access to the interior at Vishnii Volochek, the journey between the two centres taking a fortnight in spring, when the rivers were in full flood, and three weeks to a month in summer and autumn, when they were low and shallow. The infrastructural framework for the exploitation of the lands of the wooded steppe had been created. The consolidation of Russian military rule on the frontier ensured security within the area. The creation of a new transport system afforded access to markets in St Petersburg as well as Moscow.

Nor were would-be exploitants slow to seize the opportunities opened up to them, creating two spatially distinct but structurally similar agricultural regimes, to the south of Moscow and along the western banks of the Volga between Nizhnii-Novgorod and Simbirsk-Samara (map 5.4). In the former of these regions, embracing the wooded steppe in the upper reaches of the Don, the margin of estate cultivation pushed south between 1720 and 1740 into the provinces of Belgorod, Voronezh, Tambov, and Penza and a system of mixed agriculture producing livestock and grain envolved, the latter, during the years to 1762, more prominent in northern areas close to the older established towns of Tula, Kaluga, and Orel, and represented further south by a scattering of seigneurial villages intermingled with those of the Don cossacks.[29] Here settlements were widely dispersed, and being concentrated in sunken river valleys and concealed by the depth of the banks of the river below the level of the plain they were almost invisible to view. Yet in spite of their numerical insignificance these southerly settlements played a critical role in the development of the region. In the villages themselves the inhabitants lived well, their fields, gardens, and orchards yielding not only corn but also a profusion of grapes (from which was made a most delicious wine), water melons, and an abundance of fruit. Their true wealth, however, resided not in these fields and gardens but in the vast herds of cattle which grazed the plains and which, being culled of their young stock each spring, yielded a regular supply of steers to be sold at the cattle markets of Voronezh and Belgorod to those who would prepare them for the long drive north.[30] In the autumn, if not killed for local consumption, they passed on to the finishing grounds of Tula, Kaluga, and Orel where they mingled with Ukrainian beasts, manuring the land which would ultimately produce an abundance of corn. The agricultural regime in these provinces, organized on the basis of a field-grass husbandry, was peculiarly well suited to the exigencies of a mixed regime, as the heavily manured grassland, when cropped after seven years of being grazed, continued for several years to produce cereals at yields which were often as much as double those obtained in the older settled areas.[31] As earlier in the Ukraine, so also, during the years 1732–62, in the

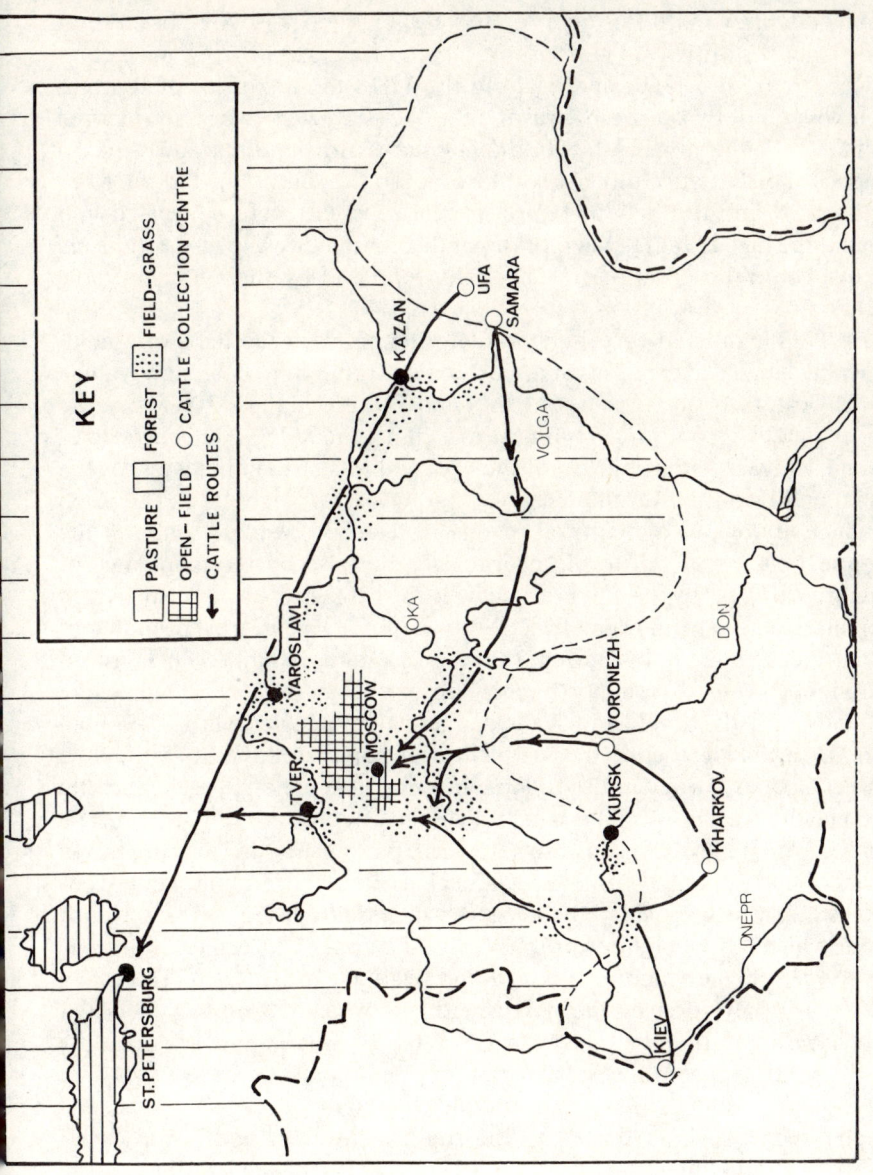

Map 5.4 Muscovy: the northern steppe and Volga regions, 1725–62

KEY

PASTURE ☐ FOREST ▨ CATTLE COLLECTION CENTRE ○

OPEN–FIELD ⊞ FIELD–GRASS ▦ CATTLE ROUTES ▸

ST.PETERSBURG

YAROSLAVL

TVER

MOSCOW

OKA

KAZAN

UFA

SAMARA

VOLGA

DON

VORONEZH

KURSK

KHARKOV

DNEPR

KIEV

valley of the Don, an extensive pastoralism in the south of the region supported an intensive agricultural regime in the north-west.

During subsequent years, from 1762 to 1788, the lands to the east of the Don were drawn into this agrarian regime. Previously isolated because of the higher costs involved in transporting produce to the Moscow market, as prices rose in the 1760s and again in the 1780s the provinces of Ryazan, Tambov, and Penza were drawn into the supply network of the old capital.[32] In the north estates enjoying access to the Tsna and Moksha shipped their produce directly to Moscow. To the south, beyond the town of Tambov and throughout Penza province, cereals were converted into vodka, the higher value–weight ratio of this commodity allowing producers to overcome the costs of transportation imposed by their unfavourable location.[33]

By 1788 high-yield cereal cultivation, organized on the basis of a field-grass husbandry, existed in an arc extending from Orel, Tula, and Kaluga in the west, through Ryazan, to Penza and Tambov in the east, each of the nodes of this agricultural regime despatching about 250,000 chetverts of grain each year, either by way of the Oka and Mtsensk, in the west, or the Tsna and Moksha in the east, to the Moscow market.[34] The high productivity of this agricultural regime rested, however, not only on the natural richness of the fertile chernozems but also on an abundance of manure ensured by its integration with an extensive pastoralism which dominated the central zone of the Don valley. Each year, west of the Don, young stock culled in the spring from the vast herds of cattle which grazed the steppe were driven north, arriving in the autumn at the finishing grounds of Tula, Orel, and Kaluga. Here they mingled with Ukrainian beasts, fertilizing the pastures which after some seven years would be broken up and sown to grain. In Ryazan a similar situation prevailed, cattle from further east playing the same role as did Ukrainian–Don beasts in the lands of the upper Oka, but further south, particularly in Tambov, it was herds of horses bought from the Kalmuck at Saratov, forming the basis of local studs, which provided the necessary manure to sustain cereal yields at levels which, although inferior to those of the upper Oka region, were 20–50 per cent higher than in Moscow and Vladimir provinces.[35] Thanks to the security afforded by the defensive lines, which ensured that martial activity, even during the bloody war of 1768–74, was confined to areas far to the south and south-west of the region, in a period of little more than fifty years a major, new, high-productivity agrarian regime had been established in regions located to the south of Moscow and in the upper reaches of the Don.

A similar process took place contemporaneously to the east in the middle Volga region.[36] North of the fortified line, extending from Penza in the west through Samara, Simbirsk, and Syzran to Ufa in the east, and around the southern fortress of Saratov, the completion of the Volga waterway system to the capital in 1732 opened up new opportunities for would-be exploitants drawn from amongst the ranks of the St Petersburg nobility. Menshikov and the Vorontsovs, who had played a major role in the seigneurialization of the Don valley, acquired estates in Nizhnii Novgorod, Simbirsk, and Saratov, the latter centre also attracting such families as the Annikovs, Dolgoruckis, Shuvalovs, and Chernishevs. All along the west bank of the Volga, extending from Saratov to Nizhnii Novgorod, cereal cultivation was established displacing the previous extensive pastoralism eastward into Orenburg, Perm, and Ufa provinces. Villages proliferated, in the north about Nizhnii Novgorod the extensive fields and rich vegetable gardens being adorned with orchards which produced an abundance of fruit. Yet here, as further west, the high productivity of the prevailing agricultural regime rested not only on the natural fertility of the soil but also on an abundance of manure ensured by its integration with the extensive animal husbandry of lands to the east. Each year young stock cut out from the eastern herds were sold at the great cattle markets of Ufa and Samara to those who would prepare them for the long drive west. Over the summer they grazed on the verdant pastures in the vicinity of the villages of the middle Volga before being driven on in the autumn towards their final destination. In the south of the region this involved their transportation to the finishing grounds of Ryazan where they were fattened for the Moscow market. In the north the process was more complex due to the much greater distances involved in transporting the animals to St Petersburg. Because of the problems of loss in weight and quality during transit no attempt was made to drive the animals directly to the capital but rather the route was fragmented into a series of interconnecting sections, each no longer than 550 km and each embracing a marketing-transport system similar to that found in the regional networks of the Don valley and the southern zones of the middle Volga region. There thus emerged in the vicinities of Nizhnii Novgorod and Yaroslavl–Kostroma regional economies where cattle, after having been bought in the autumn by the merchants of the fair towns, were overwintered by them and then sold in the spring to those who would prepare them for further transit. Each sub-element in this long-distance trade thus embraced a structure similar to that found in regional networks elsewhere, characterized by spring and autumnal peaks in commercial activity. Yet whether extending north-westward along the cattle trails to St Petersburg or confined to the lands of the middle Volga region, in the prevailing agrarian regime an extensive pastoralism to the east of the region supported an intensive agricultural regime characterized by high-yielding field and garden crops.

Between c. 1735 and 1785 an 'agricultural revolution' had taken place within the Russian countryside as high-productivity agrarian regimes were established in lands to the south, east, and north of the old capital. By the end of this period the new agrarian systems sustained population densities (16–25 persons per square kilometre) which fifty years before had been confined to the immediate vicinity of Moscow (map 5.5). The impact of these changes was not only direct, enhancing *per capita* product in the regions of agrarian change, however, but also indirect, effecting major structural changes in the rest of the economy.

Nowhere were these changes more apparent than in the new capital of St Petersburg where the linking of the Volga waterway system to the Neva transformed the provisioning of the capital.[37] Able to draw on the abundant surplus of cheap grain from the lands of the middle Volga, merchants transported increasing quantities of cereals along the new waterway system, causing the 'real' price of rye to fall dramatically in St Petersburg.

Table 5.1 'Real' rye prices, 1700–79[38] (grams of silver per 10 puds)

	St Petersburg	Moscow	Urals
1700–9		231	105
1710–9	(c. 1,000)	252	95
1720–9	129	386	114
1730–9	74	104	126
1740–9	73	139	107
1750–9	63	121	91
1760–9	61	118	71
1770–9	74	119	99

Following the years of acute shortage (1703–17/18) prices in the capital, if lower than in Moscow due to the arrival of increasing supplies of grain from Estland, were still in the 1720s higher than in such internal regions as the Urals, but fell heavily thereafter as the colonizers of the middle Volga despatched ever-increasing quantities of grain along the waterways to the capital, feeding the citizens of St Petersburg which began a period of rapid growth from the 1730s. The expanding aristocratic and commercial population, who lived permanently in the city, as well as the growing army of obrok serfs and foreigners, who were employed or provided services there, could all be amply provisioned with cheap grain allowing the city to undergo, particularly during the years *c.* 1730–60, a phase of rapid expansion.

In the lands lying between the Volga and Oka about the ancient capital of Moscow, the heartland of the old Muscovy, the changes wrought by the 'agricultural revolution' were far more complex, but symptomatic of the economic transformation taking place in these lands at this time was a fundamental alteration in methods of estate management. In the late seventeenth and early eighteenth centuries (*c.* 1680–1725) fewer than 20 per cent of the seigneurial peasants in these non-black-earth provinces had

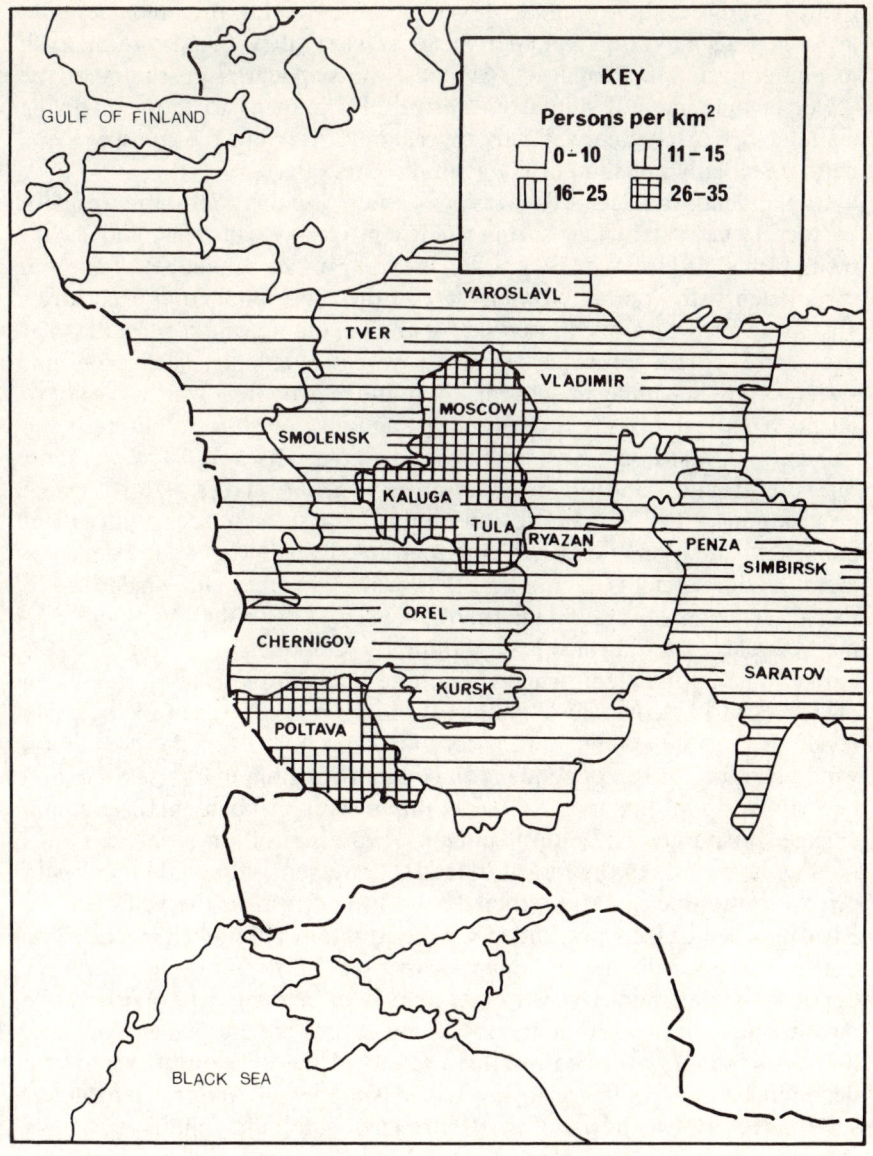

KEY

Persons per km²

0-10 11-15
16-25 26-35

Map 5.5 Russia, 1788

paid cash quitrents, while more than 80 per cent performed labour
services. By the 1780s, however, 63 per cent of the seigneurial peasants
there paid quitrent.[39] As the 'real' price of grain fell from the 1730s, many
lords were prepared to forgo the use of direct labour in cereal cultivation on
their demesnes, freeing their peasants from agricultural work and allowing

231

them to redirect their energies towards non-agricultural employments as artisans, wage labourers, and petty traders where they would earn the cash to pay quitrents to their lord. Amongst the seigneurial peasantry of the region a fundamental shift from agricultural pursuits to those involving manufactory or commercial activity was under way but the transition was neither spatially nor temporally a simple one.

In the lands adjacent to Moscow, where in the 1710s the resident aristocracy had their villages from whence provisions came easy and cheap, being brought to them by their serfs, the old ways continued; each autumn carts laden with grain plied their way to the old capital from each lord's villages to provision his household for the oncoming winter season.[40] Yet even here, in the provinces of Moscow and Vladimir, such economic systems were becoming anachronistic. In the metropolis – which was not so much 'a city of houses in mere rank and file of streets, but rather a collection of mansions, each embosomed amongst its own lawns, gardens, pleasure grounds, and the dwelling of its necessary slaves' – direct estate provisioning of the aristocratic households, and of the hordes of attendant household and obrok serfs servicing their needs, undermined the development of commercial grain markets. Elsewhere, in areas enjoying access to the Volga waterway system via the Klyazma and its tributaries, such grain markets played a central role in the life of the populace.[41] The time was already fast approaching when the arable lands in these non-black-earth provinces of Moscow and Vladimir would produce only half of the grain needs of their inhabitants and a very large proportion of the population would be more or less dependent on commercial grain markets.[42] Old and new thus stood juxtaposed. Near the metropolis peasants were still engaged in purely agricultural pursuits, working full-time on their own holdings and those of their lords to feed themselves and provide supplies to provision the households of their lords. In contrast, in the valley of the Klyazma, east of Moscow, and along the tributaries of that river stretching northward towards the Vladimir–Kostroma border, a true *kustar* as opposed to craft industry was emerging, with peasants who enjoyed the most tenuous links with the land working up imported raw materials (silk and cotton) in accord with Smithian tenets of the division of labour and depending upon commercially provided supplies of foodstuffs for their sustenance. Between these two extremes, and more characteristic of manufactory in general, were a host of peasants engaged in a wide variety of crafts – linen weaving to the north and east of Moscow; pottery manufactory in the vicinity of Gzhel near the city; and icon painting in the region about Khouy near Vladimir – who though retaining older production forms saw the balance between indigenously produced and commercially supplied grain in their diets gradually shift in favour of the latter.[43]

Elsewhere to the north in Yaroslavl and Kostroma provinces or to the east in Nizhergorod province the picture was different again, as it was also in Ryazan and Tula provinces to the south. Where direct access to imported grain supplies existed and either local or imported supplies of metal were available, as in the region to the west and north of Tula, or in the villages Pavlovo and Vorsma to the south of Nizhnii Novgorod, conditions reminiscent of the Klyazma valley emerged. A host of artisan smiths, working in accord with an acute division of labour produced a profusion of metal wares: Pavlovo, the 'Russian Sheffield', specialized in locks, knives, scissors, and surgical instruments; the Tula region, hardware and cutlery for the peasants and curiosities for travellers, one of whom, slightly later, declared, 'As soon as you arrive at the inn, a number of persons crowd the room, each bearing a sack filled with trinkets, knives, inkstands, incense pots, silk reels, scissors and corkscrews.'[44]

Yet as in Moscow and Vladimir provinces such centres were exceptional in an environment which was characterized by its own unique organizational forms arising from the direct intrusion of the 'agricultural revolution' into these lands. As has been suggested in Yaroslavl and Kostroma, Nizhergorod and Ryazan, the exigencies of provisioning the metropolitan centres of Moscow and St Petersburg with livestock had created distinctive local agrarian regimes.[45] The presence of large numbers of cattle in transit manuring the land ensured high yields not only of cereals, which rendered villagers independent of commercial grain markets, but also of such garden crops as flax, which provided them with abundant supplies of raw materials for their craft activities.[46] The local slaughtering of some of the animals for local meat consumption also provided the basis for a leather industry, the Sheremetev estate in the Bogorodskoe district, south of Nizhnii Novgorod, being noted for its leather tanning as were the villages of the upper Volga, whilst the Kimry district, across the border from Yaroslavl province in Tver, was especially renowned for its manufactory of boots.[47] For many peasants, therefore, the old forms of craft activity continued, the intrusion of the 'agricultural revolution' into their lands ensuring that increasing population densities did not deprive them of either home-produced food or raw materials.

With regard to the lands lying between the Volga and Oka about the old capital of Moscow it would be anachronistic to project early-nineteenth-century conditions back into the 1780s. *Kustari* were still exceptional, and although the balance between agricultural activity and that involving manufactory, the provision of services, and commerce was shifting heavily in favour of the latter pursuits, old production forms remained. Yet in spite of the perpetuation of archaic organizational structures, as in the other great metropolitan centre of St Petersburg, the direct and indirect effects of the 'agricultural revolution' wrought fundamental changes in the economy, bringing about major productivity gains, and leading to what might justifiably be termed a 'manufacturing-commercial revolution'.[48]

Throughout the Muscovite lands, therefore, the years *c.* 1730–88 witnessed revolutionary changes which, by raising *per capita* incomes amongst the inhabitants of the area, brought the Muscovites by 1788 to within a hair's breadth of the prosperity enjoyed by their counterparts in the western provinces of the empire for a long time past. In this latter region, as has already been indicated, intensive growth had long before given way to extensive, as within its boundaries as well as on its southern margins a rising population had been accommodated at high, but stable, levels of *per capita* income.[49] Inevitably, however, as change took place in Muscovy, these lands could not remain unaffected. During these years great herds of Ukrainian cattle continued to be driven north, but on arriving at the fattening grounds of Orel, Tula, and Kaluga, they now found animals brought from the Don region being prepared for sale on the Moscow market.[50] Accordingly as numbers increased new market outlets had to be found and new routes were driven north, passing by way of Moscow to join the eastern drove road at Tver from whence the cattle continued on to St Petersburg and Riga.[51] By the 1780s according to Coxe, in his passage along this route, he had espied 'numberless herds of [Ukrainian] oxen, moving towards Petersburg for the supply of that capital'.[52] Similarly the development of cereal cultivation to the south of the old capital, in the region of the Upper Oka, brought to an end the trade in Ukrainian grain to the Moscow region which had been active in the 1720s.[53] The rise of a powerful eastern neighbour, closing off markets in Muscovy, forced an introspective pattern to evolve in the commerce of the western provinces and created new arteries of trade which would bind together an economy itself undergoing structural change. During the years from *c.* 1718/22 to 1782 rapid population growth, particularly concentrated in the old Russian provinces of Tver and Pskov and Polish and Russian White Russia, undermined the pastoral–forest base of the economy by causing an expansion in cereal cultivation, reduced manuring levels, and adversely affected yields of grain and garden crops like flax and hemp, whilst at the same denuding the area of its forest cover. As the trade of the Baltic ports in these commodities recovered after the war, accordingly, merchants had to seek out new sources of supply.[54] In the case of flax this meant tapping supplies from the Muscovite regions along the Volga and Oka where, as has been shown, rising yields, by reducing 'real' flax prices, paved the way for increased manufactory of canvas, sailcloth, and a variety of coarse linens.[55] Surplus production, moreover, now found its way to the Baltic ports, particularly St Petersburg, where its low 'real' price ensured a steady sale. Similarly hemp supplies, though still predominantly drawn from White Russia, from the 1750s also began to arrive at the Baltic seaboard from the Ukraine, and more easterly provinces like Orel. In the case of timber, as the forest cover in the hinterland of the ports was destroyed, merchants began to cast their net further afield, extending their network during the 1730s and 1740s through White Russia into the region of the Upper Oka until in the 1750s they began to draw supplies from 'the Polish provinces bordering Turkey'. As the western provinces of the empire were thus drawn into the commercial networks of Muscovy, augmenting deficiencies in local supplies of export commodities from producers in the rapidly developing lands to the east, the balance within the regional economy shifted southward. As the northern provinces of the western economic region experienced the spread of extensive low-yield cereal cultivation the Ukraine became a cornucopia of agrarian produce, its inhabitants continuing to drive a great trade in hemp, timber products, and corn. The last product now augmented a trade which grew rapidly, if irregularly, from the 1730s, drawing also on lands further north and establishing Riga as the primary Russian grain-exporting port of the 1770s. Yet the Ukraine did not simply make up inadequacies in supplies emerging in the northern provinces of the western region. Its merchants became agents of economic change therein. The hordes of Ukranian cattle arriving at St Petersburg and the Baltic provinces had to be finished off before final sale, and a new agrarian regime evolved to accomplish this task. As has already been suggested, faced with competition on the St Petersburg market from the grain producers of the middle Volga, the Baltic German aristocracy had increasingly turned the grain from their estates into vodka.[56] A by-product of this process, the lees left over, provided an ideal concentrate food for cattle. By introducing leys into their field rotation they further augmented the availability of cattle feed. By the 1780s large numbers of cattle were being maintained each year in the Baltic provinces grazing the leys over the summer, before being fattened on concentrates in the autumn in readiness for the great winter sale, held for three days annually on the Neva ice at the end of the long fast on 24 December, when the citizens of St Petersburg bought whole carcases which would remain naturally refrigerated through the whole of the winter season.[57] The lords thus contributed considerable value to their grain and the store animals they acquired, but these animals, in producing large quantities of manure, also contributed to enhancing cereal and garden crop yields, ensuring that these lands would remain major producers of flax.

The years *c.* 1730–88 thus witnessed an increasing integration of the western provinces with the Muscovite heartland of the empire, yet the indigenes of these western lands, as a result of the rise of the eastern economy, were at the same time forced into an introspective pattern of commerce, creating new arteries of trade which bound together a changing local economy where a bi-nodal pattern was being created. Thus as economic growth of an extensive character continued in the western provinces of the empire, sustaining *per capita* incomes at high but basically stable levels to 1788, the impetus which maintained this prosperity was largely confined to the Baltic provinces and the Ukraine. In White Russia and Lithuania the first signs of an impending crisis were beginning to appear as an extensive low-yield cereal cultivation displaced the previously diversified activities of the inhabitants. As yet, however, the impact of these changes on average *per capita* incomes within the region as a whole was slight, leading to a mere 4 per cent reduction between 1762 and 1788.

Inevitably, given the brief nature of the preceding description of the imperial Russian economy during the years 1718/22–88 which has necessitated a concentration on the major regions of economic change, many of the developmental ramifications and regional subtleties of the growth process have been lost. Yet the broad outlines remain. During these years rapid economic growth in the Muscovite heartland of the empire raised *per capita* incomes there to within a hair's breadth of those prevailing in the previously more advanced western provinces and indeed raised both aggregate and *per capita* incomes in the empire as a whole to a level which would have been regarded as inconceivable at the end of Peter's reign. So rapidly did the economy grow that in spite of an eight-fold increase in the size of the monetary stock, ensured by a massive increase in the indigenous production of monetary metals, prices only increased 2.5 times and even then by far the greatest part of this increase (74 per cent) was concentrated in the 1770s and 1780s when, as the 'agricultural revolution' drew to its close, the economy experienced an acute bout of inflation. By this time, however, the economy had evolved into an infinitely complex and advanced structure.

The Russian economy, 1788

The Russian economy by 1788 was very different from the strife-torn, war-weary structure of seventy years earlier and in its new form it attracted the scrutiny of contemporary economic arithmeticians who tried to reduce it to a state of statistical simplicity. Their works, if a little statistically naive, were rooted in a deep and intimate understanding of contemporary society and provide a fascinating picture of the workings of the economy, that of the much-travelled Hermann being particularly noteworthy in its description of the Russian economy in 1788 (table 5.2).[58] Before commencing any description of the economy at this time, however, it must be noted that in his study terms like 'town' and 'country', 'industry' and 'agriculture' enjoyed a usage which was both legalistically more narrowly defined and spatially and temporally more fluid than today. Their employment must accordingly be suspect in describing an environment which was over-whelmingly 'rustic' and within which economic activity was characterized by an exceptionally high level of self-consumption amongst both lords and peasants. As much as 94 per cent of gross product was distributed through systems involving the direct provisioning of households. Only 6 per cent of wares passed through the market place. Within this diminutive 6 per cent,

Table 5.2 Russian national income, 1788: estimate of B.F.J. Hermann (millions of silver roubles, at current prices)

Product	Marketed	Total product			
Foodstuffs					
grain	(144)	144 ⎫			144
Raw materials					
grain, alcohol	15	295 ⎰	1,034 ⎱		
flax, hemp, tobacco etc.	30	595 ⎱			
cattle	58.1		1,152		
fish	15	297 ⎫			
timber	20	397 ⎬	798 ⎬	2,840	
furs	5	100 ⎭			
Minerals, salt	4.2	4.2 ⎭			
		Raw materials 2,800 ⎱			
		Manufactury[a] 40 ⎰			
Metals	9				9
Total	300				2,993
Exports	25.8				
Import	18.7				
NI		3,000			

Note: a Not calculated in Hermann but estimated here by the method employed in Golitzuin (1807) whose treatment of salt production is also utilized.

moreover, only a quarter, or 1.5 per cent of gross product, involved goods destined for, or derived from the foreign trade sector of the economy.[59] Three-quarters or 4.5 per cent of goods and services produced (133.5 million roubles at current prices) passed along the arteries of domestic commerce. In this situation few indeed were the persons who could uniquely describe themselves as urban dwellers or industrial workers in anything other than a jurisdictional sense. Most people lived within a basic economic and social unit – the peasant household – which embraced and synthesized in itself all later categories. The economic unit might have its focus in the countryside but as soon as one of its members departed to join the hordes of obrok serfs who flocked to the metropoli each year for the winter season it assumed an urban aspect. Similarly a member of such a household might till the soil and was thus involved in agriculture, but if he converted the grain produced into alcohol he was also involved in industry. Terms like 'town' and 'country', 'industry' and 'agriculture' thus did not define a person in terms of a permanent location or function but delineated a temporal aspect of a unitary activity. For this reason they were eschewed by Hermann who was concerned with the distribution of goods between what was bought and sold and what was self-consumed, encompassing all activities – agricultural, industrial, and commercial – involved in the production of a good within a valuation of the final product, and thereby abandoning any attempt to make meaningless divisions within what was a unitary whole.

Bearing these remarks in mind it is now possible to examine the nature of the economy in 1788, at the end of a phase of rapid economic growth, as it was perceived by one, particularly acute, observer. First, it is clear from his study that although the environment within which men worked might be described as 'rustic' it was very far removed from the grain monoculture of a later age. Cereals and cereal products, which had a mark-up on raw-material costs of 15–20 per cent on flour (c. 60 per cent if distributed commercially) and 375 per cent on vodka, comprised only 14.5 per cent of gross product, the net harvet before processing only 7.4 per cent.[60] In spite of its relative insignificance, however, thanks to the anachronistic pre-occupations of historians intent on perceiving the past in terms of conditions prevailing in a later age, it is this sector of the economy about which most is known. Some 32.6 million desyatini (35.9 million hectares) of sown plough land or about three-quarters (73 per cent) of the available arable produced a harvest of 156 million chetverts (327 million hectolitres) involving about 13 million 'souls' in the operation. Most of this population, 6.5 million 'souls', or half the total, each endowed with 2 desyatini of land, inhabited the northern non-black-earth (podzol) provinces but already by this time their command over land and contribution to production was small. The 12.96 million desyatini of plough land under their control comprised 40 per cent of the total, but at a seed-yield ratio of 1:3, yielded only 52.8 million chetverts of grain or one-third of the total harvest. During the years 1718/22–88 they had been eclipsed by the inhabitants of the black-soil (chernozem) lands to the south: the Ukrainians, brought within the empire during the closing years of Peter's reign, and the settlers who had migrated southward onto the rich friable black soils of the wooded steppe. The former, numbering 2.1 million 'souls' each possessing 3 desyatini of sown plough land, now contributed 20 per cent of the total harvest in spite of their small numbers, but it was the latter who now dominated the agricultural scene. Collectively, these 4.5 million 'souls' each endowed with 3 desyatini of land which would yield a return of 3.5 times the seed sown in the Volga and eastern Don regions and 4.5 in the lands to the south of Moscow, cultivated 41 per cent of the land in tillage and produced almost half of the total harvest, their high returns (represented in a seed-yield ratio of 1:4) raising the national average to 1:3:5.[61] The exploitation of rich land, utilizing a system of field-grass husbandry which ensured a heavy manuring, was slowly dragging Russian agriculture out of the low productivity levels which had characterized it at the beginning of the century when, confined to the exploitation of the poorer podzols by means of an open field regime, average decennial seed-yield ratios of 1:2.7 had not been unknown, and was ensuring a harvest which, if only contributing modestly to the prevailing level of national income, did provide the populace with a more than adequate supply of grain.

Each of the Tsarina's subjects, with 3 litres of grain available to them daily and with the prodigious quantities and enormous variety of alternative foodstuffs at their disposal, could afford to turn less than one-third of the cereals into bread, consuming about 343 g (12 oz) daily. The residual two-thirds was either fed to stock, at the rate of one-fifth of a peck per horse per day, or converted into alcohol, equivalent to 10 cl of vodka.[62] Bread, which figured so markedly in the diets of the peasantry of western Europe, who had undergone a process of 'depecoration' or growing meatlessness since the sixteenth century, played a small role in the consumption of the Russian populace, so that although endowed with similar quantities of grain as their western counterparts, most was put to other uses.

Table 5.3 Average annual *per capita* food consumption, 1788: estimate derived from B.F.J. Hermann

Product	Quantity		Value[a]
Grain, bread	1.1 chetvert	= 125 kg	2.2
alcohol	1.3 chetvert	= 38 l	9.9
Vegetables and fruits			9.8
Vegetable oil[b]	1.89 pud	= 31 kg	3.5
Honey and sugar	0.3 puds	= 4.9 kg	4.6
Flesh, fish	17.75 pud	= 290 kg	9.7
beef	10.63 pud	= 174 kg	8.6
mutton	1.22 pud	= 20 kg	1.0
Dairy produce, milk	20 vedro	= 251 l	5.1
butter	1.8 pud	= 29.5 kg	5.8
Total			60.2

Notes: a In silver roubles at current prices. b Derived from hazel nuts (Kazan province) cedar nuts (Siberia), poppy seeds and juniper (Yaroslavl), and hemp seeds (rural peasants in many areas).

During the 1780s the Russian populace lived in a very different world from their counterparts in western Europe, their dietary regime still conforming to norms which in the west had become historical abstractions.[63] Bread, far from being central to their diet, was merely a subsidiary adjunct, comprising about 3.7 per cent (by value) of their total consumption, a constituent element of those snacks of black (rye) bread and gherkins which contemporaries observed the peasantry consumed during their working day in the fields. At their main repasts, however, it was of little importance, the tables groaning under the weight of the major staple in the Russian's diet – flesh. 'On account of their numerous fast days', fish figured largely in their diet, average *per capita* consumption levels reaching the prodigious amount of 794 g (1.75 lb) a day but displaying marked inter-annual variations between those times when it comprised the only legitimately consumable form of flesh and the more normal occasions when it ranked second to meat, some 602 g (1.33 lb) of fish complementing the 706 g (1.56 lb) of beef and mutton which, either in the form of stews or roasts, provided the centrepiece of the average Russian's daily diet. In whatever form it was consumed, flesh, amounting to 1.33 kg (2.9 lb) per head per day, comprised the main element in the Russian's diet constituting almost one-third (by value) of his total food consumption. Add to this, moreover, a wide variety of other produce – oils, in the form of butter or vegetable oils amounting to c. 166 g (5.8 oz) a day; vegetables; fruit, consumed either fresh, in the form of fruit juices, or as the main element of the ubiquitous 'sweet tart which stands perpetually ready for use'; sweeteners and liquid milk, amounting to slightly more than a pint (0.687 l) a day – which collectively constituted almost half (47.8 per cent) of the average Russian's food consumption and one cannot but agree with Coxe who, familiar with the prodigious appetites of the English lower orders, noted that their Russian counterparts enjoyed 'plenty of wholesome food'.[64]

As has been noted, however, even in the terms of their diets, cereals, directly or indirectly, comprised only c. 19 per cent (by value) of the Russian populace's consumption. Yet of the other items, which made up more than four-fifths of the whole, little is known. Leaving the plough land for the garden grounds, hayfields, pastures, and forests one enters a *terra incognita* largely unexplored by the historian but, in the light of the information contained in table 5.2, of paramount importance in the economy. Both the garden grounds and hayfields, comprising c. 6 million and 13.5 million desyatini respectively, produced 'crops' worth more than the total value of the net grain harvest. The garden grounds in particular were highly productive. Each 'family farm', or tiaglo, with its average 6 statutory desyatina of sown plough land had 1 desyatina (2.7 acres) of garden ground located in the vicinity of the house or in special enclosures, which contained the peasant's hut or izba and threshing floor (0.45 acres), as well as 2.25 acres devoted to labour-intensive, high-yield crops: on average 1 acre of vegetable garden and 1 acre of orchard, 0.08 acres for flax and hemp and 0.17 acres for tobacco. Regionally varied in its production mixes, each element in this sector of the farm economy was highly productive. In terms of land, a mere 10 per cent of any acre could be made to yield 24 lb of hemp oil, 88 lb of tobacco, 2.75 litres of fruit or some 175 cabbages, produce which, if sold, would return between 2.5 and 2.75 roubles.[65] Even in terms of primary production, with gross output valued at c. 350 million roubles, this was an important sector of the economy, a relatively minor element such as fruit yielding a crop worth more than that portion of the grain harvest devoted to satisfying the population's bread requirements and generating a domestic commerce larger than the 'mighty' iron trade.[66] Where the product could be processed on the farm, as in the case of flax and hemp, yet more value could be added. Of almost 4 million families, or two-thirds of the population, involved in the production of flax and hemp, more than a quarter were involved in the fabrication of these raw materials, contributing some 1.6 million individuals to a part-time labour force whose production was, in terms of value and quantity, far greater than that of the much-vaunted manufactories.[67] The latter, concentrated in the case of linen in Moscow and Yaroslavl provinces and in the production of cordage and sailcloth in the ports of St Petersburg and Archangel, absorbed only 5 per cent of total flax and hemp production and just short of 20 per cent of that portion of the crop destined for domestic textile production, turning out, in the case of linen, c. 13.4 million arshins of low-priced broad linen cloth, 10–11 million of which was exported.[68] The former group of peasant craftsmen, working up raw materials derived from their holdings during the 'dead' winter season, deployed some 237,600 looms producing 214 million arshins of narrow linen cloth, most of which was distributed within peasant society, affording an annual *per capita* consumption of c. 7 arshins (5.5 yards) or enough material every two years to make a complete set of peasant's summer clothing: a coarse linen shirt, wide trousers, and a knee-length, double-breasted kaftan or surcoat. A clothing element was thus added to the foodstuffs and other consumables produced in this sector of the farm economy, even if it was only marginal to its total product.[69]

Activity about the villages and within their fields and garden grounds, which encompassed c. 10 per cent of the total land area west of the Urals, thus generated about one-third of the average Russian's income in the late 1780s. His true wealth did not reside here, however, but in the seemingly limitless wilderness beyond. It was by the exploitation of the pastures and wastes, forests and rivers, which extended over 90 per cent of the landscape, that he generated the greater part, two-thirds, of his income.[70]

To these lands the peasantry now brought the 'civilizing' effects of the 'agricultural revolution'. During the years 1725–96 an estimated 32 million ha of forest were cleared, about one-third in the period 1725–63 and two-thirds between 1763 and 1796, to create new arable and pasture. A burgeoning population of peasant cultivators was making inroads into the wilderness, and although their environmental impact was as yet minimal, reducing the forest cover within European Russia by a mere 6.5 per cent from 51.2 to 44.7 per cent of the total land area, their activities, plundering the natural environment, yielded them a handsome return. At an annual clearance rate of c. 636,000 ha in the 1780s such felling not only created new agricultural resources for the population, each family adding about one-tenth of a desyatina a year to its land reserves to secure its children's future, but also yielded a regular supply of timber for construction purposes and firewood, amounting in total to some 532 million cubic sazhens (40.4 million cords) worth 397 million roubles.[71] For the moment, in relation to the forests, the peasant enjoyed the best of two worlds. He could plunder their reserves at will, destroying without thought of conservation, yet his activities had so little impact on the natural environment that he could also reap the harvest of the untamed wilderness which not only surrendered up minor items like honey and beeswax, which continued to be collected from wild bees' nests set high in the trees, but also the skins of fur-bearing animals which, although now no longer enjoying an international reputation, still found a ready sale on domestic markets to 'manufacturers' of winter headgear or tailors making up the linings of woollen overcoats, goods worth some 100 million roubles a year.[72]

Similarly, whilst contemporaries deplored the deleterious impact of an encroaching civilization upon the nation's rivers and lakes, highlighting in particular the pollution caused by flax retting, the effects of this environmentally destructive activity seems to have been minimal. The rivers, lakes, and coastal waters teemed with fish. Of the Volga it was said that it had 'a great plenty of sturgeon, sterlet, citera, salmon – both red and white–saudack, perch, crawfish, carp, pike and tench'.[73] The fishermen of Archangel, Kholmogory, and Olonets landed large quantities of cod, plaice, sole, and herring. All over the country men only had to cast their nets to obtain a rich catch which collectively yielded some 8.5 million tons of fish annually for direct consumption or the production of oil, isinglass (that product 'almost peculiar to Russia' which was used to preserve eggs), and caviar – wares valued at c. 297 million roubles a year.[74] Nor was this the sole product derived from Russia's waterways and lakes. A number of salt lakes, together with brine pits, rock salt mines, and sea water refineries, produced about 0.2 million tons of salt which, distributed through government agencies, sold for 4.2 million roubles a year.[75] Whilst a burgeoning population of peasant cultivators thus nibbled at the edge of the wilderness, the forests and rivers continued to yield a prodigious harvest of wares allowing the average Russian, by his exploitation of these resources, to obtain about a quarter of his income.

Fields and gardens, forests and rivers, each provided resources which allowed the average Russian to earn an important portion of his income, but both elements of the rural economy paled in significance at the side of the third – the pastures and wastes. During the years of the 'agricultural revolution' from c. 1735 to 1785 the Russian peasant, like his Ukrainian counterpart for many years past, held the greater part of his wealth in his herds and flocks from which he obtained the most important part (40 per cent) of his income.[76] In the north, within the confines of Old Muscovy and westward into the lands of the White Russians and Lithuanians, families possessed herds similar to those found in western Europe in an earlier, more land-plentiful age, a pair of horses providing the peasant with enough draught power to work his holding, whilst a couple of milch cows yielded a regular annual supply of 'white meat' and constituted the principal breeding stock. It was their progeny which formed the basis of the population's meat supply after the requirements of herd replacement had been satisfied. Peasant herds thus normally contained eight or ten young beasts, yearling calves and heifers to replace old stock, and 2–4-year-old steers which assured the family an abundant supply of meat. Plentifully supplied with livestock, with some 60 million cattle and 13 million sheep and goats grazing the extensive northern pastures, each peasant family was assured of abundant supplies of food and raw materials, allowing each member to consume some 174 kg of beef and 20 kg of mutton a year as well as 65 lb of butter and 20 vedro/eimar (251 litres) of milk worth about 20.5 roubles, as well as 3.6 puds (59 kg) of hides and tallow which, when fabricated, yielded produce worth 16.09 roubles.[77] Collectively the peasantry obtained from their herds and flocks 1,093.9 million roubles' worth of produce which was almost entirely consumed within peasant society. Market demand was largely satisfied from elsewhere – from the vast herds (*cheredi*) of cattle which grazed the northern fringes of the vast pastures extending over the open steppe. Here, although Russian aristocratic families like the Orlovs, Razumovskii, Sharshinskii, and Vorontsov now joined the ranks of Poles who had previously dominated the animal husbandry of the area, old production forms continued in use, the characteristic production unit being the herd of 100–800 animals which remained all year on the grazings and from which, each spring, young stock 3–5 years old were culled and sold to the *prekashtshiks* who despatched them abroad or to markets in the north. Every year about 1.4 million cattle, worth 58 million roubles, were driven to market from the lands of the southern steppe.

By their exploitation of a wide range of rural resources encompassing fields and gardens, pastures and wastes, forests, rivers, and lakes, the Russian populace thus assured themselves a very high level of *per capita* income in the late 1780s amounting to about 100 roubles at current prices.

Most of their efforts were deployed to provide themselves and their principal livestock with food. As has been shown their personal food consumption was prodigious. That of their principal livestock was no less bountiful, the family's horses receiving a daily diet of one-fifth of a peck of oats, *c.* 20 lb of hay, and an indeterminate amount of concentrate feed derived from the residue left after brewing or distilling. Each person was also adequately provided with clothing, producing enough material every two years to make up a complete set of winter and summer clothes, and had abundant supplies of materials to build, heat, and light his house. The average Russian, accordingly, lived in a state of rude abundance.

Averages have only a limited analytical significance, however, either in terms of consumption or employment at prevailing low levels of commercial activity. With only *c.* 6 per cent of goods passing through the market, consumption displayed a high degree of regionality as men exploited the resources which were locally available to them to satisfy their needs. Similarly, the limited impact of the market ensured that their work in exploiting these resources would be characterized by a singular lack of

Table 5.4 Composition of average annual per capita income, 1788: estimate derived from B.F.J. Hermann (silver roubles at current prices)

| Sector | Product | | | | | | | Total |
	Food and drink		Horse fodder		Clothing		Sundries	
Plough land	Grain	12.1	Grain	2.6				14.7
Gardens	Vegetables and fruit	9.8			Linen, hemp	1.2	Tobacco 0.9	11.9
Hayfields			Hay	8.1				8.1
Pastures	Meat, butter	21.6			Leather, wool	1.5	Tallow 15.5	38.6
Forests/rivers[a]	Fish	9.7			Fur	3.4	Timber 13.3	26.4
Minerals							Salt 0.2	0.2
							Metal 0.3	0.3
Total		53.2		10.7		6.1	30.2	100.2

Note: a   excludes vegetable oils and honey not included in Herman.

functional specialization. The work experience of most of the population was encompassed within a basic economic and social unit – the peasant household – which embraced and synthesized into a unitary whole those categories of activity which only in a later age would assume separate identities. For the participants in this unit terms like 'urban' and 'rural', 'industrial', 'commercial', or 'agricultural' did not define the individual in terms of permanent location or function but only delineated a temporal aspect of his unitary activity as a peasant, involved in the life and work of a peasant household. It is, accordingly, not meaningful to impose such categories on the work patterns of most of the population.

Such terms only have relevance for a tiny group in society who could legitimately and uniquely describe themselves, in real rather than jurisdictional categories, as urban dwellers, workers or professionals, soldiers or traders. Just how large this group was is difficult to assess but one approach to this problem is by way of a consideration of the number of people who depended on commercially supplied grain for their sustenance. For Hermann such a group did not exist. All grain was consumed within the bounds of peasant society. A recent consumption-based estimate (formulated on anachronistic assumptions which would have been eschewed by contemporaries) puts the figure at 9–10 per cent of the population – 2.7 to 3 million people.[78] Setting these figures as upper and lower parameters it is clear that the number of people who could stand outside the bounds of peasant society and enjoy a functionally specialized existence was small. Just how small this group was may perhaps be gauged if an alternative method of estimation is adopted, based upon a calculation of the amounts of grain transported over long distances to meet a commercial demand. In the 1780s it may be estimated that about 1,565,500 chetverts of grain, or c. 1.4 per cent of the net harvest, entered into long-distance commerce.[79] By far the largest part of this grain (c. 42 per cent) passed by way of the complex riverine and overland transport system of the western provinces, feeding the massive export trade of Riga and the other Baltic ports (c. 323,000 chetverts) as well as the minor traffic (14,455 chetverts) which found it way to the Black Sea ports, and provisioning the capital with some 310,000 chetverts of cereals. As has been suggested, however, these latter supplies only formed part of what was necessary to provision the inhabitants of Peter's city on the Neva, whose population had been growing rapidly over the previous half-century to attain a level of some 218,000 in 1788.[80] By this time more than a quarter of the city's supplies were transported thence by way of the waterways linking the Volga to the Neva. Of *c.* 190,000 chetverts a year passing along the Volga via Liskovo and Nizhnii Novgorod almost 60 per cent was destined for St Petersburg, the rest being distributed either in the vicinity of the fair town or amongst the *kustari* resident in the valley of the Klyazma river, east of Moscow, or along the tributaries of that river stretching northward towards the Vladimir–Kostroma border. In the northern and eastern parts of Moscow province and throughout Vladimir province a large number of industrial workers had already by the 1780s become dependent upon commercial grain markets for their sustenance. Similarly in the southern part of Moscow province, extending into the north-western part of Tula province, the numbers of such people were far from insignificant, as some 50,000 chetverts of grain passed annually along the Oka, Tsna, and Moksha to provision this region. The time was indeed rapidly approaching when half of the population of Moscow and Vladimir province would be dependent upon imported grain. Already in the 1780s one-third found themselves in this position, consuming *c.* 590,000 chetverts of imported grain.[81] Further north there was the trade on the northern Dvina for export through Archangel (*c.* 56,000 chetverts) or the provisioning of the Pomor'e (*c.* 92,500 chetverts). By adding that sector of the Volga trade, an estimated 168,000 chetverts, which did not pass through Liskovo and Nizhnii Novgorod but rather was diverted along the Kama to the mining towns of the Urals, this survey of the European Russian grain trade is complete, suggesting that enough grain was transported along the major arteries of long-distance commerce to allow some 865,000 people, or 2.9 per cent of the population, to live solely on the basis of commercially supplied grain.[82]

Divorced from the agrarian regime and dependent upon the market for their sustenance, this tiny group derived their incomes predominantly from trade and manufactory. Slightly less than half (46 per cent) of their number were involved in the provision of commercial and transport services, deriving about 39.5 million roubles a year for their work servicing the requirements of a domestic and foreign commerce valued at 178 million roubles.[83] The remainder (54 per cent) or some 470,000 workers, were engaged in manufactory or industrial activity. To comprehend their work experience, however, it is necessary to sub-divide this group, for many (perhaps one-third of the whole) were not found in manufactories or industrial enterprises but laboured within the confines of workshops. Some of these artisans, craftsmen, or tradesmen were to be found in all towns but they were most numerous in St Petersburg where they were to be numbered in the thousands. The wealth and luxury of the capital and the devotion of its aristocratic inhabitants to foreign modes demanded a host of carriage-builders, watch-makers, dressmakers, milliners, embroiderers, wig-makers, bookbinders, cabinet makers, painters, mechanics, tailors, and shoemakers, many of whom were German, jewellers, French tutors, and Dutch landscape gardeners, silversmiths and goldsmiths. . . .[84] Although St Petersburg possessed a number of manufactories, it was truly a city of workshops.[85] Nor should such enterprises be considered primitive because they employed small numbers and were confined within workshop walls. They stood at the pinnacle of manufacturing activity, their labour force, highly skilled and often organized on the basis of an advanced division of labour, returning a *per capita* product 8–10 times greater than in the manufactories of the Moscow region and 2.5 times more than that of workers employed in the industrial enterprises of the Urals and the metal fabricating works of Tula.[86]

Table 5.5 Manufactories and industrial enterprises, *c.* 1788

Type of enterprise	Value of output (million roubles)	Employment (thousands)	Per capita product (roubles)
Mining and metallurgy			
Copper and iron	7.0	81.0	86
Salt	2.4	35.0	68
Manufactury			
Textiles, flax and hemp	3.1	121.3	25
wool	0.4	21.3	20
cotton and silk	0.6	30.4	20
Metal fabrication	1.8	24.4	76
Other	1.2	5.7	200
Handicraft/artisan	30.2	148.0	200
Total	46.7	467.1	100

Manufactures were similarly widely dispersed but the largest concentrations were found in the lands between the Volga and the Oka, particularly in Moscow and Vladimir provinces. This sector, moreover, encompassed a wide diversity of activities. At one extreme were the manufactories making up luxury wares, like the St Petersburg tapestry works that 'produces such excellent work that better is not to be seen from the Gobelines at Paris' or the state porcelain factory at St Petersburg, serving the Court and employing 400 workpeople, which employed similar techniques to the workshop enterprises of a similar character and enjoyed a comparable level of output per man.[87] At the opposite extreme were the textile manufactories which, both in terms of employment and value of output, dominated this sector, engrossing 85 per cent of the labour force and contributing about 58 per cent of sectoral income. These were barely distinguishable from the peasant-craft sector. Their product was sold at a broadly similar price to the peasant's wares, and although output per man-year was about 3.3 times that of the peasant this was largely due to the greater number of man-days spent per year on this activity by the full-time worker in comparison with his part-time peasant counterpart.[88] Finally, turning to the only truly industrial enterprises, the world-renowned metallurgical works of the Urals which employed 81,000 full-time workers as well as 167,000 ascribed serfs who formed an integral part of peasant society, and the *kustari* engaged in metal fabrication, these, in terms of the value of output, constituted by far the largest branch of manufacturing and industrial activity. Full-time workers therein, moreover, displayed a relatively high level of productivity, output per man-year being some 3–4 times greater than in the textile manufactures of the Moscow region.[89]

Taken as a whole, embracing all handicraft and artisanal activities, manufacturing, and industrial pursuits employing full-time workers, however, this sector of the non-peasant economy whose denizens were divorced from the agrarian regime was tiny. It encompassed only 1.5 per cent of the population and contributed a similar proportion of national income. Productivity levels, though varying greatly within the sector, were on average similar to those prevailing amongst the peasantry. Certainly, whether measured in terms of output per man-day or per man-year, the full-time workers were more efficient in the working up of raw materials than their part-time peasant counterparts but in terms of their total productivity they were not, their work yielding them a similar average *per capita* product of 100 roubles per year.[90]

Nor was the situation significantly different for those workers, comprising some 1.4 per cent of the population, whose labours in the provision of commercial services yielded them a similar *per capita* product. The world they inhabited involved an intricate pattern of exchanges transcending, but closely interwoven into, the fabric of peasant society. As has been shown, peasant society at this time was largely self-sufficient, households satisfying their requirements for food, clothing, accommodation, heating, lighting, and other sundry items by their own endeavours or by recourse to country markets encompassed within the bounds of peasant society. Only 6 per cent of produce worth 178 million roubles passed through markets of anything other than local significance. Of this, a quarter, or 1.5 per cent, involved goods worth 44 million roubles destined for, or derived from, the foreign trade sector of the economy, and three-quarters, or 4.5 per cent, of goods and services worth 134 million roubles passed along the arteries of domestic commerce.[91] Most of this latter trade, perhaps as much as 90 per cent, involved transfers of goods between peasant society and the nodal points of the non-peasant sector, allowing the inhabitants of the latter sector, who were predominantly concerned with satisfying aristocratic and state demand, to replicate the consumption patterns of their country cousins. Food (grain, alcohol, fruit, cattle, and fish), animal fodder (oats and hay), raw materials (timber and hemp, flax and wool), and a whole host of specialist items (such as Ukrainian clays used in porcelain manufacture) all flowed from peasant society to centres like St Petersburg, Moscow, and the villages of its region or the Urals mining and metallurgical communities. From the latter centres, moreover, an intra-sectoral cross-flow supplied other consumables (salt) and raw materials (metal) for manufacture as well as export. The denizens of the non-peasant sector were, accordingly, well provided with foodstuffs (allowing them to replicate peasant consumption patterns) and raw materials which, as has been shown, formed the basis of a major 'industrial' complex, but most of the goods they produced were destined not for peasant consumers but for the aristocracy, the state, or export. That element within this complex catering for state demand, initially created at the behest of Peter to fulfil his martial ambitions, was still important but many of its wares were now exported rather than consumed domestically. The shipbuilding complex of St Petersburg, the Admiralty, which at its height in the 1710s had employed as many as 7,500 workers – carpenters, sawyers, joiners, caulkers, blacksmiths, sailmakers – and which had absorbed most of the output of the ten manufactories producing 1.2 million arshins of sailcloth and linen and the ropeworks producing cordage, still played an important role as a customer for these input supplies, but they now found their major markets abroad.[92] In the case of linens and hempen wares, for instance, whilst the Admirality bought about 2 million arshins of sailcloth, or about 15 per cent of total production, most, about 80 per cent, of the manufactory's output was exported in the form of Flemish cloths, Ravenducks, and sailcloth.[93] Similarly, in the case of metals, whilst enterprises working to state orders still remained important customers for the products of the Urals and central region metallurgical works most of their wares were exported. Of a total output worth 10.8 million roubles, 7 million roubles' worth of iron or two-thirds of the whole was exported. Only one-third was fabricated in Russia and of this, as has been indicated, more than half, or 1.8–2 million roubles' worth of copper, passed to the mints to be manufactured into coins, and an indeterminate amount passed to those enterprises producing armaments which had dominated the demand for metal in Peter's reign.[94] Many of these Petrine works were still operating up to a century after their creation, and in many cases were producing more in Catherine's reign than they had in Peter's. The Tula firearms factory, which in the 1710s had manufactured 'twenty thousand muskets, ten thousand pairs of pistols, besides other iron goods' each year, was producing sixty years later, in 1777, in association with the Sestoretsk and Bryansk works, c. 42,000 rifles and 25,500 pairs of pistols; the enterprise was characterized not only by its mass production but also by a division of output between the plants, Sestoretsk producing not only complete handguns but also locks for assembly at Tula, and Bryansk supplying barrels to the latter works. Nor did this situation change over the next quarter of a century, although in the 1800s what had appeared advanced production techniques failed to impress one English visitor, who remarked of these works that 'the machinery is ill-constructed and worse preserved. Everything seemed out of order. Workmen with long beards, stood staring at each other wondering what was to be done next, while their intendants and directors were drunk or asleep.'[95]

A similar pattern may be discerned in the production of artillery pieces although in this instance the development of this branch of the industry was much more erratic. The original Petrine creations or recreations (Pushech, near Moscow remodernized in 1707, Lipetsk producing cannon for the Black Sea fleet, and Olonets and Petrozavodsk similarly supplying the Baltic fleet), together with those private works in the Urals producing cannon for the army, managed by a supreme effort to create a stock of c. 13,000 pieces during Peter's reign. But thereafter numbers remained stable until the 1760s, private works converting production to civilian markets and state enterprises, confined to repair work, either falling into decay or, as in the case of the Lipetsk works, being sold off. Following the Seven Years' War, however, the deplorable state of the artillery stock was only too clear, and it was decided that of 13,000 guns of all types at the disposal of the army 9,500 needed to be replaced. The Lipetsk work, taken in hand again, the Alexandrovsk plant at Petrozavodsk, retooled in 1772, and the Batashev iron works, which had become a major military supplier in the 1760s, thus had more than enough work to carry them through the 1780s when new demands to supply the Baltic and Black Sea fleets led to the creation of new capacity at Kronstadt–St Petersburg and Lugansk.[96] Yet in spite of the fact that these armaments works produced more than that Petrine industry which had dominated metal markets in the 1710s, they now absorbed far less than 20 per cent of the output of an industry selling most of its products abroad. Amongst those manufactures catering for the demands of the state only one – the woollen manufactory – had no export market. It had been created specifically to displace imports of cloth and *karazeia*, a woollen lining, used in military uniforms. Initially the supply of these wares had largely been monopolized by English and Prussian merchants, who had fought energetically to secure the contracts put up for tender by the War Commissariat, but slowly yet relentlessly both had been ousted by indigenous producers who from the 1750s took over the task of supplying the military and thereafter devoted the whole of the industry's output to this end.[97] A large group of those workers engaged in manufacturing and industrial activity, who drew their raw materials from, and were sustained by, that flow of goods which originated in peasant society, thus produced wares (c. 25 per cent of sectoral output) which were destined to meet state or export demand. An even more important group, employed in a small number of manufactories and a host of workshop establishments, produced goods (*c.* 68 per cent of sectoral output) which were designed to satisfy the luxury requirements of the aristocracy. Together these groups dominated the output of those workshops, manufactures, and industrial enterprises of the non-peasant sector whose labour force was sustained by the produce derived from peasant society.

Many of these luxury and other items destined for aristocratic consumption together with imports of a similar character (worth *c.* 12 million roubles) were consumed within metropolitan society. So also were many of the goods produced to meet the state's military requirements and such as were not were largely distributed through a non-commercial military supply system. Accordingly the inter-sectoral balance of trade between peasant and non-peasant society was highly active, the peasantry producing goods, either by direct labour for their lords (barshchina) or on their own initiative which when sold yielded cash to meet taxes or seigneurial quitrents (obrok), which served to fill the coffers of the state and aristocracy. This concentration of cash in lordly hands even affected the reciprocal trade from non-peasant society back into the countryside for, each year, many of those carts which had carried the produce of the lordly demesnes of the western or central agricultural regions, or had accompanied the great herds of cattle on their way north, returned bearing luxury wares, bought in the old or new capital, to furnish the rustic manor houses or administrators' residences of the provinces.

Yet in spite of the overwhelming importance of the aristocracy and government in shaping the form of exchanges between peasant society and the non-peasant/foreign-trade sector of the economy, the peasantry did benefit in some degree from participation in this network. A narrow range of goods (salt, fish, textiles, and metal wares) worth some 12 million roubles, or 7.5 per cent of this commerce, together with a host of small-value copper coins, returned to peasant society. Add to this that long-distance exchange of goods (alcohol, fruit, tobacco, meat, and fish) between regions within peasant society of somewhat greater value (*c.* 18.3 million roubles or 10 per cent of trade) and it can be appreciated that the peasantry were not entirely excluded from the benefits of market participation, having command over sufficient commercial goods to make on average a 1 per cent adjustment to their indigenously determined consumption patterns, or 5 per cent if only those participating at the fairs are taken into account.[98]

In the 1780s, therefore, the commercial sector of the economy embraced a complex web of currents and cross-currents, primarily concerned with satisfying the needs of government and aristocracy, but closely interwoven into the fabric of peasant society whose members also benefited from their involvement in the system. Organizationally, the institutional network superimposed upon this system was, moreover, no less complex. Many of the wares, particularly where they were involved in the foreign-trade sector of the network, passed directly from estates, manufactories, or industrial enterprises to their destinations. Of these goods perhaps the best known is iron, which was shipped by ironmasters like the Demidovs directly to the ports, where it was sold through the factory *kontor* to foreign merchants, and to government enterprises, only a small proportion of output being sold at the fairs.[99] It is also likely that the products of manufactories (linens, hempen wares, or leather goods) or estates (timber, grain, alcohol) were disposed of through factory or estate offices. Similarly if one considers luxury wares, imports in the 1780s were normally sold direct to customers from the showrooms attached to the merchants' residences, and it is most improbable that a similar pattern was not found amongst domestic producers.[100] Subsequent transportation of these wares within the metropolitan centres may have been the responsibility of the seller, but if they passed beyond the bounds of the cities transport was normally arranged by the buyer using estate carts. A very large proportion of total trade, perhaps as much as half, thus passed directly from supplier to customer. The remaining half passed through more normal commercial channels. In the non-black-earth provinces of the north many an obrok serf found a ready market for his produce (vegetables, fruit, milk, butter, or freshwater fish) at the numerous weekly markets servicing the needs of the non-peasant population, whilst others carted hay to be sold in the metropolitan capitals or, showing a little initiative, entered into the 'convenience food' business of those centres. Whole meals could be bought on the streets of Moscow and St Petersburg for a few kopecks and even an item like fruit played an important role in this business.[101] Stalls sold apples, fresh or preserved in kvass, baked pears spitted a dozen on a stick, oranges, dried fruit, and wild berries. Hawkers sold hot and cold drinks which were carried in bottles and jugs and poured out to customers in handsome glasses: kvass was the most popular throughout the year but lemonade, kislischi, sbitene, mead, and tea were also sold.[102] Because of the high concentration of non-peasant workers and enhanced income levels in this area such activities were handsomely rewarded, as was also the work of the numerous shopkeepers who also serviced the retail trade in this area. In Moscow alone there were some 6,450 shops as well as 199 eating houses, 162 public houses, and 79 horse-hiring establishments, whilst in the administrative centres of the *uezdi* of Moscow province there was a reported total of 902 stores encompassing both large centres like Kolomna and Serpukhov (each containing 200) and small ones like Podolsk (with thirteen) and Zvenigorod (with nine).[103] Distribution was highly polarized, however, and the two metropolitan capitals contained 80–90 per cent of the total. The northern black-earth provinces and particularly the capitals and their hinterlands were especially well provided with retailing outlets provisioning the non-peasant inhabitants of the area, their trade, overlapping the direct sales system already described, engrossing one-third of the total and leaving little place for fairs (like the great winter mart for the sale of animals' carcases held on the frozen Neva) in the supply system of the area.[104] Fairs essentially serviced the needs of the rural communities, providing markets where goods destined for the non-peasant/foreign-trade sector of the economy could be sold and produce of that sector as well as commodities involved in intra-sectoral trade could be bought. In the north-western provinces they were a focal point for merchants to acquire flax and hemp for export and to dispose of imported salt and fish to peasant customers.[105] The situation was not significantly different in the Ukraine, although the commodities exported from this region included not only hemp and woodland products destined for the Baltic ports but also cattle for northern consumers and wool for northern manufactories. Each spring, young stock were driven to the Kharkov fairs (Kreshchenskaya, Prokrovskaya, and Uspenskaya held on 6 January–6 February) or to those of Kiev (Kontraktnaya Kievskaya) to be sold on to those who would rear them over the summer. In the autumn they were assembled into herds and, accompanied by oxcarts bearing other produce of the Ukraine, were driven north, some of the stock being sold at Bryansk to those who would finish them off for the Moscow market, and the remainder continuing on via Kaluga, Moscow, and Tver to St Petersburg and Riga. During the winter these traders returned, bearing not only the staple import commodities – salt and fish – but also large quantities of foreign silver coin. Stopping *en route* to convert the balances they held at Moscow into northern manufactures they subsequently resumed their journey south, returning to Kharkov in time for the spring fair. During the spring and summer, as the cattle acquired by rearers were made ready for their voyage north the following autumn, the wares brought from the north were distributed – at the Maslianskaya fair in Romny (17–30 February); Elizavetgrad (21–9 April); the Voznesenskaya fair at Romny (20 July–1 August); Karkhov again (15 August–1 September); and Krolevets (14–26 September). The traders then returned for the autumn cattle fair at Kharkov (1 October–1 November) and the cycle began once more. The pattern repeated itself again, commodities destined for the non-peasant/foreign-trade sector to the north (cattle, hemp, wax, potash) exchanging against wares derived from that sector (imported salt and fish, small quantities of textiles and metal wares) creating, because of the high income levels of the region, an especially high density of fairs.[106] Within the Muscovite lands, outside the non-black-earth provinces of the north which have already been considered, the situation was very different. The 'foreign-trade' sector of this commercial network was well catered for with major fairs at Makar'evskaya, near Nizhnii Novgorod, which apart from its role in the China trade was also a major focal point importing raw materials (Urals iron and copper) to the manufactories of the Moscow region and exporting their wares (textiles and metal goods). Irbitz serviced the trade to the Siberian steppe and Astrakhan played a central role in the Persian and central Asian trade, but 'domestic commerce' within the region was poorly developed. The direct trade in estate produce northward undermined the 'export' trade from the region, restricting the fairs, most notable Voronezh and Belgorod, to handling the cattle trade of the area. These centres, moreover, also handled most of the diminutive reciprocal 'import' trade into the area. It was here that fish (obtained at Tula and Kaluga from Ukrainian traders passing south), salt (obtained from Lake Elton, near Saratov, or Perm), and other wares were sold, many of the latter being brought by those 'carriers of Woronetz' who

> go every three years to Tobolsky in Siberia which is a rendezvous for all caravans bound to Kiatka, on the frontiers of China. . . . From Siberia they bring furs; from Kiatka, Chinese merchandise of all sorts, as tea, raw and manufactured silks, porcelain and precious stones. Thus laden, many of them set forth for Francfort, and bring back muslin cambric, silks, the porcelain of Saxony and the manufactures of England.[107]

Their commerce was essentially a transit trade, but both here and at Kiev it is not unlikely that some portion of the wares was sold to local consumers at the fairs. Yet this was the icing on the cake, as the basic 'import' goods sold at the fairs were old staples – fish, salt, and small quantities of textiles and metal wares – and the lack of a major 'export' staple other than cattle ensured that the central agricultural region had the lowest density of fairs in the empire.[108] Taken as a whole, however, whether concerned with the sale of goods to the non-peasant/foreign-trade sector of the economy or with the distribution of a narrow range of commodities derived from that sector or from peasant suppliers engaged in intra-sectoral exchanges, the fairs, which had grown rapidly in number during the previous half-century, engrossed a considerable volume of trade, their turnover amounting in the late 1780s and early 1790s to c. 64 million roubles or 36 per cent of the whole.[109]

Table 5.6 Estimated trade at fairs, c. 1788

Commodity	Value (million of roubles)
Cattle	39.5
Fish, salt	10.9
Flax, hemp, linen	9.0
Furs	3.1
Metal wares	0.7
Sundries	0.8
Total	64.0

Taken as a whole that group engaged in commercial activity, the denizens of estate or factory sales departments, the staffs of foreign merchant houses, the second corporation merchants active at local markets, and those guild members engaged in retailing, as well as the first corporation merchants with a minimum capital of 10,000 roubles who dominated activity at the fairs, was numerically tiny, comprising some 395,000 persons or 1.4 per cent of the population. By their endeavours, however, they did earn a similar average *per capita* income to those engaged in 'industrial' pursuits – 100 roubles per year.[110]

During the years 1718/22–88, therefore, rapid economic growth in the Muscovite heartland of the empire raised *per capita* incomes there to within a hair's breadth of those prevailing in the previously more advanced western provinces and enhanced both aggregate and *per capita* incomes within the empire as a whole until in 1788 the average Russian, with a *per capita* income (at current prices) of 100 roubles a year, was, in real terms, 50 per cent better off than in 1718/22 and 85 per cent richer than his counterpart in 1700 who had been subject to the disastrous impact of Peter's martial ambitions. This phase of rapid economic growth had not been associated, however, with major structural changes in the economy involving the transfer of resources from a low-productivity agricultural sector to a high-productivity urban-industrial one. In so far as manpower had been transferred from the ranks of the peasantry to man state-sponsored manufactories or industrial enterprises it actually retarded growth and only the development of high-productivity artisanal activities, largely in response to aristocratic demand, neutralized the impact of ill-considered state policies, ensuring that average *per capita* incomes in the urban-industrial sector, tiny as it was, would equal those of the peasantry. At best the role of the so-called 'modern' sector, including workshops and manufactories producing luxury wares, was passive in the growth process, the small number of people employed therein producing no more than if they had remained within the ranks of the peasantry. At worst, excluding the production of luxury wares, its impact was a negative one. In either case however, because of the small numbers involved, the urban-industrial sector exercised only a minimal influence on growth. Growth came from within peasant society not from without. By reorganizing the time-allocatory patterns of family members who were deployed within a spatially extended network in order to exploit the greater abundance of resources available to them, patriarchs, during the years 1718/22–88, had created new work opportunities for family members, thereby enhancing not only their individual incomes but also that of the household – and society. Growth had been a product of both better organization of production and greater resource availability. Accordingly, when in the 1780s resources began to be fully utilized and the balance between them disturbed it seemed that an age was drawing to a close, raising fears amongst the populace which were in no way assuaged as the economy became subject to an acute bout of inflation.

Crisis and renewed growth, 1788–1807

In the event, however, the populace's fears were not realized as during the next twenty years, 1788–1807, after a severe crisis, the economy continued to grow, ensuring that *per capita* incomes were largely maintained.[111]

Table 5.7 Russian economic growth, 1788–1807: estimates of B.F.J. Hermann and I. Golitzuin (millions of silver roubles, at current prices)

Producer	1788	1802/3	1807
Foodstuffs			
grain	144	109	195
Raw materials			
grain, flax, hemp			
cattle, fish, timber,			
furs, salt etc.	2,800	3,802	4,787
Manufactury	40	79	79
Metals	9	3	3
Total	2,993	3,993	5,964
Trade balance	+ 7	+ 7	+ 6
NI	3,000	4,000	5,070
$\frac{NI}{Pop}$ Current prices	100	103	124
constant 1788			
prices	100	87	105

With the successful conclusion of the war of 1788–92 peace was brought, under Russian tutelage, to the Black Sea steppe and the Crimea and the resource base of the economy, which in 1788 had appeared confined, was thrown wide open. Reserves of arable, pasture, forest, and waste all increased. The fashioning of these resources, however, took place in a very different environment to that of the 1780s. The prevailing 'agricultural crisis' of the years 1788–1802 ensured that growth would take place within a very different structural framework.

The 'agrarian crisis', 1788–1802

As a result of the new annexation potential arable reserves increased by about 50 per cent during the years 1788–1807 so that at 1788 levels of cultivation the proportion of available land utilized for tillage was reduced from 73 to 48 per cent.[112] Both peasants and lords thus once more had room to manoeuvre. Their response to the new resource availability was highly conditioned, however, by the prevailing crisis in this sector of the economy. Initially this crisis had been masked by the effects of the market

255

disruptions of 1786–9 and 1794–5 occasioned by government intervention in supply systems servicing the commercial grain trade, levying a tax in kind in the black-earth and southern regions of the empire, during and in the immediate aftermath of the Russo-Turkish war of 1787–92. As a result of this measure, by which 1.5 million chetverts of grain were taken annually at sub-market prices, the commercial grain trade was denuded of supplies and prices rose, but underlying this market crisis, and not entirely unassociated with it, was a much more serious production crisis.[113] In 1786/7, 1795, and 1805 famine swept the land, initially affecting the central and southern agricultural regions but subsequently extending its grip over the north and west. Climatic deterioration, coupled with a developing ecological imbalance within the agrarian sector, had begun to undermine the agricultural economy.[114] First to feel the impact of these changes during the years 1786–92/6 were cereal producers in the Ukraine and central Russia (1786–8) to be followed shortly by their counterparts in the middle Volga region (1795–6) where an overextension of the arable element in the prevailing system of field-grass husbandry made cultivation particularly susceptible to any failure of precipitation.[115] Much more serious, however, was the subsequent crisis during the years 1792/6–1802 which affected the north and west. Here, as has been shown, the western and north-western agricultural regions were particularly badly affected as rapid demographic growth concentrated in the Old Russian provinces of Novgorod, Tver, and Pskov and Polish and Russian White Russia undermined the pastoral–forest base of the economy by causing an expansion of cereal cultivation, reducing manuring levels, and adversely affecting yields of grain and garden crops like flax and hemp, whilst at the same time denuding the area of its forest cover.[116] The provinces of the old Muscovite heartland were also affected by the developing ecological crisis, although its impact on incomes was in part alleviated as an increasing proportion of the population turned to commercial pursuits and manufactory, becoming dependent upon imported grain for their sustenance. Underlying a series of acute commercial crises affecting grain markets there emerged, during the years 1788–1802, an acute production crisis in the agricultural sector of the Russian economy, arising from a disturbance in the balance between agricultural activity and that involving the exploitation of pastoral and woodland resources, which caused a disastrous decline in cereal yields and reduced the size of the annual harvest until in 1802 it was 23 per cent below the gross level of the 1780s (*c*. 120 million chetverts) and net almost 40 per cent below (*c*. 68 million chetverts).[117]

Inevitably a crisis of such magnitude could not but affect those lords and peasants who contemporaneously were making decisions about the deployment of the resources which had newly been made available to them. It should be noted, however, that in formulating their strategy the question of food shortages was not of central importance. Even at the much reduced levels of grain availability, amounting on average in 1802 to 1.76 chetverts *per capita*, there were more than enough cereals to satisfy the population's bread-grain requirements, 1.1 chetverts *per capita*, in a situation where their intake of staple foods – meat and fish – was in no way diminished. What they were forced to abandon during the crisis was animal fodder and drink-grain, which had previously comprised 70 per cent of the harvest. For the human population this introduced an element of insecurity into their lives, for the malt which had been prepared for brewing or distilling and which had previously been available to tide them over a bad harvest in the form of malt bread, was no longer at their disposal. For the animal population, and particularly the draught horses, the situation was nothing short of catastrophic. The peasant could no longer afford to feed them oats and with the decline of brewing/distilling they were now denied the concentrates derived from the residues of these processes.[118] Hay might still be available but lacking a regular intake of energy foods they no longer had the strength to undertake heavy work, like breaking the virgin sod which was an integral part of the field-grass system of husbandry. It was these problems upon which men pondered, and in attempting to resolve them they sought new methods to ensure a supply of animal feed commensurate with the work requirements of the draught stock and to alleviate the risks inherent in harvest fluctuations. Old consumption-allocation methods thus gave way to new production-organization ones, resulting in a rebirth of the open-field system. The introduction of regular fallowing in a three-course rotation, even if returning a 15-per-cent-lower yield (from the crop land) than could be obtained from the field-grass system operating in optimally balanced conditions, at least ensured a 26-per-cent-better return than could be obtained from the latter system in its contemporaneously debased state, thereby enhancing grain availability to a level, which if lower than in the 1780s was higher than in 1802, and allowed the draught stock a modest diet of energy food. Nor did they require more, for the new system of husbandry entailed no heavy work comparable with the breaking of the sward in the old methods. The institution of a repartitional system of landholding, moreover, went some way to optimizing the balance between population and resources, whilst the 'scattering' of strips about the arable provided that 'insurance' against harvest failure which had previously been embodied in the drink-grain of the earlier consumption-allocatory system.[119] Even as the 'agricultural crisis' deepened during the years 1788–1802, surveyors, working to lords' instructions, set about reshaping the landscape, incorporating the new reserves of arable into the open-field regime to accommodate the need for fallow land and thereby once again increasing the proportion of available land utilized for tillage to *c.* 68 per cent.[120]

By 1807 the process of reorganization was far from complete, old and new systems continuing to intermingle for many years to come, but the broad outlines of the new order were clear. Over a vast area of the trans-Pontine plain, encompassing the central black-earth and non-black-earth agricultural regions, the land of the middle and nether Volga and White Russia, open-field systems dominated the agricultural regime, engrossing two-thirds (24.7 million desyatini) of the contemporaneously tilled cropland. Almost 11 million 'souls', each endowed with about 2 desyatini of cropland in the western part of this area and 2.4 in the more arid, eastern Volga lands, worked to an entirely new labour discipline, their toil in the fields being rewarded with yields 13–17 per cent lower than twenty years before. Only on the periphery, in the northern and north-western or lake agricultural zones, the pre-Ural area, the Ukraine, and the newly-annexed steppe, did the old ways continue. In the forested, thinly populated, and infertile northern provinces of Archangel, Olonets, and Vologda, and the neighbouring provinces of Novgorod, Kostroma, Vyatka, and Perm, slash-and-burn tillage continued in use, the peasantry taking advantage of the short growing season to raise a crop in a forest clearing. They grew cereals and flax mainly, the ash-enriched soil giving good and sometimes spectacular yields. The field was cropped continuously for two to eight years, depending on its fertility, and then was allowed to revert. In the Ukraine and the newly-acquired steppe, field-grass husbandry remained the norm.[121] Such irregular cropping patterns, which still encompassed one-fifth of the cropland (8 million desyatini), required, however, abundant amounts of pasture and woodland to operate effectively and in 1807 such conditions existed only in the far north and the open steppe. Elsewhere, in the lake region and the left-bank Ukraine, they were rapidly disappearing, and although the system here still yielded returns higher than in neighbouring open-field regions (thereby providing no incentive for conversion) they were in no way comparable with those obtained where the system operated optimally. In the north a seed-yield ratio of 1:3.34 was still normative and in the steppe returns of 1:4.58 could be obtained. Only in these latter areas, therefore, were the agricultural traditions of the 1780s still preserved in 1807. Elsewhere the agricultural system operated on the basis of either the 'new' open-field system or a debased form of field-grass husbandry returning yields which were respectively 13–17 and 9–12 per cent below those obtained in the 1780s. Over most of Russia the introduction of new techniques had thus stabilized production at levels somewhat lower than those of the 1780s.[122] The one exception to this rule was in the Baltic provinces and Smolensk where, as has been suggested, the introduction of a rotational system incorporating root crops and leys to form the basis of a mixed husbandry regime raised crop yields to as much as 1:5.1, but as yet the impact of this on total production was insignificant.

Far more important was a structural realignment of landholding within the agrarian regime, for if the introduction of the open-field system had raised and stabilized yields at a level 26 per cent higher than in the crisis year of 1802, the incorporation of populations working richer soils into the empire, as a result of the second and third partitions of Poland, contributed a further 8 per cent to raising average yields, so that in 1807 they were only 10 per cent below the level of the 1780s. Within the political boundaries of 1788, although the landscape underwent fundamental changes during the subsequent twenty years, the balance of cropland between the northern podzol zone and the southern black-earth lands remained largely unchanged. As a result of the introduction of the open-field system the area of land in tillage (but not the cropland) had been extended in the central non-black-earth region, White Russia, the Volga lands, and the central black-soil region, incorporating arable land which had remained unutilized and under grass into the agricultural regime to make provision for the fallow necessary to the working of the new system. In those areas, accordingly, cereals and fallow displaced grass so that in the northern reaches of the Don, for instance, Clarke, passing across the lands of the wooded steppe where cattle had grazed in the 1780s, could report thirty years later that 'you travel for miles and miles and see nothing but corn', and fallow, but in spite of this extension of the land area in tillage little else had changed.[123] The distribution of cropland between the podzol and chernozem zones of the empire was in 1807 much the same as it had been in 1788, minor changes in land use affecting neither the overall distribution of land between regions nor average intra-regional yields. Thus in the north, whilst the amount of cropland diminished in the northern region and expanded in the central non-black-earth, lake, and eastern White Russian regions the net increment to the total during the years 1788–1807 was 940,000 desyatini or 6 per cent of the whole; and as the land abandoned was higher yielding than that taken into cultivation the impact on production was largely nullified. Similarly in the south whilst the extension of cultivation in the nether Volga region brought some 0.99 million desyatini under the plough this was counterbalanced by the abandonment of 1.01 million acres of higher-yielding land in the central black-earth region, the Ukraine and New Russia. Overall, within the political boundaries of 1788, in terms of either the distribution of crop land or relative interregional yield levels, nothing much had changed during the years 1788–1807. Only the incorporation of the former Polish territories of Minsk, Vilna, and Grodno in the north and Volynia and Podolia in the south affected average yield levels. Here the annexation of 2.72 and 4.06 million desyatini of crop land yielding 40 and 18 per cent higher returns than in neighbouring Russian regions respectively had the effect of raising average yields within the empire by some 5 per cent whilst leaving the productive balance between north and south unchanged. In 1807, as in 1788, 40 per cent of the crop land located in the twenty-one non-black-earth provinces of the empire contributed one-third of the harvest whilst 60 per cent located in the twenty-four black-earth provinces contributed two-thirds of a harvest which gross (158.6 million chetverts) barely exceeded that of 1788, and net (107.8 million chetverts) was 3 per cent down, affording the population a considerably reduced (− 28 per cent) average *per capita* grain supply, amounting to 2.65 chetverts a year.[124] In the context of the contemporary 'agricultural crisis', moreover, even this had only been achieved by the introduction of 'new' technologies, represented by the open-field system of Muscovy and White Russia and the multi-course rotations of the Baltic provinces, and by the incorporation of relatively high-yield Polish agricultural regimes within the bounds of an extended empire. During the years 1788–1807 Russian agriculture had been transformed, the traditions of an earlier more productive age only being preserved on the agricultural frontier and new systems dominating the agricultural heartland, but this retarded rather than halted the decline of the agricultural sector within the economy so that in 1807 the net harvest before processing comprised only 5 per cent of gross national product in comparison with 7.4 per cent twenty years earlier.[125]

If, however, agricultural production stabilized and the agricultural sector declined in importance within the economy such was not the fate of the grain trade during these years. It actually increased, rising from *c.* 1.5 million chetverts or 1.4 per cent of the net harvest in 1788 to in excess of 5 million chetverts or 4.5 per cent in the 1800s.[126] Lords, taking advantage of the agricultural reorganization to delineate more precisely the extent of their demesnes and to define their peasants' obligations, were now able to compensate for declining incomes by increasing their share of total agricultural output and, pushing more and more grain onto the market, in the process effected a complete restructuring of the commercial grain trade (map 5.6). The previously dominant riverine and overland transport network of the western provinces was now eclipsed, for although its trade doubled during these years its share of the total grain trade declined from 42 to 26 per cent. Moreover, all of the 1.31 million chetverts which now passed through this system was destined for the Baltic ports, feeding an export trade which, thanks to the destruction of Danzig's commerce as a result of the draconian tariff policies of the Prussian Crown, increased more than 4.5 times between 1788 and 1807.[127] The region's former role as the major supplier of St Petersburg now ceased. That city, which at this time contributed about 190,000 chetverts or 12 per cent of Russia's Baltic grain trade, had during the previous twenty years become entirely dependent on supplies transported thence by way of the waterways linking the Volga and the Neva. The impact of metropolitan demand upon this trade along the Volga, however, had much more to do with this displacement effect than with changes in aggregate demand, for during these years the population of St Petersburg barely increased, numbers only rising from 218,000 in 1789 to 220,000 or 230,000 in the 1800s.[128] Even so,

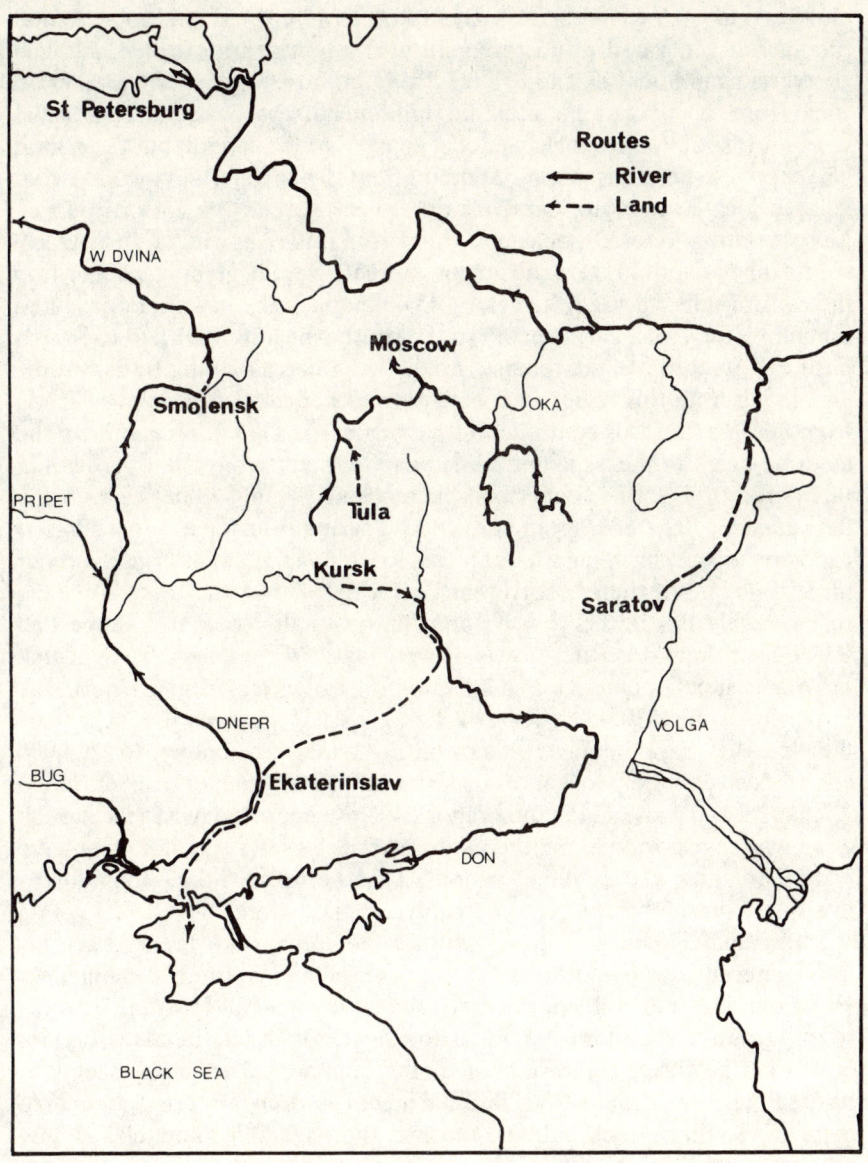

Map 5.6   Grain trade, 1800

as sole purveyors of grain to the capital those merchants engaged in this commerce along the Volga now had to acquire *c.* 650,000 chetverts of grain each year in place of the 114,000 required twenty years earlier. Accordingly, they were forced to cast their net wider, encouraging an increase in the production of, and trade in, grain from the lands of the middle and nether Volga.[129] By 1807 one-third of the incremental production generated in this area since 1788 entered into trade, 225,000 chetverts from Simbirsk and 658,000 from Saratov, bringing the total grain trade from the Volga region to 1.26 million chetverts or 25 per cent of the whole. Most of this grain (52 per cent) passed to the capital, but the region also supplied the Urals mining communities (15 per cent) and the industrial villages located in Vladimir, Nizhnii Novgorod, and the eastern part of Moscow provinces (33 per cent), this latter trade showing a particularly spectacular, five-fold, rate of growth. Cereals were beginning to pour into the central non-black-earth region of the empire, the main supply-source remaining the black-earth region to the south which by 1807 had risen to a position of absolute supremacy in the commercial grain trade, traffic emanating from this region encompassing 43 per cent of the whole. Trade from here northwards continued to pass from the Oka basin and from the lands adjacent to the Tsna and Moksha rivers, but as the rate of growth in this waterborne traffic began to slacken, rising by only 90 per cent during the years *c.* 1788–1807, grain began to be carted from Tula, some 536,000 chetverts a year, bringing the total trade to almost 1.5 million chetverts or three times more than twenty years before. Nor was the trade from the region solely northward for the annexation of the southern steppe had opened up markets there which were largely supplied from Kursk province, supplemented by small quantities of grain from Ekaterinslav province, some 630,000 chetverts being carted southwards each year. Finally, the trade along the northern Dvina, amounting to 297,000 chetverts a year or 6 per cent of the total, supplied the export trade through Archangel and provisioned the Pomor'e. These figures suggest that enough grain was transported along the major arteries of long-distance commerce in 1807 to allow *c.* 2.2 million people or 5.5 per cent of the population to live solely on the basis of commercially supplied grain.

Unfortunately, because of the failure of Golitzuin to break down his 'raw-material' sector into its component elements in the manner of Hermann, it is impossible to analyse the employment and earning patterns of this group in the same manner as for the 1780s. As in the case of grain, however, the foreign-trade sector of the commercial economy seems to have grown more rapidly than national income so that if internal commerce grew at a similar rate to national income, then overall commercial activity rose in value from *c.* 178 million roubles to 357 million roubles, a rate directly comparable with the growth of manufactory during these years.[130] If this was the case then the distribution of employment between the two sectors probably remained unchanged during the years 1788–1807; at the latter date slightly less than half the labour force (47 per cent or 1.04 million people) were employed in the provision of commercial and transport services and derived 75.9 million roubles a year for their work servicing the requirements of a domestic and foreign commerce valued at

357 million roubles or 73 roubles *per capita*. The remainder, or some 1.16 million people comprising 53 per cent of the group, were engaged in manufacturing, industrial, or artisanal pursuits. Their numbers had increased markedly during the previous twenty years, 250 per cent, but the regional incidence of this increase was highly varied. As has been suggested, employment in both the capital and the mining and metallurgical communities of the Urals stagnated during these years.[131] In the north it was only in the lands between the Oka and Volga, that is in the central non-black-earth, or perhaps now more appropriately central-industrial, region, that employment increased. Here, as has been indicated, grain imports increased, rising from *c*. 591,000 chetverts a year in 1788 to almost 2 million chetverts in 1807, allowing more and more of the population of these provinces to sever their links with agriculture and obtain supplies for their sustenance through the market. By the latter date slightly more than a quarter of the *whole* population of the region could be so supplied, in Moscow and Vladimir provinces the proportion rising to half and elsewhere, in Kaluga, Yaroslavl, Kostroma, Nizhni Novgorod, Tver, and the extra-regional province of Tula, the figure averaging 10 per cent.[132] Amongst those members of this group who worked in 'industrial' pursuits and who numbered about 930,000 persons, some, but certainly not more than half, found employment in manufactories, many of which had been established during the 1790s and 1800s. Almost equally numerous were the members of another group, perhaps more characteristic of the new age, who found no place in the 'industrial' structure of the 1780s. These comprised those members of the peasantry who, through lack of land, had crossed the divide which separated them from the full-time workers. They stood betwixt and between. Still adhering to peasant values, they now derived most of their income from their craft activities and derived most of their sustenance from the market place. Together these two groups comprised perhaps 64 per cent of the whole but it is about the first, those engaged in manufactory, numbering about 483,600, that most is known. As in the 1780s, textiles still dominated this sector of the economy, hemp and flax, employing about 281,000 workers either in centralized establishments (20 per cent) or dispersed through the countryside (80 per cent), and producing, in the case of linen, *c*. 30 million arshins of broadcloth a year of which 22.7 million (75 per cent) were exported, maintaining its supreme position. Woollen-cloth production, much more centralized, provided work for another 36,900, and silk and cotton for a further 63,500. In all, in the 1800s, the textile industry provided employment for some 381,400 people or 79 per cent of those engaged in manufactory. It had grown massively during the last twenty years as existing establishments took on more men and new ones were established creating yet more employment opportunities, but in spite of doubling its workforce it was losing ground. The most rapidly growing sector of manufactory was now metal fabrication

which over the previous twenty years had almost quadrupled the size of its workforce. As exports of iron had diminished, largely as a result of tariffs imposed in Britain to protect its inefficient new coke-fuelled industry, more and more of the produce of the Urals industry had been sold at home, employment in domestic metal fabrication increasing from 24,400 to 95,100 in the 1800s when it comprised 20 per cent of the manufacturing labour force. As employment in those manufactories producing luxury wares stagnated, rising by only 25 per cent during these years, and textile manufacturers struggled to double their labour force in the face of domestic competition it was thus metal fabricating which led in creating new work opportunities in the manufacturing sector.[133] Starting from a small base, however, its impact was limited and although this sector employed almost 2.5 times more people in the 1800s than twenty years earlier, or some 483,600, it was now rivalled by the 262,000 or so impoverished peasants who sold their woollen and linen cloths at the country fairs of Tula, Ryazan, Moscow, Vladimir, Yaroslavl, Tver, Kostroma, Kaluga, Vologda, and Nizhegorod provinces. Nor were their wares confined to intra-gubernial sales. At the end of the eighteenth and the beginning of the nineteenth centuries in Tula province alone more than 10 million arshins of linen cloth were produced by this group annually for despatch to other provinces, whilst at the great fair of Nizhnii Novgorod these wares found a ready market amongst those merchants who would sell them to the cossacks of Voronezh, Penza, and the Don region or their counterparts in Siberia.[134]

The years 1788–1807 had thus witnessed momentous changes amongst that population who were employed full-time in artisanal, manufacturing, industrial, commercial, or transportation activities. Nor, on the whole, were these changes for the better. Partly as a result of a shift from high-productivity artisanal or manufacturing activity in the production of luxury wares towards low productivity manufactory of textiles and metal wares, and partly as a result of the emergence of a group of impoverished peasant-craftsmen producing low-grade textiles, incomes within this sector had fallen, until in 1807, with an average *per capita* product of 73 roubles a year, full-time 'industrial' and commercial workers were almost 40 per cent worse off than their counterparts twenty years before.

Pausing for a moment, therefore, to consider the implications of the analysis so far undertaken, it is clear that the story is one of unrelenting decline. Beyond the bounds of peasant society, for the 2.2 million people who had severed their links with the agrarian sector and had become dependent on full-time work of an 'industrial' or commercial character to earn their livings, incomes fell by 38 per cent during the years 1788–1807. For the 38.4 million people remaining in peasant society an even more severe crisis had affected incomes derived from their exploitation of the plough lands and garden grounds. In spite of a major reorganization within the agricultural sector of their economy, average *per capita* material product derived therefrom declined by some 26 per cent. In value terms the decline was nothing short of catastrophic – 52 per cent. With the virtual destruction of fruit growing in the north as a result of deteriorating climatic conditions, incomes from the garden grounds were similarly affected, falling by 40 per cent.[135] The economy was undergoing a severe crisis during the years 1788–1807 which was only compensated for by an increased exploitation of the pastures and wastes, woodlands and rivers which was of sufficient magnitude not only to restore the average Russian's income to the level of the 1780s but even to enhance it by 6 per cent.

Southern annexation and renewed economic growth, 1792–1807

In order to investigate this dynamic element in the contemporary Russian economy it is necessary, however, to focus not only on a new sector – the pastures and forests – but also on a new area – the recently annexed southern steppe. Agricultural reorganization in the central black-earth region had displaced the extensive pastoralism of the northern Don southwards into lands containing most of those reserves of pasture and forest which after *c.* 1795 served to augment total Russian supplies of these resources by 21 and 5 per cent respectively.[136] Accordingly, whilst in the northern lands peasant herds were probably largely maintained, in the context of the agricultural reorganization it is unlikely that much potential for growth existed there. Only in the south could herders expand production unfettered by land resource constraints.

As has already been shown, links were rapidly formed after annexation in 1793 between the dominant Russian economy and the diminutive Tartar one, which since the climatic amelioration beginning in the 1780s had assumed a stable form, thereby permitting curious visitors a unique opportunity to observe the latter economy before its destruction at Russian hands. Many seized this opportunity, and the picture they painted is a fascinating one.[137] In the Perekop and the northern part of the peninsula, which during the mid-century had been occupied by Tartar shepherds who each possessed flocks numbering 1,000–5,000 sheep, pastoralism gradually gave way to tillage, a belt of cereal cultivation extending as far south as the mountains of the Crimea and as far east as Kaffa (Feodosiya) where the thin, gravelly soils gave only a meagre yield of corn but provided grazing for thousands of sheep.[138] Along the eastern coast and on the Kerch peninsula cattle rearing predominated, whilst in the southern coastal valleys, sunny, well-watered, and protected from the northerly winds, orchards abounded, luxuriant with fruit trees, nut trees, and vineyards (map 5.7).[139] A complex, stable pattern of husbandry was thus established in the Crimea which displaced, within the confines of the peninsula, the previously extensive and highly volatile nomadic system, and provided the

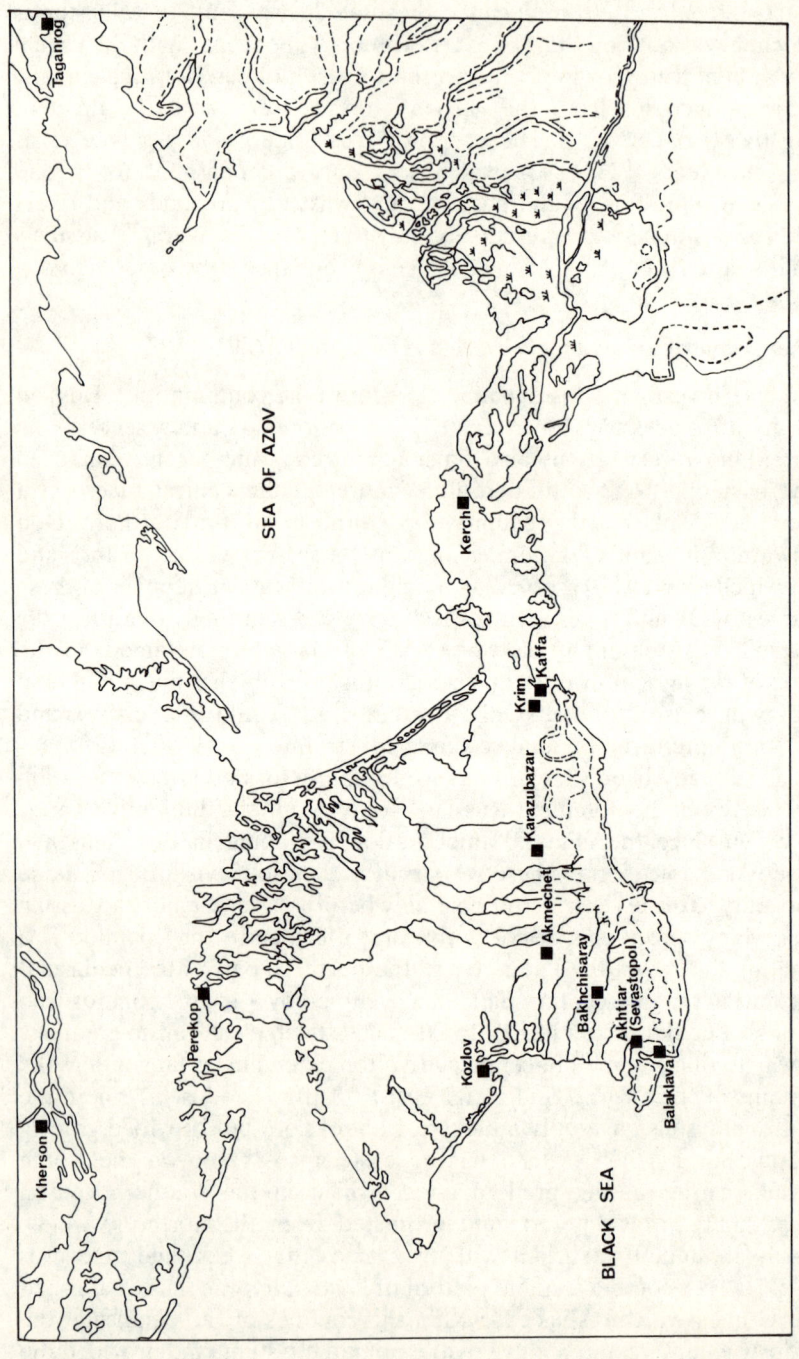

Map 5.7 Crimea and Black Sea steppe

basis for a high level of domestic consumption and a commerce which, in spite of political frontiers and route realignments occasioned by the Napoleonic wars, continued to be focused on Constantinople.[140] All along the northern Black Sea littoral from Ochakov to Taganrog and amongst the ports of the peninsula commercial activity was directed towards the Ottoman capital, as agricultural produce – clarified butter or ghee and wheat – was exchanged for Turkish manufactures, Greek wines, and Aegean fruit. Typical was Kaffa (Feodosiya) whose trade consisted 'chiefly in the export of wheat; besides which barley, salt and a few manufactures in iron and woollens. The imports are more numerous: Greek wines, Turkish stuffs, silks, and other manufactures, raw cotton and a few copper utensils.'[141] But Kozlov (Evpatoriya), Taganrog, and Ochakov were all broadly similar and Kerch differed only in the importance of clarified butter or ghee amongst its exports.[142]

The initial links between Russia and the Khanate involved the urban sector of the southern economy, for, unlike in the northern lands where, outside of the capital, towns were little more than rural settlements endowed with grandiose jurisdictional privileges, here they enjoyed a distinct and separate economic identity from the countryside. Whether in territory under Turkish or Cossack rule, prior to the conquest they had encompassed the residences of a military caste and their attendant craftsmen, supplied with grain from the indigenes' military fiefs. Following the Russian annexation this group had fled and their pashaliks and spahiliks had fallen into decay so that the incoming Russian military personnel who took their place became dependent upon imported grain for their survival. As has been shown this resulted in the development of a major grain trade, as the central black-earth province of Kursk, supplemented with small amounts of cereals from Ekaterinslav, began to export some 630,000 chetverts of cereals southwards each year.[143] In the east this trade passed by way of the Don, which afforded 'an easy intercourse with the Black Sea. Every year, vessels go laden to Tscherkaskoy with corn; and they accomplish their voyage in about two months. In winter (the traders) receive merchandise by sledges from the Crimea and Turkey.'[144] Their destination was the settlements at the mouth of the river, where prior to the bloody war of 1787–92 Azov had reigned supreme but where now this old fortress town, reduced to a settlement of some fifty huts, had been displaced by Novo Cherkask (c. 15,000 population), the island capital of the Don cossacks which, if 'not so grand as Venice it somewhat resembles that city. The entrance is by broad canals, which intersect it in all parts. On either side wooden houses built on piles, appear to float upon the water', and shops 'kept chiefly by Greeks. They contain the produce of Turkey and Greece, as pearls, cloth, shawls, tobacco, fruit.'[145] Azov had also been displaced by Nakhichevan (c. 2,000 population), an Armenian colony again with some 400 shops and 'neat and comfortable houses, many of which are limestone and covered with tiles', Rostov, the new garrison town; and Taganrog.[146] In the west a similar trade by cart was directed to the Crimea, each vehicle carrying 'one or two, even two and half tons of' wares, corn in their passage south and salt, brandy, and wool on their return north.[147] Their destination was the old Tartar cities of the peninsula – Karazubazar (population 8,000), situated in the richest agricultural region of the peninsula and an important centre for the production of wares made from sheep and goatskins; Bakhchisaray (population 5,000), the old capital; and Akmechet (population 3,000), in a valley of the northern foothills – and the new Russian settlements – Odessa, which already by 1804 had 2,900 houses and 15,000 people, and Sevastopol, with perhaps as large a population of sailors, soldiers, and shipwrights – which were already in the 1800s displacing the Tartar cities and ports.[148] To these centres now flowed an enormous grain trade, provisioning the towns (c. 140,000 chetverts), but also feeding a rapidly growing export trade (c. 490,000 chetverts) which was now superimposed upon the indigenous commerce of the region.[149]

Even simply considering the direct impact of annexation, as measured in terms of the northern production pattern already examined, it is clear that the incorporation of the indigenous population of the south into the empire served to raise average *per capita* incomes there. In terms of cereal production, as has already been indicated, the incorporation of the highly productive agricultural regimes of the former Polish and Turkish territories made a significant contribution to raising raising average national yields to even their debased levels of 1807.[150] Add to this the impact of southern fruit production, urban craft activity, and extractive work, and the effects were not inconsiderable. Fruit grew in abundance throughout the area. Indeed, such was the profusion within which the indigenous population lived that a quarter of a century later, when trade links had been forged with the north, southern fruit drawn from these supplies completely displaced northern in the consumption of the populace, and fruit brandies, already passing northwards in the 1800s, began to challenge vodka in alcohol consumption.[151] Craft activity, oriented towards satisfying the luxury requirements of the native elite, valuably supplemented that high-productivity sector which was losing ground in the north. Extractive work, most notably salt production from the crusts of the Crimean salt lakes, added yet another element, enhancing Russian production by almost a quarter.[152] Quantification of the direct impact of annexation is inevitably difficult but even the most conservative 'guesstimate' would suggest that, setting aside the impact of the southern agricultural regimes, it was sufficient to raise average *per capita* incomes within the empire by some 5 per cent.[153]

Even more important were the indirect effects arising from a minor 'revolution' in animal husbandry as the southern steppe lands were engrossed within the Russian agrarian regime. Stabilization of conditions in the south and the establishment of Russian hegemony over the open steppe transformed conditions, allowing a basic restructuring of the herding economy as the frontier was pushed far to the south-east. By the 1800s Little Poland, the province encompassing the south-western agricultural region recently annexed from Poland, together with the rest of the Polish and Russian Ukraine remained the principal area of cattle rearing, but, by that date, agricultural reorganization confined this branch of husbandry in the north to Kharkov, Kursk, and Orel in the western part of the zone and to Kazan, Ufa, and Saratov in the eastern part. In the centre agriculture displaced animal husbandry southwards and eastwards where both the Don cossacks and the Khirghiz, Kalmucks, Bashkirs, and other Tartar peoples now furnished large numbers of beasts to the rest of the empire. The Don cossacks, in particular, occupied themselves with cattle herding, utilizing the excellent pastures of the fertile steppe and the abundant herbages which grew on the river banks. The short duration and temperate nature of the winter in this region particularly favoured cattle rearing and it was a poor cossack who did not possess 50–200 beasts. Further to the south-east cattle were less prominent in the pastoral economy, forming only a small part of the herds and flocks of the nomadic people whose primary preoccupation was with horses and sheep. The Kirghiz raised fine poled-cattle, but only the poorest of the Kalmucks kept oxen, the rich concentrating on the breeding of horses and the maintenance of flocks which remained all year on the steppe. Amongst the Nogais cattle constituted the principal form of wealth, but these were a poor people and anyone possessing 500 beasts was deemed a very rich man indeed.[154] Peace had brought a new security to the steppe, and cattle rearing, once confined to the northern fringes of the area, now extended far to the south and east, to the shores of the Caspian and beyond, resulting in a fundamental restructuring of the industry. Southern animal husbandry, generating an annual income of 1,142 million roubles, now rivalled that of the north where a peasantry, still maintaining the size of their herds, earned some 1,289 million roubles a year.[155]

The 'export' sector of this southern industry continued to be dominated by the products of the 'Ukraine', which still enjoyed a reputation for the sweetness of their flesh, but as one rather acute observer noted, the inhabitants of this region were now covering beasts from other areas, drawn through an extensive supply network and sold under their name.[156] The trail extended far to the south and east. Across Russia's Asiatic frontiers, stretching from the Caspian to Semipalatinsk, some 60,000–85,000 sheep and 3,000–5,000 cattle were driven annually from the lands of the Kirghiz Horde to Orenburg, joining the great route west. Thereafter passing westwards *en route* to the Ukraine these herds were joined by others drawing animals from the lands of the Kalmuck on the lower Volga and from amongst the Don cossacks to create a burgeoning flood of livestock to provision both domestic and foreign markets. Shipments of livestock across the frontier, which had amounted to no more than 22,000 beasts in the 1770s, increased rapidly, attaining 40,000 animals worth 1.445 million roubles a year when trade was uninterrupted in the 1800s.[157] Even more spectacular was the growth of the traffic northwards, some 3.5 million animals being despatched annually to feed that rapidly increasing population who were entirely dependent on the market for supplies. The impact of this commerce on the northern economy was considerable, for these animals yielded not only sufficient meat for that population's consumption but also a supply of hides, of which two-thirds were exported (sustaining a commerce worth 3 million roubles a year) and one-third provided raw materials for a manufactory producing wares worth 1.5 million roubles a year.[158] Yet this was merely the tip of the iceberg, the 'export' of livestock probably generating no more than 12 per cent of sectoral income. Vastly greater numbers of animals, possibly as many as 25 million a year, found their way to centres specializing in the production of tallow: Kolomna, Tula, Voronezh, and Rostov. At Voronezh, a whole sector of the town was

> covered by storehouses, cauldrons and tubs, for the production of grease, which is a great article of trade here, and which they send to England and America in vast quantities. The stench of the bones and horns of the animals, slaughtered for the purpose of obtaining grease, made the spot absolutely intolerable.[159]

On the basis of this noxious activity *c.* 75 million puds (1.2 million tons) of tallow worth 367.5 million roubles were produced each year of which 2.6 per cent was exported (1.97 million puds worth 9.6 million roubles) and the remainder sold at home affording the population an annual average supply of 1.8 puds (*c.* 65 lb) per head. As a by-product of these operations, moreover, the inhabitants of the central black-earth region, whose animal husbandry had suffered badly from the effects of agricultural reorganization, were afforded a meat consumption far greater than was enjoyed by their counterparts further north and which was surpassed only by that of the denizens of the open steppe.[160] Once again, therefore, the south had been called upon to redress decline further north, the emergence of a formidable animal husbandry on the open steppe contributing to raise average *per capita* incomes within the empire by about 29 per cent.

Crisis and renewed growth, 1788–1807: an overview

The years 1788–1807 thus witnessed fundamental changes within the Russian economy. That portion of the population living within the lands bounded by the frontiers of 1788 experienced an acute crisis which encompassed both peasant and non-peasant society. Agricultural yields declined disastrously (− 15 per cent) and as the extent of sown cropland did not increase this was reflected in a parallel decline in the size of the gross harvest and a more severe one in the net harvest (− 23 per cent) resulting in a disastrous collapse in 'real' *per capita* sectoral product (− 52 per cent). In the context of the contemporary 'agricultural crisis', moreover, even stabilization at this level had only been achieved by the introduction of 'new' technologies, represented by the open fields of Muscovy and White Russia and the multi-course rotations of the Baltic provinces. During these years Russian agriculture had been transformed, the agrarian traditions of an earlier and more productive age were now preserved only on the agricultural frontier, and new systems dominated the agricultural heartland, but this process of agricultural innovation had retarded rather than halted the decline of the agricultural sector and had left the population of these lands almost two-thirds worse off in terms of 'real' *per capita* cereal product in 1807 than it had been twenty years earlier. With the virtual destruction of northern fruit growing, as a result of deteriorating climatic conditions, and a severe depression in flax and hemp production, income from the garden grounds was similarly affected, falling by about 40 per cent in 'real' terms, so that taken as a whole peasant incomes derived from plough lands and garden grounds halved during the years 1788–1807. Nor was the non-peasant economy immune from the effects of the crisis. The 2.2 million people who had severed their links with the agrarian sector and had become dependent upon full-time work of an 'industrial' or commercial character to earn their livings experienced a 38 per cent decline in 'real' earnings. Only within the pastoral and woodland economies were the effects of the crisis not felt, and even here the illusion, if not the reality, of crisis persisted. Within the pastoral economy, whilst herds were largely maintained and the supply of animal products may even have increased slightly, on a *per capita* basis there was at least the illusion of decline as falling prices, resulting from southern competition, ensured that average *per capita* sectoral product *at current prices* was down by some 8 per cent. A similar impression of crisis pervaded the woodland economy, for although the forest cover remained largely intact allowing the peasant to continue his exploitation of the wilderness, agricultural reorganization leading to an extension of the arable, if not the cropland, led to large-scale peasant felling, abundant timber supplies, and a marked depletion of reserves in the vicinity of the villages. This caused one contemporary to note that

with all her wealth in forests, Russia, contains districts that are totally destitute of timber and fuel; and even in the governments where these necessaries of life were lately in abundance, the increasing population and industry have made the decline of them sensibly felt. The immense

> consumption of wood in a territory where it is necessary for eight to ten months of the year to provide against the cold, and where all of the habitations in town and country are constructed of timber, rises in the same proportion in which the number of people increases.'[161]

Even as Tooke's book left the presses in 1799 the peasantry were destroying the sylvan heritage in the vicinity of their villages, precipitating the contemporary ecological crisis but at the same time creating abundant supplies of firewood and lumber for construction. In terms of current production, therefore, decline was largely illusory and the woodland economy like its pastoral counterpart continued to buoy up an otherwise flagging northern economy. Their stabilizing influence ensured that 'real' regional *per capita* product fell only by about 20 per cent in spite of catastrophic decline in both the non-peasant sector (− 38 per cent) of the economy and the agricultural/horticultural element of the peasant sector (− 52.5 per cent). Even so, economic activity within the lands encompassed by the frontiers of 1788 yielded a 'real' average *per capita* product which in 1807 was still 20 per cent down on the level of 1788.

Only the annexation of new territories allowed average *per capita* product to once more grow, attaining a level in 1807 about 6 per cent higher than in 1788. The incorporation of the relatively productive agricultural regimes of the former Polish territories annexed as a result of the second and third partitions not only augmented the amount of cropland within the empire (and the population dependent upon it) but also raised average national yields by some 5 per cent and average *per capita* product therein by 1.5 per cent.[162] Even more important was the direct and indirect impact of annexation of the open steppe lands of southern Russia. The 1.94 million inhabitants of New Russia, including the 230,000 Don cossacks and 35,000 kalmucks occupying their territory with whom the cossacks had intermarried, the 74,000 Black Sea cossacks uprooted from their homes on the Dnepr in 1783 and resettled in the Kuban, and the 385,000 Russians and Tartars living in the Crimea, all lived in conditions of relative abundance.[163] Extensive fruit growing throughout the area added an important element to their diets which was increasingly denied their contemporaries in the north. Craft activity, concentrated amongst the relatively numerous urban population of the area and oriented towards satisfying the luxury requirements of the native elites, valuably supplemented the output of that high-productivity sector which was losing ground in the north. Extractive work, most notably salt production from the crusts of the salt lakes of the Perekop and Crimea, further augmented indigenous incomes and provided a major 'export' commodity to be despatched northward from the region. In total, even when measured in terms of northern production patterns, *per capita* product was much higher here than further north and contributed to raising average *per capita* product within the empire by some 5 per cent. The direct effect of annexation was thus by no means insignificant in raising average *per capita* product within the empire. It was, however, totally eclipsed by the indirect effects arising from the establishment of Russian hegemony over the steppe. The establishment of a formidable animal husbandry in the area not only totally transformed the structure of the Russian pastoral economy but also contributed to raising average *per capita* product within the empire by almost a quarter to 68 roubles (in constant 1762 prices) per head or 6 per cent more than in 1788 and double the level attained in the 1700s.

The years 1788–1807 had thus witnessed a fundamental transformation of the Russian economy as, against the background of an acute crisis in the north, the development of the recently-annexed southern lands once more established the nation on the path of economic growth. Resulting from (and symptomatic of) these changes in the productive base of Russian society was an equally fundamental transformation of the commercial sector of the economy. That intricate pattern of exchanges transcending, but closely interwoven into, the fabric of peasant society which had characterized domestic commerce in the 1780s underwent a process of extreme simplification, assuming a simple bi-nodal structure of exchanges between peasant society and the non-peasant/foreign-trade sector of the economy, with the foreign-trade element in the latter complex gaining ground on the domestic element. Within this bilateral exchange, moreover, long-distance trade in the basic staples (grain and cattle) increased at the expense of the shorter-distance traffic encompassing the wide variety of wares which had dominated commerce in the 1780s.[164]

In 1807 trade flowed exclusively along that bilateral axis between peasant and non-peasant society. That system of commodity exchanges between regions within peasant society, which had allowed individuals within that sector to make up to 5 per cent adjustments to their indigenously-determined consumption patterns, had, in the context of the developing agrarian crisis, disappeared. Within the remaining commerce directed exclusively to the non-peasant sector, moreover, there had been an extreme simplification in the range of goods marketed. Food, animal fodder, and raw materials still flowed from peasant society to the nodal points of the non-peasant sector but its inhabitants, greatly increased in numbers but much poorer than twenty years earlier, now relied much more on the basic staples – meat and grain – to make up their diets. These items which had made up 55 per cent of their food consumption now comprised almost 90 per cent. That wide variety of foods – vegetables, fruit, alcohol, freshwater fish, milk, and butter – which had added interest to their gargantuan repasts in the 1780s were disappearing from the market. They remained adequately fed but now enjoyed only the most monotonous of diets as the cattle and grain trades established their ascendency in the provisioning systems of this sector.

These changes, which were transforming consumption patterns within the domestic element of the non-peasant/foreign trade sector of the economy, moreover, left their imprint on the foreign-trade element. Amongst Russian exports that collection of commodities associated with flax and hemp production – raw hemp and flax, linseed, hemp seed and linseed oil, sailcloth, canvas, ropes and cordage, linen and linen goods – continued to dominate, but they were losing ground, their share in total exports falling from half to one-third. By 1807 their position was being challenged by the products of the new trading system: grain, which increased its share in total exports from under 3 per cent to 17 per cent, and animal products (tallow and candles, live animals and butter, leather and leather goods), which increased their share of the whole from c. 12 per cent to 23 per cent, tallow becoming the *single* most important commodity in the export trade. Not only did these rapidly expanding trades displace the traditional raw-material exports of the nation, however, but they also pushed into a very poor third place those manufactures and industrial products which had enjoyed such an important position amongst exports twenty years earlier.[165]

This subordination of industrial wares in the export trade only reflected, however, the relative stagnation of that sector in which so many of the non-peasant population were employed. Here very little had changed during the years 1788–1807. Within the production framework of the 1780s a large group of those workers engaged in manufacturing and industrial activities, who drew their raw materials from, and were sustained by, the produce of the peasant sector, continued to produce wares (c. 23 per cent of sectoral output) which were destined to meet state or export demand. An even more important group of workers, employed in a small number of manufactories and a host of workshop establishments, produced goods (62 per cent of sectoral output) which were designed to meet the requirements of the aristocracy for luxury wares. Together these groups continued to dominate the output of the workshops, manufactories, and industrial enterprises of the non-peasant sector of the economy just as they had in the 1780s. Many of these wares were either distributed through non-commercial military supply systems or were destined for consumption within the aristocratic households of the metropolitan centres, but now producers catering for the latter market began to suffer from the impact of a rising flood of luxury imports which over the years 1788–1807 increased some 2.5 times, attaining a level of c. 33 million roubles a year. Many of these wares (coffee, wine, and exotic produce like sandalwood and incense, pearls and gemstones) were complementary to rather than competitive with domestic produce. Others (most notably fruit and sugar) filled gaps created by emergent shortages in domestic supply. In the case of fruit, declining northern production during the years 1780–1810 left the market dependent on either the St Petersburg importer or an increasing number of glasshouse producers. In part this new domestic response to the problems of fruit supply was a direct result of commercial pressures: a small group of foreigners, mostly Dutch, established forcing houses in St Petersburg and sold their produce on the market. More important, however, was the contribution of the aristocracy who built large numbers of hothouses in the vicinity of Moscow and St Petersburg. The production of these latter groups was not competitive with imports, however, but complementary to them.[166] Similarly in the case of sugar, as domestic supplies of honey declined as a result of deforestation, imports increased, but here there was an element of competition as the products of the cane-sugar refineries, which had been established in St Petersburg (1718) and Moscow (1723) and which now processed some 320,000 roubles' worth of raw sugar, were swamped by imported supplies worth some 4.8 million roubles.[167] On the whole, however, these wares, valued at *c.* 15 million roubles or about 45 per cent of the luxury import trade, posed no threat to domestic producers. They merely extended the range of goods available to aristocratic consumers, allowing them to set such tables as might on occasion (at least in the provinces) evoke the not uncritical comments of visitors like Martha Wilmot, who reported of Moscow society that

> many a bad dinner have I made from the fatigue of being offered fifty or sixty different dishes by servants who came one after the other and flourished ready carved fish, flesh, fowl, vegetables, fruit, soups of fish etc., etc., in their turn before your eyes; wines, liqueurs etc., etc., in their turn.

In St Petersburg, however, they did nothing but excite admiration, one visitor at least remarking with awe that he had seen at his hosts' table

> Sterlet from the Volga, veal from Archangel, mutton from Astrakhan, beef from the Ukraine and pheasants from Hungary and Bohemia. The common wines are claret, Burgundy and Champagne; and I never tasted English beer and porter in greater perfection and abundance.'[168]

Imported foodstuffs thus merely added to the already rich variety of produce available to aristocratic consumers, and when distributed amongst the lower orders they served only to embellish an otherwise unexciting diet, posing no threat to the position of domestic producers. Similarly, 'traditional' imports of manufactures (woollens, silks, and metal goods) as well as the newer products of the western European 'industrial revolution' proffered no significant challenge to domestic manufacturers of these wares. Indeed, if anything, they were losing ground at this time. The largest single item in this category remained textiles, worth *c.* 10 million roubles or about 30 per cent of 'luxury' imports, and in this group of wares imports of woollens held up best, controlling some 85 per cent of the market, the Dutch and French supplying the personal requirements of the aristocracy and the English clothing their servants, but long before importers had been forced to cede the army and navy contracts to the Russians.[169] During the years 1788–1807 indigenes had moreover proved themselves more than able to hold their own in the growing market for cottons and silks, securing two-thirds of sales of these products against the foreigners.[170] Perhaps most spectacular, however, was the success of the rapidly growing metal-fabricating sector of Russian manufacturing which confined foreign imports to a mere 4 per cent of the market. If imported foodstuffs were thus complementary to, rather than competitive with, Russian produce, and imports of 'traditional' manufactures or the newer products of the western European 'industrial revolution' like cottons could make little headway in the market, such was not the case with that wide variety of luxury manufactures which, during the years 1788–1807, constituted the fastest-growing sector of the import trade, attaining a level of c. 8 million roubles a year or about a quarter of total luxury imports by the 1800s.[171] A huge variety of metal wares, made from iron, brass, and copper, many of them beautifully crafted versions of everyday items, glasses of different types and sizes, furniture of all kinds, clocks, wallpaper, carpets, and materials for curtains, bed linen, and table linens, silver cutlery and plate, porcelain and fine glassware, were all imported to adorn the houses of the rich. Luxury fabrics for clothing, accessories and trimmings in accord with the dictates of fashion, jewellery, and watches similarly were imported to clothe the wealthy denizens of the capital who also sought from abroad elegant carriages with well-bred horses accoutred with handmade harness, and, to indulge their leisure pursuits, sporting guns, fishing rods, playing cards, gaming pieces, scientific instruments, *objets d'art* – as well as bloodstock for hunting and racing, and dogs for sporting purposes. These imported wares were not solely items of current manufacture, for successive rulers and members of their courts availed themselves of the services of the newly established western-European auction houses to acquire antiques such as paintings, sculptures, and books. Commencing its phase of rapid expansion in the 1760s this trade had still been an affair of little import in the 1780s. Judging by the British experience at that time these wares comprised no more than 5 per cent of imports which, if extrapolated to the total Russian trade, would suggest a commerce amounting to something less than a million roubles a year. Alongside the output of that group of manufactories and workshop enterprises producing luxury wares which had evolved in the Russian metropolitan centres during the preceding fifty years, amounting to c. 31.4 million roubles, this was small beer. Twenty years later, in the 1800s, the situation was very different. Imports had risen eight-fold, domestic production only by a quarter, and the flood of imported high-quality wares had forced domestic producers into the fabrication of either the more lowly items required by St Petersburg society or wares fit only to satisfy the less eclectic tastes of the provincial nobility. As in the 1780s, therefore, carts, which brought demesnal produce to the capitals, returned to the provinces bearing luxury wares but now, in the 1800s, the goods they bore back to the rustic manor houses and administrators' residences in the provinces were the homemade aristocratic junk which was no longer acceptable to polite St Petersburg society. Aristocratic and state demand thus continued to dominate the activities of those inhabitants of the non-peasant sector involved in artisanal, manufacturing, or industrial pursuits but, within the production framework of 1788, luxury-good production lost ground as domestic manufacturers and craftsmen making these wares succumbed to foreign competition.

Indeed, during the closing years of the eighteenth century and the beginning of the nineteenth, this sector of non-peasant society was assailed on all sides, for not only did producers of luxury wares lose ground to foreign imports but also manufacturers of textiles failed to take advantage of fundamental changes taking place in peasant society. During the years 1788–1807, salt producers and manufacturers of metal goods continued to sell their wares to the peasantry, goods destined for this market very slightly increasing their share of sectoral output (from 10.7 to 13.25 per cent), and a somewhat diminished flow of small-value copper coins passed to peasant society; but textile manufacturers made no more headway there than they had in the 1780s. Yet, during those years, a fundamental change was taking place amongst the peasantry, whose importance can hardly be underestimated for it signalled the first breach in the self-sufficiency of that sector. The peasantry had always bought one or two items, like metals or salt, which they had been incapable of producing, but the overwhelmingly greater part of their needs, for food, clothing, accommodation, heating, and lighting, they satisfied by their own endeavours or by participation in that commercial network spanning, but encompassed within, peasant society. Now they abandoned the production of one of these elements – clothing – selling the raw materials (wool and flax) they had previously worked up, and buying the finished product from denizens of the non-peasant sector. It was not, however, to the manufacturers of these wares that they turned for these fabrics but to that group of impoverished peasant-workers, who stood outside the bounds of the production structure of the 1780s and who now sold some 10 million roubles' worth of their goods to the peasantry, or some 12.5 per cent of sectoral output.

The years 1788–1807 had thus witnessed a fundamental transformation of the commercial sector of the Russian economy, resulting from and symptomatic of equally fundamental changes within its productive base. That intricate multilateral pattern of exchanges, transcending but closely interwoven into the fabric of peasant society, which had characterized commercial activity in the 1780s underwent an extreme simplification. That system of inter-peasant exchanges disappeared and trade became exclusively concentrated on that bilateral exchange between peasant society and the non-peasant/foreign-trade sector of the economy. Within this exchange, moreover, the range of goods was drastically thinned, as short-range commerce gave way to long. The denizens of the non-peasant sector, apart from those engaged in the manufactories which, thanks to protection, held their own against imports on domestic markets (textiles, metal fabricating, and salt production), accordingly experienced an attenuation of their consumption which was reflected in a reduction in their 'real' earnings. Activity within the sector, still overwhelmingly concentrated on satisfying the demands of the aristocracy and government, was experiencing acute difficulties which were in no way relieved by the emergence of a new element catering for peasant demand. Overall peasant purchases had declined with the reduction in their incomes (− 55 per cent per head in 'real' terms), and although that proportion bought from non-peasant workers had increased the latter could only vend the textiles, which now found a major place in peasant purchases, by drastically cutting prices, which doomed producers to the most appalling of lives.

These changes, which reflected the developing crisis in the north, were also mirrored in the organizational structure of the commercial sector as both the direct commerce and petty trade of the market and retail stores lost ground to the fairs. Least affected was direct commerce, whose share in total trade declined marginally (to 47 per cent), largely as a result of the diversion of iron supplies to the Nizhnii Novgorod fair to be sold to manufacturers. Far more significant was the decline of the northern markets and stores which suffered from the effects of the acute thinning of short-range trade. Their business, even including those goods involved in the overlap with the direct trade, fell from one-third to a quarter of total commerce, and amongst the wares displayed on stalls or the shelves of stores imports figured much more prominently than before. Both the direct trade and petty commerce, however, declined. Only the fairs increased their share of commercial activity, their numbers increasing both as a result of this displacement effect and as a result of the overall growth in trade.[172]

Table 5.8 Estimated trade at fairs, *c.* 1807

Commodity	Value (millions of roubles)
'Exports'	
Cattle	126.0
Flax, hemp	15.4
'Imports'	
Textiles	10.0
Fish, salt	6.8
Metal wares	5.6
Furs	0.8
Total	164.6

Servicing the needs of the rural communities, they continued to provide a market for the disposal of goods destined for the non-peasant/foreign-trade sector of the economy, though as the intersectoral balance of trade swung against the peasantry this aspect of activity increased disproportionately. 'Imports' into the rural communities correspondingly declined in relative importance, but within this collection of goods, now acquired exclusively from the non-peasant sector, textiles now reigned supreme. To accommodate this new 'import' trade, moreover, a new system of fairs had been created during the years 1788–1807 which was superimposed upon the old. In each of the main production centres of the 'new textile industry' – Tula, Ryazan, Moscow, Vladimir, Yaroslavl, Tver, Kostroma, Kaluga, Vologda, and Nizhnii Novgorod – new fairs had been born, to which the peasant-workers brought their low-grade fabrics and from which these wares were distributed through the existing trade network. Thus not only was there an overall increase in the number of fairs but there was also a major structural realignment in their distribution as the northern non-black-earth provinces increased their share of the total, creating a much more even spread of fairs over the country.[173]

The years 1788–1807 thus witnessed a considerable simplification of the structural and organizational patterns of Russian commerce. They also saw an acute rationalization in the range of goods carried along the major arteries of internal trade as the basic staples – grain and cattle – increased their share of total commerce from 24 to 42 per cent. Any assessment of the efficiency of the new commercial system must focus, accordingly, on these commodities and upon their trading networks which during these years were extended to incorporate the newly annexed, rich lands of the south.[174]

Table 5.9 Regional grain availability, *c.* 1788 (chetverts per head)

Region	Before trade	After trade
Grain export regions		
Central black-earth	8.04	8.03
Middle Volga	8.05	7.90
Nether Volga, South, pre-Ural	3.55	3.43
West	3.20	3.11
Grain import regions		
Lake	2.71	1.96
Central non-black-earth	2.68	2.81
North and North pre-Ural	2.65	2.71

In the case of grain during the 1780s the regional availability of grain was widely differentiated, ranging from *c.* 8 chetverts a head in the rich and fertile lands of the central agricultural and middle Volga regions to 2.7–3 in the grain-deficient regions of the north. The principal cause of this wide variation resided simply in the amount and fertility of the land an individual possessed. Trade played only a minor role in eliminating surpluses or alleviating shortages. In the lands of the Volga and Dnepr, trade reduced individual surpluses by only 2–3 per cent and so great was production in the central agricultural region that it had virtually no effect on *per capita* availability. Nor was its impact very much greater in the grain-deficient regions of the north. The inhabitants of the northern regions and the central non-black-earth region, through their access to the market, were able to augment local supplies by some 2–5 per cent. Only in the capital and its region was the situation different as trade along the Volga–Neva riverine system allowed the population to increase supplies by almost 10 per cent.[175]

Table 5.10 Regional grain availability, *c.* 1807 (chetverts per head)

Region	Before trade	After trade
Grain export regions		
Central black-earth	4.15	3.84
Nether Volga	3.65	2.96
West	3.30	2.96
Middle Volga	3.10	1.96
Grain import regions		
Central 'industrial'	2.22	2.56
Lake	2.13	2.46
North	1.92	1.94
New Russia	1.31	1.63

Twenty years later, in the 1800s, the situation was very different. Regional variations in grain availability were enormously reduced, a feature reflected also in the increasing convergence of regional grain prices.[176] Yet, as will be appreciated from an examination of tables 5.9 and 5.10, this owed much more to production changes than market ones. As a result of an equalization in the size of holdings and a disastrous decline in yields in the richer central agricultural regions of the empire the degree of variation about the on-farm mean was reduced by one-third. Trade served to reduce this by a further 13 per cent. Certainly trade was beginning to have an impact, particularly within the trading network of the 1780s, and merchants were starting to make serious inroads into the harvests of exporting regions. Least affected were the older areas of settlement – the central black-earth (7.5 per cent) and middle Volga (5 per cent) – and most, the new lands of the nether Volga (19 per cent) and the agricultural regimes of the ex-Polish territories in the west (10 per cent). Consumers, especially in the regions of the two capitals, were able to augment local supplies significantly (15 per cent) by recourse to the market. These changes, however, did not extend to the lands of the periphery. The system had no way of coping with the developing crisis in the north where climatic deterioration led to a retreat in cereal cultivation. If the denizens of these lands were *relatively* slightly better off it merely reflected the fact that everyone else was considerably worse off. Similarly in the newly annexed lands of the south, normal commercial mechanisms could do little to relieve the acute cereal deficiency occasioned by the decline in cereal cultivation in the aftermath of the war. Extraordinary measures, such as had been taken in 1794–5, were the only answer but these could not be sustained, and into the 1800s New Russia remained the most acutely grain-short region of the empire.[177] The years 1788–1807 had thus witnessed a considerable improvement in the operation of the grain market, but this was largely confined to the commercial system operating within the trade network of the 1780s. Perhaps not surprisingly, these improvements did not extend to the trade between the north and the newly annexed territories of the south.

A similiar situation prevailed, moreover, in the cattle trades. The degree of variation in meat consumption in the north was not great: non-peasant consumers, dependent upon supplies of animals from the south, acquired *c.* 1.64 beasts per head annually; their peasant counterparts obtained 1.16 culled from their herds. Both were eclipsed, however, by the inhabitants of the central agricultural region who, thanks to the mass slaughtering of animals for tallow, had at their disposal the meat of 3.5 animals per head, and by the denizens of the eastern steppe, with 10 beasts per person per year to kill. Thanks to severe market imperfections arising from an inadequate development of north–south trade, the Don cossacks could still preserve the lifestyle of their forebears of whom it had been said that they 'sow little corn . . . nor do they eat much bread, roots or herbs; their chief diet being fish, flesh and fruits'.[178] Their kalmuck neighbours could still subsist on a diet of raw horse flesh and kumiss.[179]

Trade had failed to reduce the marked differences between north and south. *Per capita* product in the 1800s was characterized by high interzonal variations. Yet great as they were, these variations existed about a very high mean. Russia, by that time, was a very rich country indeed.

Russian economic growth, 1718/22–1807. International comparisons

Within the brief space of some ninety years, and within contemporary boundaries, Russia had experienced an almost five-fold increase in national income. As a result of Peter's annexations of the Swedish Baltic provinces and the Ukraine, and rapid economic growth in the Muscovite heartland of the empire, average *per capita* incomes rose by some 85 per cent between 1718/22 and 1788, attaining a level of 63 silver roubles a head. Subsequent annexations of the Turkish Tartar lands of the south and their development, further increased *per capita* incomes by some 8 per cent until in 1807/9 they attained a level of 68 roubles per head – or double that enjoyed by the inhabitants of Peter's war-torn kingdom a century earlier.

In these circumstances it is hardly surprising that in spite of a massive increase (times 12.5) in the size of the Russian monetary stock, ensured by an equally massive increase in the indigenous production of monetary metals, prices were only 2.25 times higher in c. 1807 than they had been in c. 1700 and even then by far the greatest part (83 per cent) was concentrated in the years 1775–95 as growth within the Muscovite heartland of the empire first faltered and then gave way to decline, injecting an acute bout of inflation in the economy. By this time, however, international conditions had dramatically changed as a flood of South and Central American silver pouring into western Europe pushed up prices there to such a level that the Russian export boom continued unabated in the late 1790s and 1800s. At this point, as Russian prices stabilized and then fell with the recovery of the economy, exports rocketed upwards and silver poured into the realm, augmenting declining domestic production and providing sufficient coin to satisfy an increasing transactions demand associated with renewed growth. Rapid Russian economic growth thus ensured that the massive quantities of monetary metals produced domestically in the period 1730–95 were all deployed within the economy and that from 1795 to 1807 the nation drew on international specie markets to satisfy an increasing transactions demand associated with renewed growth.

That poor, strife-ridden nation of Peter's reign, whose denizens enjoyed a *per capita* income only two-thirds that of the English and three-quarters that of the French had by 1807 come of age.[180] During the years 1718/22–88 the gap between Russia and the then most powerful nations of western Europe had rapidly closed until the average Russian was only 15 per cent poorer than his French counterpart and less than 5 per cent worse off than the average Englishman (figure 5.2). Nor did this position weaken during the subsequent years 1788–1807 when, as a result of the decline in French incomes during the period of the French revolutionary wars, Britain assumed economic leadership in western Europe. In 1807 the average British citizen was barely richer than his Russian counterpart. The two nations – Britain and Russia – stood at the very top of the European national-income league table.

Lest this conclusion surprises the reader, however, it should be noted here that it would not have surprised contemporaries. Even English visitors to the Tsar's realm, who were accustomed to view the world with their customary insular arrogance, were impressed at its denizens' wealth. Of polite society Sir James Harris was constrained to remark that, 'prepared as I was for the magnificence of this court, yet it exceeds in everything my ideas', whilst John Carr simply noted that 'the princely magnificence in which some of the Russian nobility live is prodigious'.[181] Nor was aristocratic display simply a product of peasant impoverishment.

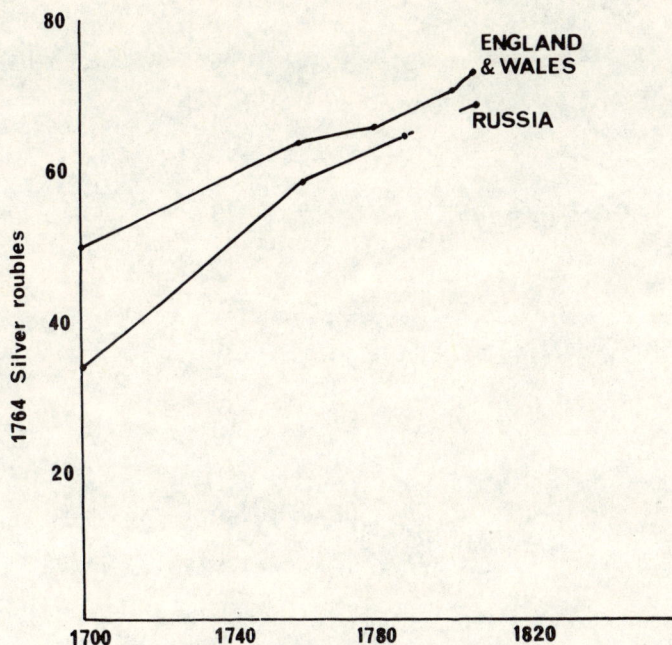

Figure 5.2   Russia and England: estimated *per capita* national
income, 1700–1807 (constant 1762 prices)

As has already been noted Coxe could remark in the 1780s that he thought
the peasants 'well clothed, comfortably lodged' and provided with 'plenty of
wholesome food', and his cry was taken up twenty years later by Martha
Wilmot who asserted that 'those who imagine the Russian peasantry sunk
in sloth and misery imagine a strange falsehood'.[182] Indeed a slightly later
writer remarked that she 'had several opportunities of remarking the
plenty and comfort of the Russian boor' and that 'certain it is, that the
Russian peasant is happier and has fewer wants unsatisfied than the
peasantry of that country whose liberty is her boast'.[183] A century of rapid
economic growth had made both Russian aristocrats and peasants the envy
of visitors from the allegedly more civilized west.

Part Three

Russian Silver and International Specie Markets in the Eighteenth Century

Chapter Six

Conclusion. Russian silver and international specie markets in the eighteenth century

From about 1570 international precious-metal markets were totally transformed as a new group of Central and South American producers utilizing a new technology – the amalgamation process – completely altered production relationships within the industry. A new resource base, encompassing high-grade silver haloids (argentite and cerargyrite) almost unknown in Europe but abundantly available in the New World, was opened up. Metal extraction rates were enhanced and for a quarter of a millennium from 1570 to 1820 a flood of precious metals poured forth from the Americas totally dominating world bullion markets. In creating this new Central and South American industrial complex, moreover, a new supply network to channel mercury to the major mining centres was created, and during the years 1570–1820 the secular trend, measured between long-cycle peaks in *c.* 1623 and 1798, in the supply of the liquid metal was ever upwards and its price, in both nominal and real terms, was down. During the inter-cyclical trough which extended from *c.* 1623–1775 there was some contraction in supplies of mercury to the mines and enhancement in its price but neither were significant. Throughout most of the history of the amalgamation process, which as far as Central and South America was concerned spanned the long period 1570–1820, precious-metal producers enjoyed an increasing supply of mercury at falling long-term prices which allowed them to extend the margin of exploitation increasingly to lower-grade ores, thereby increasing both output and exports of precious metals (figure 6.1).

If during the period 1570–1820 Central and South American precious-metal production and exports underwent a major secular boom, interrupted only in the period 1670–1720 when activity was displaced to west Africa, however, the industries producing these metals at the same time underwent a process of major structural change. Falling productivity in the silver industry, associated with increasingly deep mining, declining ore yields, and the excavation of an increasing proportion of minerals which were intransigent to working by the amalgamation process and had to be smelted and cupelled, coupled with rising prices for the products of

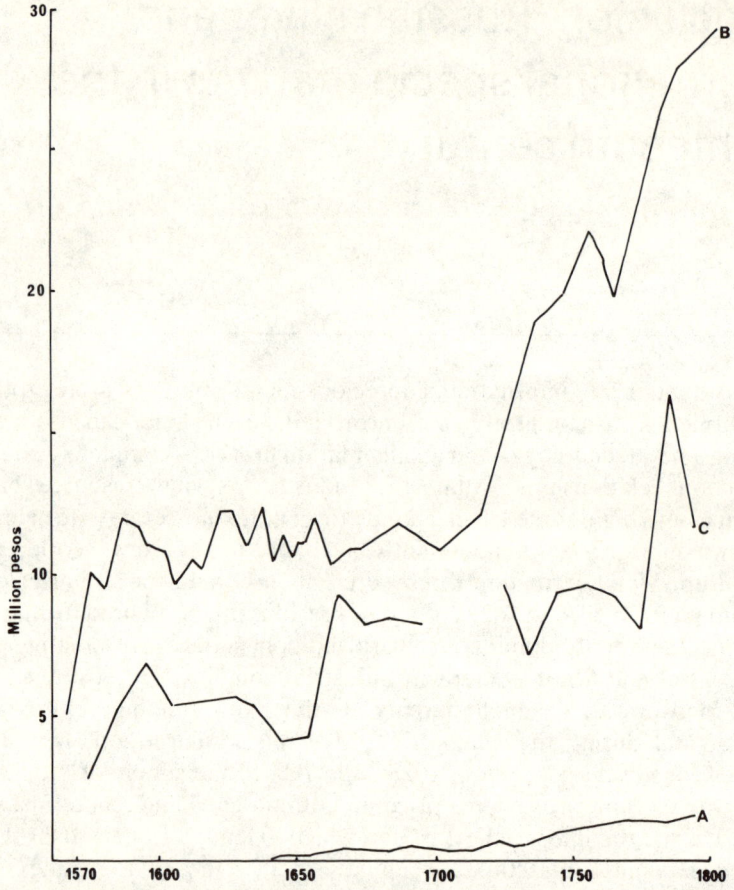

Figure 6.1 International specie production and trade

A   European output
B   Latin American and West African output
C   Latin American-Spanish trade

the low-productivity gold industry, created by about 1625 an uneasy equilibrium between the two sectors of the precious-metal industry. By that date the overwhelming importance of silver – that 'common drudge' of the sixteenth-century world monetary system – had become a thing of the past. During subsequent 'crises' in c. 1625–55 and 1670–1760, mercury supplies were redeployed beyond the base of the silver industry in accord with relative intersectoral productivity levels, causing gold for the first time to pose a significant challenge to the dominance of silver, increasing its share of total specie production from 27 per cent during the years 1625–55

288

to 47 per cent in the period 1670–1760 and establishing a trend which would ultimately be realized in the aftermath of the Central and South American mining crisis of 1810–30 when, borne upon an industrial diaspora, the gold industry was transported first to Russia and then to the lands of the Pacific basin (California, Alaska, and Australia) and established its absolute supremacy on world specie markets contributing almost 80 per cent of total supplies in the 1850s.

The 'crises' of 1625–55 and 1670–1760 thus marked the beginnings of a process of major structural change within the industries supplying gold and silver to world markets, which could not but affect European markets for specie and the monetary systems dependent upon them. Diminishing silver flows from the Americas, creating shortages and enhancing free-market prices, caused a diminishing flow of silver for coining as mint prices lagged. As supplies of gold increased on the other hand and prices fell, more and more of that metal was minted. A *de facto* silver mono-metallism thus gave way to a true bi-metallic system embracing both metals. This tendency, already apparent during the first 'crisis' of 1625–55, became particularly pronounced during the second, although initially during the years 1670–1720, as European traders scrabbled for available supplies of west African gold, mintmasters were forced to enhance the mint price of both metals in an attempt to alleviate overall short-term supply deficiencies. By 1720, however, the crisis was past, and as monetary stocks once more began to grow all over Europe, an acute shortage of silver coins prevailed and the market for that metal was characterized by an acute instability, merchants adopting the relatively stable gold as the basis for their transactions. Only in the north, amongst the lands bordering the Baltic, was the situation different. Here, although a few gold issues were minted, silver retained its importance. The reason lay in the availability of abundant local supplies of the white metal deriving from a buoyant, broadly based, European industry, for if the 'crises' of 1625–55 and 1670–1760 had led to an increasing dominance of gold in European monetary systems by creating acute shortages of silver it had opened up new opportunities for European producers of that metal.

As the massive inflow of Central and South American silver had poured onto European specie markets in the years after 1570 the pre-existing European industry, based upon the *Saigerprozess*, had been utterly destroyed. The delicate price equilibrium between product (copper and silver) and input (lead) markets, upon which the fortunes of European silver producers depended, was seriously disturbed. As silver prices fell, relative to both input and by-product prices, the industry underwent an internal metamorphosis which totally altered its market position. Enhanced copper prices relative to silver increasingly made that metal the primary object of exploitation, whilst the declining importance of silver in total returns ultimately made the refining of the copper to obtain that metal unviable and rendered the use of lead in that stage of production unnecessary. The close symbiotic relationship which for a century, 1470–1570, had bound together lead and copper/silver producers was in 1570 split asunder. The focus of silver production shifted elsewhere – to Central and South America. The erstwhile European *Saigerhändler* was transformed into a copper producer who now had to compete with other producers without the benefit of cross-subsidization of his wares from silver sales. The lead producer lost his major market and now had to vend his wares to new customers. From 1570, therefore, as the European silver producer was subject to an endemic crisis and activity was displaced to the Americas, copper and lead production evolved independently within autonomous market structures.

Nor was the enhancement in silver prices during the 'crisis' of 1625–55 sufficient in itself to create conditions conducive to a revival of European production on the basis of the prevailing technology. On that occasion, silver prices had risen enormously but increasing lead prices, due to a major crisis in the English industry, and a 'glutted' copper market, as Swedish output peaked, had limited the *Saigerhändler* in their ability to respond to the new situation and during these years they made little or no contribution to relieving prevailing shortages. Indeed, the enhancement in lead prices had provided an impetus to the deployment of an even older technology, leading to the exploitation of argentiferous lead by cupelation. High and rising primary (silver) and by-product (lead) prices thus, on this occasion, created a speculative atmosphere encouraging men to venture fortunes on the re-opening of old workings or the establishment of new ones. They met, however, with singularly little success. At best, even under conditions of excessively high lead prices, such as existed during the years 1630–70, the production of silver by the cupelation of European argentiferous lead yielded only marginal supplies of the white metal to the market. That dominance of Central and South American silver on world specie markets, first established in 1570, remained undisturbed.

Only during the second 'crisis' from 1670 to 1760 did this situation change, when thanks to a favourable conjuncture in both product and input market prices, there was a revival in the European production of that metal. Once more the *Saigerhändler* of central and east Europe, the heirs to the technological traditions of the period 1470–1570, established their ascendency amongst indigenous suppliers of European silver on the basis of rising copper and silver prices and falling lead prices. The shortfall in Central and South American silver production after 1670, as has been suggested, enhanced prices of that metal. Prices of copper similarly rose after the collapse of Swedish production in the mid-1650s and rocketed upwards after the cessation of Japanese imports in the 1670s. Lead prices, on the other hand, plummeted as the English industry entered upon an expansionary phase at the same time, flooding European markets with abundant and cheap supplies of the metal. For the first time in more than a century a favourable price conjuncture advantaged central European producers and they rapidly seized their new opportunities. From 1670 the central European argentiferous copper industries once more resumed the pattern of yield-related production cycles which had been interrupted during the previous century and production expanded rapidly. Initially, against a background of declining Central and South American silver production, central European production concentrated in the mines of Saxony and Oberharz increased during the years of the first cycle (1670–1740) to a peak in c. 1725, pouring forth a flood of copper and silver which, passing by way of the Elbe to Hamburg and Lübeck, was distributed either south to the Low Countries or east, the latter flow allowing Lübeck merchants to increase their share of the lucrative Russia trade at the expense of the Dutch and English. Product markets were transformed. The Hanseatic towns ran an adverse balance of trade and exporting silver created a distinct monetary zone in the north. The integration of the German *Saigerhändler* into the trading system of the western commercial metropolis of Amsterdam in 1701, moreover, bestowed on western European consumers the benefits that their central European counterparts had enjoyed for two decades. From 1701–25 copper prices fell steadily, interrupted only by Swedish hostilities in the 1710s, to equilibriate ultimately at a level which, if higher than in the early 1650s was still sufficiently low to ensure market dominance in those areas where the products of the Harz forests and the Erzgebirge were not excluded by tariffs. A fortunate conjuncture of prices in silver, copper, and lead markets during the years 1670–1725 had thus created the conditions for the establishment of a broadly based central European metallurgical complex which formed the focus of an important European-metal supply system. Over the same period production of silver from this complex rapidly increased, reaching a level of 20,500 kg a year or 4 per cent of total European silver supply, before a downswing in the output of the Oberharz mines, which began slowly in 1725 and gradually gained momentum in the years to 1740, again caused a crisis, raising silver and copper prices, initiating a new industrial diaspora, and leading to the beginning of the second great European production cycle which ran its course during the years 1740–80 largely on the basis of Hungarian production.

In the new location, production once more increased but intersectoral changes in the balance of mining activity altered product market structures and the intrusion of increasing quantities of Central and South American silver onto European markets after 1760 resulted in the cyclical upswing aborting before the European industry had attained peak output. Silver production, which had steadily declined between 1725 and 1740 leading to a displacement of that metal on Netherlands markets by gold and precipitating a series of acute monetary crises in the Baltic region, once again from *c*. 1740 began to increase, rising to 36,700 kg of silver a year in the early 1760s when the European industry held a 7 per cent market share, before a lowering of silver prices caused output to fall to 31,000–32,000 kg a year. During the second production cycle European producers had thus, at least initially, increased their market dominance. Such was not the case with respect to copper. Here as Harz production declined there was no compensatory increase elsewhere, and prices doubled between 1725 and 1760. The industry was thus already in a weak state when the inflow of American silver dealt it a final death blow; rising copper prices shifting it into a phase of rapidly diminishing returns as the margin of exploitation extended into lower- and lower-grade ores and falling silver and rising lead prices finally undermined its position.

Against a background of declining Central and South American silver production and in conditions of product and input price equilibrium the years 1670–1760 had thus witnessed a revival in the fortunes of the central European *Saigerhändler*, production of silver steadily increasing between cyclical peaks in *c*. 1725 and 1760 and the central European industry equally steadily increasing its market share.

Even as the central European production cycles ran their course, however, events were taking place far to the east which would cast a shadow over the whole European scene as new mines, opened up in the lands of the Russian Empire, rose in the ascendant. Initially discovered during the first production cycle of the European industry, the mines of Nerchinsk (opened in 1704) and Kolyvan–Voskressensk (opened in 1729), together with the Uralian gold mines of Beresov (opened in 1754), contributed an ever-increasing production of precious metals to the second great production cycle until, at the height of the boom in the early 1770s with a combined output worth 2.2 million roubles, equivalent to 40,000 kg of silver a year, the mines produced collectively more of the precious metals than all of the rest of Europe put together (figure 6.1). Contemporaneously, moreover, the Russian copper industry underwent a parallel process of expansion, until, with an output of *c*. 3,500 tons a year in the 1760s, it ranked second to none in the world. For some sixty years, *c*. 1735–95, therefore, whilst specie continued to be imported into the recently annexed Baltic provinces, a flood of both base metals (copper) and precious metals (silver and gold) derived from indigenous mining and metallurgical enterprises, passed to Russian mints, rendering the Muscovite heartland of the empire independent of international supplies of specie. Imports of precious metals were displaced and Russian monetary stocks were rapidly augmented on the basis of domestic production, creating potentially strong inflationary pressures within the economy and threatening to transform the nation from being a net importer to an exporter of specie. In the event, however, a 'revolution' within the Russian countryside, and associated transformation of internal transport systems, prevented this occurring by reducing 'real' export prices. The export boom continued, and as imports rose Russia experienced a passive balance of trade during the years *c*. 1735–95. The full impact of increasing monetary supplies during these years was felt within the domestic economy which acted as a sponge absorbing the burgeoning flood of coins which poured forth from the mints in order to meet a rapidly growing transactions demand arising from rapid economic growth which transformed the poor, strife-ridden nation of Peter's reign into one of the richest nations in Europe.

The 'crisis years' 1670–1780 had thus witnessed revolutionary changes in eastern Europe as Russia's precious-metal industries had risen to a position of European supremacy and its base-metal industries – copper and iron – had come to dominate world production. These changes, however, had paled into insignificance alongside the more general transformation of the Russian economy which had at the same time undergone a process of rapid economic growth which converted the poor, strife-ridden nation of Peter's reign into a nation which stood at the very top of the European national income league table, its citizens enjoying a *per capita* income comparable with that of Britain.

Neither phenomenon was, however, anything more than ephemeral. In the case of both base- and precious-metal industries the 1780s saw the beginnings of a process of decline. In the case of precious-metal production, plagued by both resource-depletion problems and ill-advised government intervention in labour markets, output stagnated even as the Old World industry as a whole was once again eclipsed by that of the Americas. Similarly with respect to Russian copper and iron production, whilst output stagnated during the years 1780–1835, these industries, excluded from foreign markets, were eclipsed – this time by Britain. Most significant of all, however, were the first signs in the 1780s that all was not well with the Russian economy. As growth gave way to stagnation and then, between 1788 and 1795, to decline, famine stalked the land and the economy was subject to an acute bout of inflation. Both subsequently subsided as, from 1795 to 1809, territorial expansion laid the foundations for renewed growth, but from *c*. 1810 national product began slowly to decline, *per capita* product rapidly falling until *c*. 1835 (figure 6.2). The promise of the mid-eighteenth century, when Russia had risen to a position of some importance on the world stage, had turned sour. After *c*. 1780 both national product and the output of her metallurgical industries had stagnated, *per capita* product falling rapidly to 1835.

After a phase during *c*. 1670–1780 when precious-metal production had been displaced from the Americas to Europe and when economic activity within Europe had been displaced eastward to Russia, the *status quo ante* 1670 had been restored. From 1780 to 1835 Central and South American specie again dominated monetary systems within an international economy in which Britain again reigned supreme.

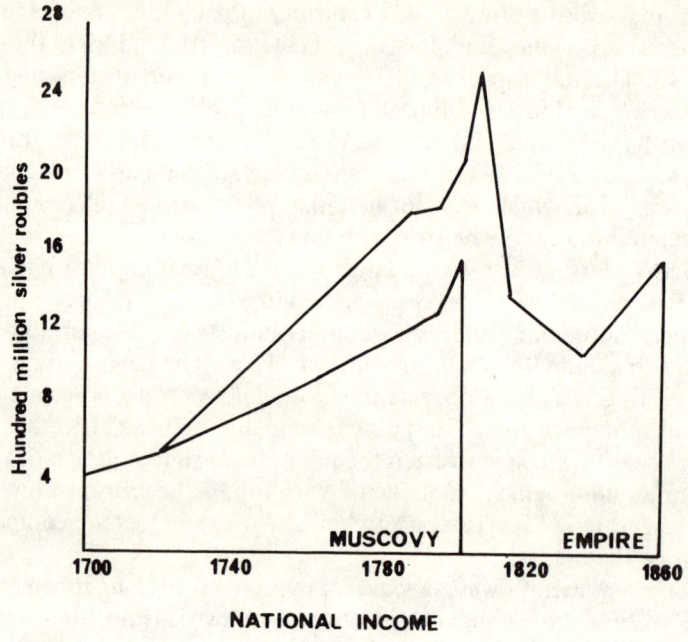

NATIONAL INCOME

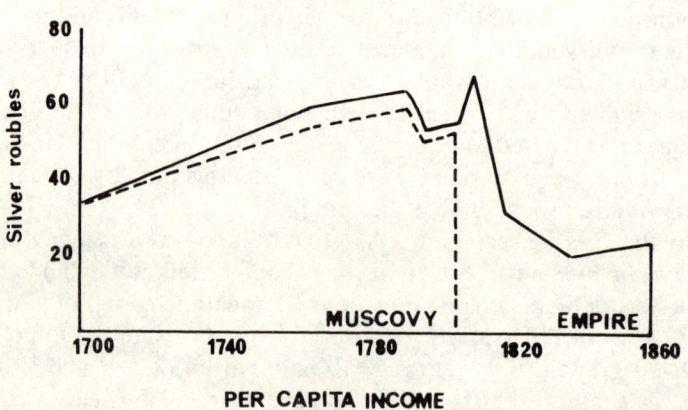

PER CAPITA INCOME

Figure 6.2 Estimated Russian national income, 1700–1860 (constant 1762 prices)

Appendices

Russia. Precious-metal and copper production

The relative obscurity into which the history of Russian precious-metal production in the eighteenth century has lapsed arises largely from Adolf Soetbeer's treatment of this subject in his classic *Edelmetall – Produktion und Werthverhältniss zwischen Gold und Silber Seit der Entdeckung Amerika's bis zur Gegenwart* (Gotha: Erganzungsheft No 57 zu 'Petermann's Mitteilungen, 1879), section VII, pp. 38–42. Restricted to the use of German translations of mid-nineteenth-century Russian official statistical abstracts or contemporaneous secondary works based upon these materials, the greater part of his essay (pp. 39–41) involves a simple presentation of the available (often contradictory) production data for the years 1815–76 and (pp. 38–9) a description of the main mining regions in the 1870s. As far as the eighteenth century is concerned he does no more than devote three short paragraphs (p. 38 col. 2 and p. 39 col. 1) to describing the main workings before providing estimates (section C, Rekapitulation, p. 41) of eighteenth-century production. How these were derived is not revealed, but in the absence of the presentation of any direct evidence they seem to represent arbitrary backward linear time projections superimposed upon W. Jacob's estimates (in his *Russlands, Australiens und Californiens Gold–Produktion*, St Petersburg, 1864) of total gold and silver production for the years 1704–1810, which are even incompatible with the isolated production data presented by the author (p. 39) in his description of the Altai mines. Unfortunately although these projections lack any substance they continue to be quoted even though most recent writers have shifted their attention to a miscellaneous collection of late-eighteenth-century works containing abstracts derived from contemporary official records.[1]

These works which all present data in varying formats, degrees of completeness, accuracy and clarity from what are clearly official records or abstracts published from them are: H. Storch, *Tableau historique et statistique de l'empire de Russie à la fin du dix-huitième siècle* (Basle and Paris, 2 vols, 1801), notes et pieces justificatives, pp. 54–5, or the German translation of this work published in Leipzig, 1803, *Supplementband*, table 2.[2] W. Coxe, *Travels into Poland, Russia, Sweden and Denmark.*

Interspersed with Historical Relations and Political Enquiries (London, 2 vols, 1784), pp. 283–6.[3] B.F.J. Hermann, 'Sur l'exploitation des mines de l'empire de Russie', *Nova Acta Academiae Scientiarum Imperialis Petropolitanae*, 11 (1798), pp. 418–33.[4] To fully comprehend them, however, it is necessary to understand something of the administration which created the documentation upon which they are based.

The basic unit within the mining administration of the eighteenth century was the individual mine or smelting establishment which encompassed all operatives at work in that particular location. In the case of a mine this involved all pickmen (*Berghauer, Bergairov*), apprentices (*uchenik*) and children (*gruben, yuni*), as well as auxiliary mineworkers, technicians, and craftsmen who were subject to a foreman (*Steiger, Shsheigerov*) and his assistants (*Unter-Steiger, Unsher'-Shsheigerov*) who organized the operatives into work gangs. At the smelting establishment the ore carriers, washers, charcoal-makers, as well as the bewildering array of workmen at the smelting and refining hearths, were similarly subject to the authority of an under-furnace master (*Unterhüttenmeister, Unter Gershshen Meister*).[5] Neither of these senior workers at this primary level of administration, however, enjoyed officer status and neither contributed to the creation of the administrative paper work.

This was reserved for those located on the next rung of the hierarchical administrative ladder – the denizens of the district office. Each of the great mining and metallurgical complexes, which dominated Russian precious-metal production in the eighteenth century, was divided into districts, there being six at both Kolyvan and Nerchinsk in the years after 1770, each engrossing a number of mines and smelting establishments.[6] The denizens of the district office comprised two officials and their subordinates. First, there was the chief of mining operations (*Schichtmeister*), who enjoyed the status of a lieutenant of engineers or artillery, was in charge of all materials, and was expected to verify the presence of all operatives at their place of work and to assess the quantities of ore raised as well as fixing wages. He was expected, accordingly to keep records of:

   (i) The number of workers and their wages;
   (ii) the amount of ore raised;
   (iii) the stock of materials and state of equipment on hand;

at each of the mines under his control, and to despatch at three-monthly intervals copies of these accounts to the principal mining office of the complex. Second, there was the master smelter (*Hüttenmeister*), also of lieutenant's rank, who was in charge of all smelting operations in the district and who, in collaboration with the chief of mining operations assisted by the ore driver and assayer, complied a record of

   (iv) The amounts of metal produced and the yield of the ores smelted.

Subordinate to these officials was the assayer (*Probirstchik*) whose work with his team at the district laboratory provided necessary data for both of his superiors. In relation to the production side of the enterprise, which is of primary concern to us here, copies of four basic sets of records were thus kept at the district office, duplicates of which were sent to the principal office of the mining region (*Bergamt-Nachal'stva*).[7]

The territorial limits of the mining regions underwent certain changes in the eighteenth century in line with reforms of gubernatorial jurisdiction, but they retained their independence from the authority of the governors throughout that period, one or other of the district offices always assuming a primacy over the others.[8] In the trans-Baikal mining region the principal office was always located at Nerchinsk, whilst that of the Altai mining region was first (1729–49) situated at Kolyvan before being moved to Barnaul.[9] Here the chief administration of the mining region was located under the supreme authority of a commander (*Nach'alnik*), enjoying the ranks of general and councillor, whose staff at Kolyvan in the late eighteenth century comprised two chief directors of mines (*Oberbergmeister*, of Lt Colonel's rank) with general technical and managerial responsibilities and a number of specialists. In relation to mining operations there were three directors of mines (*Bergmeister*, of Major's rank) with four mine jurors (*Berggeschvören*, of Lieutenant's rank) under their direction who assisted them and transmitted periodic reports on the mines based upon personal inspection rather than the testimony of subordinates. Smelting operations were under the direction of three directors of furnaces (*Hüttenverwalter*, also of Major's rank), assisted by sub-directors (*Hütten-verwalter*, of Captain's rank) whilst the care of the jurisdiction's forests was the responsibility of a master forester (*Forstmeister*, of Captain's rank). Assisting each of the mining, smelting, and forestry officials were also two surveyors (*Markscheider*, also of Captain's rank) whose duties involved making technical drawings of mines and furnaces and undertaking surveys of lands and forests. These then were the executive officers of the jurisdiction who were supported by a chancery, comprising two secretaries, an office manager, and fifty-five clerks and copyists, responsible for all the documentation of the mines which fell into three categories.[10] First, the chancery was an office of record, receiving from the district offices and storing the accounts recording operations at each mine and smelting establishment. Second, it was the repository of all those documents generated by the executive action of the chief administration: the reports compiled by the mine jurors which served as a check on the accounts of district officials; the synoptic accounts which provided an overview of operations; and the plans and surveys drawn up by the surveyors' office. Third, it was the recipient of instructions and requests from the central administration in St Petersburg, keeping a record of all instructions (ukazi) and compiling all reports required by the officials of the central administration. During the eighteenth century records of all aspects of operations within the jurisdiction were thus kept in the chancery of the principal mining office, the basic four series of accounts despatched thence from the district offices and the summaries compiled from them there being augmented, on the basis of executive action, by other documents of an independent, non-derivative character:

(v) reports on the mines, their production equipment and labour forces;
(vi) Plans and surveys.

From 1828, this unitary structure was split into two parts with the creation of a mining section, concerned with the general administration of the enterprise, and a section on technical affairs, created to service the administrative needs of the groups of staff-mining engineers sent out from St Petersburg at this time to reform the Altai works, but the basic pattern of documentation remained.[11] Throughout the period when the mines were in the government's hands, namely from 1704 in the case of Nerchinsk and 1746 with regard to Kolyvan, a basic stability characterized the administration within the mining jurisdictions.

Such was not the case with regard to the central administration at St Petersburg to which the local mining offices were subject. Prior to the creation of the mining College (*Berg Kollegium*) in 1719 mines were under the authority of diverse organizations whose competence was not precisely defined. Foreign experts, recruited to work in the mines and smelting establishments, were under the jurisdiction of the Office of Embassies (*Posolski Prikaz*). The plants within which they worked were either arbitrarily assigned to specific government departments like the Department of Artillery (*Pushkarski Prikaz*) or the Public Treasury (*Prikaz bolshoi kazni*) as need dictated, or, as in the case of Nerchinsk, left to the tender mercies of the local governor or to individual entrepreneurs who were subject to his authority.[12] Only from 1719, with Peter's creation of the Mining College by the ukaz of the 10 December of that year, was this situation regularized.[13] Responsibility for all matters concerning mining and metallurgy in both state and private works was now centralized in the college whose powers were strictly defined. With regard to gold and silver production, at that time solely represented by the Nerchinsk works in the hands of one Sibiriakov who had first discovered the deposits, this involved not only subjecting the works to the terms which were generally applied (clauses I–VII, XI, XIV–XVII) to enterprises but also the establishment of very specific regulations authorizing:

(i) the provision of loans and indemnities and exempting the works from fiscal and military obligations (clauses VIII–X);
(ii) the state to exercise pre-emptive rights within a month of extraction

at prices which were equitable to the entrepreneur whilst ensuring the avoidance of losses at the mint (clauses XII–XIII).[14]

As initially conceived, therefore, production of precious metals was to remain in private hands but was to be subject to special encouragement by the state, which could exercise pre-emptive rights over production. Before a year was out, however, this situation was brought to an end and the college took over the direct administration of the mines by an ukaz of 19 August 1720 establishing the future pattern of local administration with the appointment of a commander (*Nach'alnik*) responsible for general management, one cabinet-councillor (*Kabinet-Kyr'era*), Ilya Golenishchev-Kutuzov, and a principal mining officer (*Oberbergmeister*), a Swedish prisoner-of-war, Peter Dames, in charge of operations within a specific mining jurisdiction (the Nerchinsk *Bergamt/Nachal'stva*).[15] Henceforth, until 1782, although never formally codified in any of the surviving general mining ordinances, the state exercised its authority to seize any mine within which silver and gold was discovered and to establish an administration directly accountable to the centre similar to that created at Nerchinsk in 1720. As far as the Nerchinsk officials were concerned this meant their direct subordination to the college and their independence from successive directors of Siberian mines and works at Ekaterinburg from 1720 until the mines became entangled in that web of senatorial concern and bureaucratic peculation which in 1736 led to their being attached to the Ekaterinburg Chancellery.[16]

The reasons behind this reorganization of mine administration in 1736 remain obscure but certainly they may in part be understood as a response to a genuine concern in government circles about the endemic losses being made by state plants during the mining crises of 1730–4/5 and 1738/9.[17] Year after year during these crises successive commissions laboured to inquire into the viability of the state works.[18] As early as 1730, against a background of the emergence of grotesque over-capacity in the copper industry and a severe market crisis in the iron industry, the utility of the state works was made the subject of investigation of both a Senate commission and one established by the College of Finance, the former with the remit to examine the financial viability of the enterprises since 1720, the latter their condition in 1730. After extensive deliberations, each produced a detailed financial statement of the profitability of the works, revealing a steady fall in returns from an average of 100,000 roubles a year before 1730 for the seven state copper plants to 38,250 roubles in 1730 at the prevailing price of 4 roubles per pud when the state ironworks yielded a profit of *c.* 31,500 roubles. These findings were then put before Henin at the Ekaterinburg Chancellery and the College of Mines for consideration. Their reports, submitted in the course of 1731, revealed that whether 'privatization' was carried through or not the critical issue was the prevailing level of prices, particularly of copper. As consensus about the sale of the plants was thus reached the focus of discussion shifted to a consideration of price support and of the viability of the private sector as potential purchasers of the works, leading, with respect to the first point, to the embroilment of the commissions with the affairs of the contemporary commission into the circulation of false money (*Monetnaya Komisiya*) of 1731. Perhaps not surprisingly in relation to the question of 'privatization' the members of the monetary commission again came to the conclusion that the state works should be sold off but their conclusion, following the logic of both Hennin and the college, that this was only possible if the state undertook to buy the copper produced at the enhanced price of 6 roubles a pud (in Moscow, 6.50r at St Petersburg) effectively sabotaged the work of the commission whose remit was to devise ways of calling down the lightweight copper issues then dominating the coinage.[19] Accordingly, unable to resolve the irreconcilable objectives of maintaining a viable copper industry and of calling down the copper coinage, the commission was doomed. By recommending a rise in copper prices it ensured the rejection of its report by the Empress who in May 1733 created a new commission under councellor Golovkin with a new remit. Once again the Golovkin commission was to investigate the condition of the Treasury works, this time on the basis of information provided by the Commerce College, but now new questions were raised as to whether potential buyers of state plants were in a position to operate them *without* raising copper prices.

In relation to the first line of inquiry the commission reported in January 1734 suggesting that the works were indeed making a very small profit, partly as a result of unfavourable market conditions but also because of overmanning and bureaucratic inertia, and that they should be sold off but not before the organization of the works had been totally reformed. With regard to the second line of inquiry the evidence assembled was more ambiguous. Certainly they found that market conditions were unfavourable for both copper and iron producers, but when they considered the flood of accusations levelled at private entrepreneurs, involving fiscal fraud or illegal dealings, which had accumulated at the Mining College during the 1720s and early 1730s, they began to have doubts as to whether the low level of fiscal returns from the private sector reflected the unprofitability of enterprises therein or simple fraud. Accordingly, parallel to their investigation into the condition of the state works undertaken in the latter part of 1733, a second vast inquiry was instituted into the affairs of the leading private entrepreneur, Akinfi Demidov. To accomplish this task Demidov was forbidden to leave Moscow and envoys were sent to investigate his Urals, Moscow, and Tula works. For almost two years they laboured to establish the profitability of the Demidov metallurgical empire, ransacking factory offices and interrogating peasants until in April 1735 they were ready to submit their reports to the Empress. On receipt of these, she set up yet another commission, this time under the presidency of Shafirov, head of the Commerce College, to consider the findings. Its report clearly indicated that the Demidovs' non-payment of their obligations to the state arose not from any lack of profits but was due simply to fraud, and that Akinfi owed 800,000 roubles. After a quinquennium of debate, and incidentally in conditions of rising copper and silver prices during 1734/5, it thus seemed clear that customers existed to take over the state works and that they should be sold off but not before they had been totally reformed.[20]

It is thus against the background of a consistent policy of 'privatization', formulated in the context of the 1730–4/5 mining crisis, that the 1735/6 administrative reorganization of state works must be understood. New mines, like the silver mines of Arkhangelsk opened up in 1735 or the rich iron deposits of Blagodat discovered near Verkhe–Tura in Verkhoturie province in the same year, were to be sold off immediately. Existing state enterprises were to be grouped under the Ekaterinburg Chancellery and subjected to a thoroughgoing reform before being put up for sale. The instrument of reform could not, however, be the existing mining administration which the Golovkin commission had declared was subject to bureaucratic inertia. Accordingly, even as the jurisdictional authority of the Ekaterinburg Chancellery was extended over previously independent enterprises like the Nerchinsk mines that administration as a whole was displaced by a new one organized about a new central authority – the General Directory of Mines – which replaced the Mining College in 1735.[21] The whole system of mining administration was in a state of flux. Already two years earlier in 1734 the pattern for the future was becoming clear as the cabinet constantly thwarted the activities of Tatishev, the head of the Ekaterinburg Chancellery. Now in 1736 he was displaced. Nominally retaining his title of Director of Urals Mines, he was moved to Ufa, ostensibly to effect a reorganization of south Uralian production in the wake of the pacification of the region. In reality, his tenure as chief administrator of the Orenburg region during the years 1737–9 was merely a prelude to his final disgrace at the hands of Brigadier Aksakov.[22] From 1736 the Ekaterinburg administration was thus left in place but rendered leaderless. The focus of power had shifted elsewhere, to the directory, placed under the control of the erstwhile superintendent (*Berghauptmann*) of Saxon mines, Baron Schoenberg, and staffed by the group of German mining specialists who had accompanied him on his journey from Freiberg to Russia in 1736. These men were to be the agents of 'reform', directly accountable to Schoenberg and with wide, if ill-defined, powers they were to bring a new rationality to the management of the state's mining and metallurgical enterprises.

Initially, from 1736 to 1738, their work was largely investigative, the circulation of a general seventeen-point questionnaire concerning the state of the industry being backed up by the despatch of investigative missions manned by German staff officers. Then, in 1738/9, the real work of the directory began, but not before members of the old administration attempted to block its programme of 'reform'.[23] To pave the way for change on 31 May 1738 an ukaz created yet another commission, the third since 1733, again under the direction of Shafirov, to prepare a policy document outlining the main features of 'reform'. On the basis of information gathered by Schoenberg's agents, it reported, perhaps not surprisingly in the prevailing crisis conditions, in favour of 'privatization' of the state's works, but reserved its position on the new mines discovered in 1735. The reasons for its hesitancy lay in its receipt of a report from Tatishev (compiled in the boom year 1735) describing the great riches of the mines and calling into question the veracity of much of the evidence placed before them. To resolve their dilemma the members of the commission directly contacted the Ekaterinburg Chancellery for information. The response they received was clear: it was impossible to verify Tatishev's figures and to their knowledge the Blagodat works (in the crisis year 1738/9) was making no profit whatsoever. Yet in spite of the evidence placed before them they persisted in their position forcing the cabinet to exclude the '1735 mines' from the general decision to 'privatize' the state works until Schoenberg had personally appeared before the commission to provide information on the matter. His arguments, as might be expected, stressed the importance of introducing 'superior' German management in place of the incumbent 'simple and incompetent' Russian management whose retention 'would be a source of numerous difficulties'. In response to the director's overpowering logic, or perhaps only in respect of his authority, the members of the commission fell into line. The amended report passed to the Empress and, on her acceptance of it, a new two-part manifesto was published on 3 March 1739 providing the basis for 'reform'.[24] With respect to the controversial mines, which had been discovered in 1735, these were immediately sold off, their assets passing to a specially created 'Mining Company' (*Berg-Kompaniya*). In relation to the second objective, namely the 'reform' of the state works as a prelude to their sale, new mining regulations were drawn up and a new mining administration created to implement them. Within little more than a week of the formulation of the manifesto, on 11 March 1739, about forty German staff officers of the directory were despatched via Perm to Ekaterinburg . On arriving at the end of July, after a four-and-a-half-month trip, eleven of their number continued their journey to Nerchinsk and the rest turned to accomplishing their task of 'reforming' of the Urals works.[25] For three years (1739–41) two separate administrations existed within the state mining enterprises: the members of the Ekaterinburg Chancellery responsible for the day-to-day running of the state works and the teams of German 'trouble shooters', directly responsible to Schoenberg and independent of the local administration, who, pursuing a peripatetic course set about establishing new working practices in the state plants.

Unlike earlier, however, the competence of these officials extended only to production matters. For three years, 1738/9–1740/1 marketing was shorn from their grasp and put into the hands of the 'Mining Company', which paid producers the prevailing market price of 58k. a pud for iron, *on a guaranteed basis*, and, engrossing all state output, became the sole party negotiating sales with either government departments or foreign merchants. In spite of the high guaranteed price paid to producers, this does not seem to have been an unhappy arrangement, for during the years of its contract the company seems to have been able to maintain high levels of sales, far higher than was the case under English concessionaries before or after. For almost half a decade mining administration thus underwent major changes against a background of recurrent economic crisis. It is not of primary importance here whether in these changes true administrative virtue resided in either 'arrogant' Germans or 'ignorant' Russians. What is important is that the administrative tradition established in 1720 was, for good or ill, interrupted. A new administrative model had been created but its life was a short one. The whole structure was swept away in that wave of xenophobia which engulfed administrative circles following the fall of Biron and the arrest and deportation of Schoenberg. In 1741 the General Directory of Mines was suppressed and the Mining College reinstituted. The mines of Blagodat and Arkhangelsk, which had passed to the 'Mining Company', were resumed. The cadres of German technicians were dispersed but not before they were brought under the authority of the Mining College and the local administrations dependent upon it.[26]

From 1741 to 1782 the Mining College once more reigned supreme. On hearing of the discovery of gold and silver – at Voits in 1744, Kolyvan–Voskressensk in 1746, and Beresov in 1754 – its officials once again seized the mines for the Crown and once again instituted a new mining regime, similar to that created at Nerchinsk in 1720.[27]

Theoretically, Catherine II's mining regulations promulgated in 1782 altered this situation completely as property rights were extended to all prospectors discovering gold or silver subject only to their delivery of a tithe of production to the Crown. Yet much to the astonishment of Storch writing at the end of the century this clause remained a dead letter, for since its enactment he reported that 'nobody has begun to exploit silver or gold mines and no individual possesses such mines'.[28] Production of precious metals remained concentrated in mines discovered before 1782 whose officials, as a result of the enactment of that year, merely experienced a change in management at the top as they passed from the jurisdiction of the Mining College to that of the Cabinet. Henceforth, from 1782 until 1796, all local mining officials in charge of mines producing precious metals, who continued to operate much as before, were responsible to a special sub-committee of the Cabinet presided over by a major-general and comprising two councillors assisted by the chancery staff attached to the committee.[29]

Under Paul (1796–1801) they once more passed under the control of the Mining College but not for long. Even before his successor suppressed that institution in 1805, transferring most plants to the newly created Ministry of Finance, he had once again (in 1801) placed the mines of gold and silver within the jurisdiction of the Cabinet.[30]

Throughout the eighteenth century, therefore, local mining officials existed in a relatively stable administrative structure extending from the individual mine or smelting establishment through the district offices to the head office of the jurisdiction. Each element within this local administration contributed to providing the officials at the head office with an overall picture of the working of the enterprise, both as it was perceived from the bottom, embodied in the particular accounts of the district officers (documentary classes i–iv) synopsized in the chancery of the head office, and as it was perceived from the top through the periodic reports of the mine jurors and the plans drawn up by the surveyors' office (documentary classes v–vi). In their relations with the central administration at St Petersburg, however, the officials of the local chancelleries were confronted with an administrative body which both structurally and in terms of its personnel was constantly changing. Fortunately, however, because of the functional relationship existing between the two the central administration had access to such information as it might require to make up for deficiencies in its own archives resulting from the process of change. In relation to non-state enterprises, the central authority essentially regulated prospecting activity and licensed works as well as resolving conflicts within the industry. Much of the documentation associated with these activities was accordingly generated at the centre and had no counterpart in the field offices of the private entrepreneurs; it was thus irreplaceable if lost. Such was not the case with respect to state works where it acted in the capacity of senior management for the enterprise. Moreover, in carrying out this function it was hampered by the fact that, save for the years 1735–41, it had no direct access to information concerning the operation of its plants and was forced always to operate through its sub-agencies whose records, as has been suggested, existed in a long-term stable administrative environment. Whether autonomously generated from below, or created in response to orders from above, therefore, the documentation stored in the chancery of the central administration tended to replicate or synopsize that held in the local chancelleries which provided the necessary information for the executive decisions made by the central board and embodied in administrative orders.

An ordered and complex series of archives at district, *nachal'stva*, and national levels thus contained comprehensive information concerning the operations of those state plants producing precious metals. Sadly, judging by the investigations of Pavlenko, Danilevskii, and Karpenko concerning the Kolyvan–Voskressensk works, very little of this documentation now survives.[31] The records of the head office of the mining region, now preserved in the State Archive of the Altai Krai (*Gosudarstvennogo arkhiva Altaiskogo kraya = GAAK*), encompass the archive of the chancellery of the Kolyvan–Voskressensk mining region (ibid., fond 1) as well as those of the nineteenth-century mining section (ibid., fond 2) and section for technical affairs (ibid., fond 3). In addition there is a miscellaneous group of documents (ibid., fond 50) which seems to represent the archive of the surveyor's department. The modern archive thus seems ordered in the same form as the archives of the eighteenth- or nineteenth-century mining office but as Karpenko notes 'not a great number of documents survive' and those that do are somewhat arbitrarily located. Moreover, loss has been selective. That vast body of registers containing the particular accounts of the mines or the synopses made from them is represented only by a number of workers' lists (fond 1, dok. 219.1759; dok. 660a.1778; dok. 89.1798 as well as a misplaced list of 1850, fond 1 dok. 480; fond 2, dok. 387 1828–31; dok. 88. 1850) and one or two production registers (fond 1, dok. 25. 1789–97; fond 2, dok. 38. 1833). Much better preserved are those reports, compiled at the behest of members of the chancellery, on technical aspects of the mines and smelters or on conditions amongst the workforce, and the appendent plans and surveys drawn up by the surveyor's office which after 1834 were augmented by materials collected by the group of staff mining engineers under general Saint-Al'degond sent to the Altai to reform the works. Certainly such reports and appendent plans, like those concerning the Gurev (fond 1, dok. 684 misplaced and fond 2, dok. 2709 & 480 sd., fond 3, dok. 1490. 1839; fond 50, dok. 4534. 1848) and Loktev smelters (fond 2, dok. 1392. 1850) or the Zmeinogorsk (fond 50, dok. 1432. 1772; ibid., dok. 1371. 1799; fond 2, dok. 38 sd), Salairsk and Zyryanov mines, comprising documentary categories v–vi above, are of considerable technical interest but the loss of the particular accounts of these, and other works, documentary categories i–iv above, is to be especially regretted.

Nor are the records preserved in the archives of the central administrations – the Mining College, located in the Central State Archive of Ancient Acts in Moscow (*Tsentral'nogo gosudarstvennogo arkhiva drevnikh aktov = TsGADA*, fond 271), and the Cabinet Office, dealing with imperial enterprises in the Altai, located in the Central State Historical Archive in Leningrad (*Tsentral'nogo gosudarstvennogo istoricheskogo arkhiva v Leningrade = TsGIA-L*, fond 468) – capable of remedying this deficiency, although in relation to the synoptic reports compiled in the principal mining offices a considerable number survive as well as documents compiled from them in the chanceries of the central administrations. Few of these documents received or compiled at the Mining College are now preserved in its archive but some idea of its former glory can be seen from the special report compiled in 1747/8 by its officials for the Senate and preserved in its archive subsequently to be published by: N.I. Pavlenko, 'Materiali o razvitii ural'skoi promyshlennosti v 20–40kh godakh XVIII v.', *Istoricheskii Arkhiv*, IX (1953), pp. 175–282. TsGADA f. Senata, dok. 1510 fo 370–399^v.

Far greater numbers exist in the archive of the Cabinet office which contains both statistical abstracts, estimates, reports, and information from the chancery of the Kolyvan–Voskressensk mining office and that of the Altai mining okrug and documented details, assembled by the chancery of the Cabinet office, on the Cabinet plants and mines, their production, expenses, details of the labour force, and their wages. Yet judging by the use Soviet scholars have made of these documents certainly too few exist to construct a production series for any of the mining enterprises.

It is thus particularly fortunate that there has come down to us the works of two late-eighteenth-century mining officials – H.M. Renovantz, *Oberbergmeister* of the Kolyvan mining region, inspector and teacher in mining sciences at the imperial Mining School in St Petersburg, and B.F.J. Hermann who held the same office from 1785 to 1796 before being moved to Ekaterinburg – who indulged a profound interest in the history and contemporary condition of their industry.

These works are: H.M. Renovantz, *Mineralogisch – geographische und andere vermischte Nachrichten von den Altaischen Gerburgen russische kanserlichen Anteil* (Reval, 1788), and the Russian translation of this work, *Mineralogicheskie, geographicheski i drugie smeshanie izvestiya o Altaiskikh gorakh* (St Petersburg, 1792). B.F.J. Hermann, *Versuch einer mineralogischen Beschreibung der uralischen Erzgeburges*. 2 Bd (Berlin and Stettin, 1789), published after his second tour of duty, 1785–9.[32] He subsequently made a further tour of duty, 1790–6, and on returning published: *Mineralogische Reisen in Sibirien vom Jahr 1783 bis 1796* (St Petersburg, 2 vols, 1797–8), and *Sochineniya o sibirskikh rudnikakh' i zavodakh'* (St Petersburg, 3 vols, 1797–1801) incorporating materials collected in 1796/7 updating his earlier data published in 1790. After his appointment at Ekaterinburg he published *Die Wichtigkeit des russischen Bergbaues* (St Petersburg, 1810), and *Istoricheskoi nachertanie gornogo proizvodstva v rossiiskoi imperii* (Ekaterinburg, 1810), of which only one volume was published relating the history of the industry to 1740.

During their terms of office at Kolyvan–Voskressensk as senior mining officials they clearly had access to a chancery archive which had complete sets of particular accounts relating to specific mines and smelting establishments stretching back in time to the 1740s. Hermann, moreover, on an inspection tour of the Nerchinsk works in the early 1790s found a similar set of records at the principal mining office at Staroi Nerchinsk. It is these materials which provide the primary source for their surveys of the workings which often include transcripts taken from the registers.

What was clearly not available to them at this time were the synoptic tables of aggregate production of unrefined silver (blicksilver), Renovantz not including this data at all and Hermann when first compiling his general study, which was published in 1790, being forced to rely on the somewhat inaccurate published data used by Coxe.[33] Possibly only in 1789, on his return from his second tour of duty, did Hermann obtain access to this information, preserved in the chancery of the mining section of the Cabinet office, introducing it as a late insertion in his 1790 book. Subsequent visits to that archive, moreover, allowed him to update this data in 1796/7 and in the 1800s for inclusion in works published in 1797/8, which are the sources of Storch's information, and 1810. Only in the 1800s, however, does he seem to have gained access to the archive of the procurator general's office allowing him to produce the complete tables of unrefined *and* refined output published in 1810.

The basis of the information concerning aggregate production levels within the enterprises comprising the Russian precious metal industries was thus only from 1789 the original registers relating to state works preserved at St Petersburg which provided Hermann with comprehensive data on these enterprises whilst they were in the state's hands (viz. Nerchinsk from 1704, Kolyvan from 1746, and the Ekaterinburg gold works from 1754).[34] What is not in these archives, and thus was not available to him in St Petersburg, was information concerning the level of production at Kolyvan–Voskressensk when that enterprise was in the hands of the Demidovs (viz. 1726–46). Nor have modern Soviet historians been any more successful in filling this gap by locating the archive of the Demidovs' principal mining office at Kolyvan in spite of the exhaustive investigation into the records of the family by the late Professor B.B. Kafengauz.[35] Both he and Z.G. Karpenko were, accordingly, forced back onto the use of documents generated by the state either during investigations into the family's affairs in 1733–5, culminating in the Shafirov commission, and 1745–6, constituting the Beyer commission, or, during the period when the confiscated works were in the state's hands in 1735–7, to describe the history of the mines in this period. Nor is this absence of the records of the Demidovs' principal mining office from amongst the family papers entirely surprising, for the whole archive passed intact to the chancery of the Kolyvan *nachal'stva* when the works were taken over by the state in 1746 and, like so many other records lodged there, the archive has since been lost without trace. In the 1780s, however, the papers were still located in the chancery of the Kolyvan–Voskressensk *nachal'stva*, at that time situated in Barnaul and once again they are revealed to us through the eyes and by the pens of Renovantz, Hermann, and others who were afforded access to them. Thanks to these indefatigable workers in the cause of science, who whiled away their spare time at Barnaul perusing the 'ancient' papers preserved in the chancery, and on returning to the capital preferred to rummage through the papers of the Cabinet office rather than to indulge in the more sensuous pleasures of the city, therefore, it is now possible to construct complete series for the production of gold and silver in Russia during the eighteenth century, saving only the output of the short-lived Arkhangelsk mines.

Silver production

Nerchinsk (table A1.1)

Most silver at Nerchinsk was obtained from argentiferous lead ores excavated from mines located in the valley of the Arguin (table A1.1–2 cols 1–2). The extraction of the metal involved the reduction of the ore and the production of an argentiferous lead described as 'fertile' (i.e. *viplavleno iz serebryanikh rud svintsa, nazivaemogo sirogo, is kotorogo otdelyetcya serebro*). This was then placed in a cupelation hearth and subjected to an oxidizing blast which converted the greater part of the lead into lithage leaving in the hearth a residual deposit of hearth lead (*herdblei-gert*) and the 'unclean' silver (blicksilver) still containing traces of other metals (i.e. *ostavshago ot razdeleniya serebra gletu i gertu, soderazhavshago v sebe svinets*) and requiring further refining (cols 3–4). The lithage was then resmelted producing a lead described as 'sterile' (i.e. *is togo viplavleno svintsa chistogo*) whilst the silver was transported to St Petersburg. Here it passed out of the jurisdiction of the mining authorities and was delivered to the laboratory for the separation of gold and silver in that city which was under the jurisdiction of the procurator general, from which refined gold and silver was delivered to the mint (cols 5–8).

Table A1.1 Nerchinsk silver and gold production, 1704–1809 (in puds and funts)

	(1) Ore smelted	(2)	(3) Blicksilver	(4)	(5) Silver	(6) Refined	(7) Gold	(8)
1704			0	1				
5			1	22				
6			3	19				
7			5	7				
8			5	26				
9			2	3				
1710			8	3				
1			8	14				
2			11	4				
3			11	26				
4			11	30				

Appendix 1 Precious-metal and copper production

	(1) Ore smelted	(2)	(3) Blicksilver	(4)	(5) Refined Silver	(6)	(7) Refined Gold	(8)
5			2	16				
6			12	3				
7			15	13				
8			10	9				
9			5	6	280	7 (1704–1747)	1	4
1720			4	1				
1	35,296	0	5	32				
2	46,974	0	10	4				
3	47,480	0	7	17				
4	64,864	0	6	10				
5	21,944	0	3	9				
6	10,852	0	1	15				
7	4,008	0	0	4				
8	5,422	0	0	23				
9	5,115	0	1	21				
1730	115	0	0	35				
1	nothing		nil					
2	smelted		nil					
3	560	0	nil					
4	1,667	0	0	28				
5	2,000	0	1	14				
6	2,680	0	2	24				
7	8,280	0	4	4				
8	8,300	0	3	19				
9	15,161	0	7	24				
1740	41,099	0	13	17				
1	39,882	0	12	20				
2	119,608	0	9	32				
3	78,087	0	15	0				
4	91,349	0	14	37				
5	36,147	0	16	39				
6	167,275	0	16	3				
7	112,429	0	35	2				
8	166,579	0	71	6				
9	189,799	0	82	7				
1750	116,540	0	81	3				
1	50,860	0	39	27				
2	77,182	0	51	7	1343	12	5	11
3	3,290	0	100	25	(1748–59)			
4	96,000	0	51	35				
5	115,600	0	139	16				
6	341,894	0	126	16				
7	413,000	0	100	2				
8	410,450	0	134	7				
9	463,253	0	137	7				
1760	606,366	0	149	14			0	18
1	487,164	0	143	14			0	15
2	628,707	0	176	3	1770	34	0	27
3	765,177	0	306	0	(1760–5)		1	14
4	1,147,450	0	399	9			1	39
5	1,249,600	0	298	1			1	7
6	1,110,010	0	320	30	286	18	1	4
7	1,007,430	0	435	35	401	0	1	25

Table A1.1 (*cont.*):

(1) Ore smelted	(2)	(3) Blicksilver	(4)	(5) Silver	(6) Refined	(7) Gold	(8)
8	952,640	0	343	31	310	2	1	7
9	565,324	0	312	21	300	6	1	3
1770	950,324	0	414	26	384	36	1	17
1	1,521,985	0	470	6	425	34	0	32
2	1,699,020	0	503	36	452	20	0	34
3	1,843,176	0	523	28	480	21	1	17
4	1,763,980	0	619	2	574	34	0	36
5	912,092	0	539	14	500	18	2	4
6	914,006	0	398	39	382	11	1	8
7	1,513,771	0	323	22	303	12	1	17
8	1,912,659	0	380	39	353	16	2	8
9	1,977,626	0	349	16	317	29	1	10
1780	1,926,012	0	452	28	415	39	1	8
1	1,597,273	0	396	28	368	27	1	9
2	1,919,080	0	470	37	423	21	1	18
1783	1,813,535	0	507	37	442	10	0	39
4	1,858,457	0	457	17	422	17	1	25
5	1,333,406	0	288	36	271	19	1	4
6	1,251,373	0	340	31	318	36	0	33
7	1,184,794	0	320	8	305	26	0	37
8	961,084	0	281	34	260	14	0	37
9	925,399	0	314	6	274	3	0	34
1790	912,498	0	218	37	191	5	1	1
1	975,748	0	278	1	242	27	1	2
2	858,001	0	263	0	229	3	0	27
3	791,889	0	237	36	206	7	0	29
4	688,437	0	256	3	225	0	0	31
5	632,444	0	255	4	224	37	0	21
6	677,752	0	251	29	236	2	0	36
7	636,690	0	251	13	237	11	0	34
8	948,572	0	335	6	317	24	1	0
9	984,366	0	325	20	304	1	1	1
1800	961,403	0	222	1	210	5	0	37
1	995,427	0	240	15	224	22	0	36
2	908,264	0	203	39	193	13	0	30
3	790,767	0	202	8	192	3	0	31
4	931,752	0	152	5	214	4	1	2
5	1,005,136	0	180	33	245	9	1	2
6	825,954	0	175	26	231	7	0	35
7	855,005	0	195	29	255	33	0	33
8	800,376	0	177	14	233	4	0	28
9	774,390	0	184	0	235	6	0	31

Source: Hermann (1810).
Note: From 1804 much of the silver-bearing ('fertile') lead smelted remained unrefined as a backlog of blicksilver despatched to St Petersburg was cleared. Silver-bearing lead. Stocks of unrefined metal:

1804	34,542
1805	36,795
1806	25,432
1807	25,012
1808	25,002
1809	25,006

Kolyvan–Voskressensk (table A1.2)

Production in the Altai evolved in a quite different manner from in the Nerchinsk mines. Initially, from 1729 to 1731/2, the works, located on the banks of Lake Kolyvan, were equipped only with smelting and refining hearths producing respectively black and refined copper (see table A1.6). Nor did this situation change significantly with the rebuilding of the plant after the disastrous fires of 1731/2. As new mines were opened up and a new smelting works established at Barnaul in 1740 the primary focus of activity remained the production of copper, smelting capacity remaining at five hearths and refining at two of Kolyvan from 1729–31/2 and 1734–9 before there was an extension of capacity to seven smelting and nine refining hearths at Kolyvan and two smelting and four refining hearths at Barnaul during the years 1740–6. These smelters and refining hearths, during the Demidov period (1726–46) were largely employed in processing high-grade copper ores capable of yielding 8–25 per cent unrefined, 7–8 per cent refined copper, but so poor in silver, 0.15–0.34 zolotniks per pud, that the metal was not worth extracting.

At this time only the ores of the Voskressensk mine, which were poor in copper, were found in 1733 to be rich in silver, yielding up 1.5 zolotniks of silver per pud of ore. Whilst rumours concerning the precious-metal content of these ores continued to circulate within and outside the enterprise, however, it was not until the 1740s that supply problems were resolved and capacity built to extract the silver from the copper with the creation of two *Saiger* hearths at Kolyvan. Henceforth, after roasting and

Table A1.2 Kolyvan–Voskressensk silver and gold production, 1743–1809 (in puds and funts)

	(1) Ore smelted	(2)	(3) Blicksilver	(4)	(5) Silver	(6) Refined	(7) Gold	(8)
1743	3,000	0	1	6				
4								
5	50,000	0	44	6	41	19	0	15
6								
7	212,427	0	237	13	205	4	3	37
8								
9	204,294	0	309	22	268	28	6	21
1750	142,122	0	209	29	179	11	4	18
1	231,406	0	366	34	309	27	7	24
2	150,691	0	304	25	266	10	8	15
3	281,254	0	310	26	264	15	10	12
4	209,058	0	334	19	280	3	10	14
5	253,511	0	303	21	269	36	12	3
6	305,809	0	321	17	290	28	10	19
7	154,514	0	222	30	202	24	7	1
8	346,255	0	264	2	237	21	9	16

Table A1.2 (*cont.*):

	(1) Ore smelted	(2)	(3) Blicksilver	(4)	(5) Silver	(6) Refined	(7) Gold	(8)
9	336,549	0	273	10	248	24	8	28
1760	171,091	0	264	22	244	14	8	31
1	560,014	0	333	20	314	36	8	17
2	219,015	0	405	27	372	4	12	23
3	402,244	0	499	24	450	31	17	24
4	454,352	0	421	9	381	18	17	0
5	531,899	0	575	34	520	39	21	5
6	544,065	0	767	14	678	13	27	14
7	465,576	0	779	19	716	1	24	33
1768	547,645	0	741	22	690	25	21	0
1769	707,680	0	809	18	761	15	26	19
1770	1,111,304	0	1,013	11	950	29	33	31
1	1,510,491	0	1,268	19	1,189	9	39	39
2	1,026,204	0	1,277	34	1,182	29	50	31
3	1,200,638	0	1,181	12	1,103	25	42	28
4	1,375,792	0	1,146	24	1,064	32	45	29
5	1,377,096	0	1,059	32	996	14	38	34
6	1,607,229	0	1,027	20	996	12	37	23
7	1,537,147	0	1,035	26	980	27	33	17
8	1,482,299	0	913	0	863	2	27	34
9	1,654,767	0	809	11	763	6	24	31
1780	1,793,518	0	802	30	755	32	26	2
1	1,136,745	0	546	14	507	17	13	36
2	938,038	0	400	13	375	2	12	32
3	1,485,364	0	730	24	675	9	22	14
4	1,134,645	0	517	27	475	37	14	37
5	1,850,733	0	600	21	543	34	13	38
6	1,743,068	0	771	20	698	0	20	14
7	1,621,193	0	775	18	707	13	20	19
8	2,028,691	0	868	13	793	7	22	1
9	2,196,539	0	1,050	9	962	16	24	29
1790	2,177,081	0	1,054	33	968	25	20	35
1	1,996,583	0	1,052	25	968	24	19	20
2	2,067,079	0	1,022	38	946	15	18	38
3	1,886,396	0	1,011	22	932	18	22	37
4	1,913,158	0	1,042	6	958	15	22	14
5	2,357,737	0	1,018	10	935	15	20	38
6	2,483,114	0	1,026	14	950	17	18	26
7	2,918,513	0	1,064	5	983	9	20	9
8	2,821,142	0	1,069	20	991	6	20	20
9	2,941,064	0	1,100	13	1,029	19	21	1
1800	2,946,187	0	1,130	16	1,034	26	20	7
1	3,162,549	0	1,151	17	1,065	15	19	6
2	3,078,571	0	1,170	39	1,076	17	23	32
3	3,170,198	0	1,204	32	1,099	30	21	24
4	3,365,732	0	1,170	28	1,043	21	22	12
5	3,671,979	0	1,150	14	1,038	33	22	8
6	3,543,011	0	1,151	23	1,064	12	22	25
7	3,248,267	0	1,088	31	991	37	20	27
8	3,662,663	0	1,145	35	1,035	7	23	5
9	4,208,306	0	1,010	7	909	11	21	29

Table A1.3 Summary of available statistics of silver production, 1704–89 (in puds and funts)

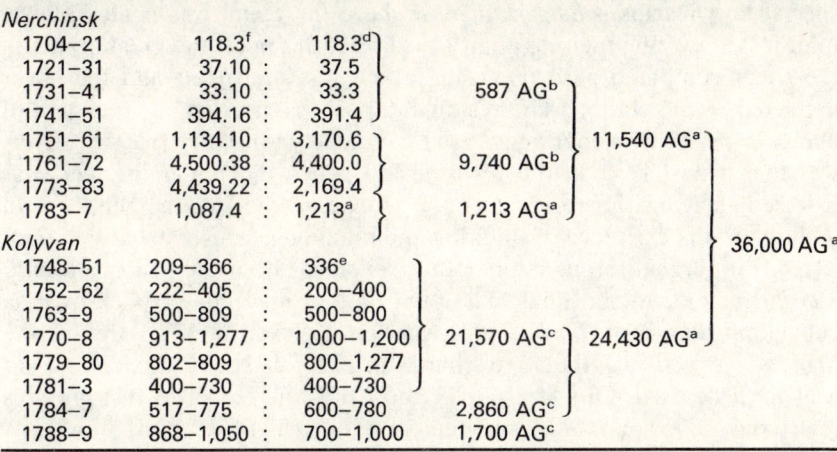

Nerchinsk					
1704–21	118.3[f] :	118.3[d]			
1721–31	37.10 :	37.5			
1731–41	33.10 :	33.3	587 AG[b]		
1741–51	394.16 :	391.4			
1751–61	1,134.10 :	3,170.6		11,540 AG[a]	
1761–72	4,500.38 :	4,400.0	9,740 AG[b]		
1773–83	4,439.22 :	2,169.4			
1783–7	1,087.4 :	1,213[a]	1,213 AG[a]		
Kolyvan					36,000 AG[a]
1748–51	209–366 :	336[e]			
1752–62	222–405 :	200–400			
1763–9	500–809 :	500–800			
1770–8	913–1,277 :	1,000–1,200	21,570 AG[c]	24,430 AG[a]	
1779–80	802–809 :	800–1,277			
1781–3	400–730 :	400–730			
1784–7	517–775 :	600–780	2,860 AG[c]		
1788–9	868–1,050 :	700–1,000	1,700 AG[c]		

Sources and key:

[a] H. Storch, *Tableau historique et statistique de l'Empire de Russie à la fin du dix-huitième siècle* (Basle & Paris, 2 vols, 1801).

Total	1704–88	36,000 pud ag,	ibid., II p. 380
Kolyvan	1745–88	24,430 pud ag,	ibid., II, p. 370
therefore Nerchinsk	1704–88	11,540 pud ag,	derived from above
but Nerchinsk	1704–83	10,327 pud ag,	ibid., II, p. 380n
therefore Nerchinsk	1783–8	1,213 pud ag,	derived from above

[b] ibid., p. 380n.

[c] ibid., p. 370n.

[d] W. Coxe, *Travels into Poland, Russia, Sweden and Denmark. Interspersed with Historical Relations and Political Enquiries* (London, 2 vols, 1784), p. 286. Note that the continuous series of decennial total only extends to 1761. The 'decade' 1761–72 is represented by data for three years (1767, 1771–2) only, although Coxe felt that 'the average produce may be estimated at 16,000 lbs (= 400 Puds)'.

[e] ibid., p. 284. This data is presented by each of the authors in the form of minimum–maximum output figures and is organized by each of them on a differing chronological basis. The series displayed here for the period 1752–78 are from Coxe, p. 284, and 1770–1789 Storch, p. 379n, which is also reproduced in B.F.J. Hermann, *Statistische Schilderung von Russland in Rucksicht auf Bevoelkerung Landesbeschaffenheit, Naturprodukte, Landwirthschaft, Bergbau, Manufakturen und Handel* (St Petersburg and Leipzig, 1790), pp. 321–2, but is followed by a non-derivative and perhaps latterly inserted entry:

1748–51	General Beyer	200–366 puds
1752–69	Lt General Poroschin	222–809 puds
1765–9		600–800 puds
1770–80	Lt General Irmann	800–1,277 puds
1780–5	Lt General Muller	400–730 puds
1786–7	Privy Councillor and Knight Kaschka	750–80 puds
1788		870 puds
1789		1000 puds

which seem to be derived from materials collected for his studies published in 1797–8.

[f] B.F.J. Hermann, *Die Wichtigkeit des russischen Bergbaues* (St Petersburg, 1810). Kolyvan: table 8aa (1745–80), ibid., bb. (1780–1809). Nerchinsk, table 9cc (1704–37), ibid., dd (1738–72), ibid., ee (1773–1806), ibid. ff (1806–9).

smelting the ore yielded up an argentiferous copper which was subsequently leaded. This involved the mixing of silver-rich 'black' copper with lead in special low hearths – *Saigerhütten* – where the metal was heated slowly until just above the melting point of lead. At this point the greater part of the silver combined with the lead, which was drawn off and the silver extracted by cupelation. The residual deposit, comprising an amalgam of about 75 per cent copper and 25 per cent lead with slight traces of silver, was then transferred to the *Daarofen* in which the remaining lead was oxidized, leaving the residual copper to be smelted in the refinery. In addition to the existence of smelting and refining hearths, the extraction of silver from argentiferous copper thus involved the use of *Saiger*-hearth, *Daarofen*, and, in the final extraction of the 'unclean' or blicksilver, a cupelation hearth or *Treibofen*, created at Kolyvan in 1743. Before the plant was seized into the state's hands in 1746, however, this equipment was but little used. Of some 42,042 puds (675 tons) of silver-bearing ores raised only 3,000 puds were smelted, yielding in 1743–4 1 pud, 6 funts, 84 zolotniks of silver.[36] During the Demidov epoch from 1726 to 1746, therefore, the primary focus of production was copper, only very diminutive supplies of silver being produced.

Such was not the case when the enterprise passed into the state's hands. Copper sales were suppressed and from 1746 until the establishment of copper minting at Susunsk in 1764 the metal made as a by-product of silver production, which could not be sold locally or used at the works, was stockpiled. Silver production, on the basis of ores (table A1.2, cols 1–2) increasingly drawn from the newly-discovered Schlangenberg (Zmeinogorsk) mine, however, rapidly increased, the extraction of the white metal at a fast proliferating number of works yielding up quantities of 'unclean' or blicksilver (cols 3–4) which twice a year were despatched to St Petersburg. Here, like the Nerchinsk blicksilver, it passed out of the jurisdiction of the mining authorities and was delivered to the imperial laboratory for refining and for the separation of the gold and silver in that city which was under the jurisdiction of the procurator general, from which refined gold and silver was delivered to the mint (cols 5–8).

Gold production

In compiling statistics of Russian gold production in the eighteenth century it must be noted that by far the largest amount of that metal produced was derived from the auriferous blicksilver despatched from Nerchinsk and Kolyvan for separation and refining at the imperial laboratory at St Petersburg (tables A1.1–2 above). Other than this gold supplies derived from intermittent production at Voits and continuous and large-scale production at Ekaterinburg.

Voits, Olonets province

The most ancient gold mine located between Lake Onega and the White Sea, its chief product was a violet pyritical copper mixed with quartz containing gold, which was either washed from the mine or found in pieces. The mine was initially worked for copper and only in 1744 was gold discovered.

Table A1.4 Voits gold production, 1744–94 (in puds and funts of 'legatur', or unrefined gold)

1744–68	1.22	1775	0.16	1780	0.08
		1776	0.18	1781	0.03
1772	0.04	1777	0.22	1782	0.08
1773	0.11	1778	0.08	1783	0.01
1774	0.11	1779	0.15		
				1794	0.02

Source: V.V. Danilevskii, *Russkoe zoloto* (Moscow, 1959), pp. 35–9.

Ekaterinburg gold works

The principal Urals gold fields comprised the Beresov and Chussovsk works in close proximity to Ekaterinburg, and there were lesser workings at Miask and Neviansk. At the former centre, which was by far the most important, the deposit comprised an ore, commonly in a cubic form in a quartz matrix, from which the gold was obtained by washing.[37] This involved the collected ore (table A.5, cols 1–2) being poured into long troughs and water added. The ore was then beaten with iron stampers, the water which continually flowed into the trough carrying off the fine powder over the washing tables which were laid out like slightly inclined terraces below the trough. Many of the heavier grains of gold fell into the interstices of the double iron bottom of the stamping trough and were collected from time to time. As was usual with poor sands, like those of Beresov yielding c. 0.17–1.02 oz/ton, they were often raked upward on the benches with a wooden rake. This constituted the first straining. The richest part of the strained ore, which remained upon the upper benches, was then well washed again in large receptacles as its weight prevented it being carried off, but the poorer and finer part was again exposed to the atmosphere and washed a second time on little tables to which water was brought through pipes which could be directed in many directions. Various methods were then employed to effect the final separation of the gold from the quartz. Until the 1790s it was usual either to continue washing the ore or to refine it with lead, the fine grains of gold/quartz being heated with the base metal until the gold was taken up by the lead, which was then ready for separation by cupelation. From the 1790s, however, experiments were

Table A1.5 Ekaterinburg gold production, 1754–1809 (in puds and funts)

	(1)	(2)	(3)	(4)	(5)	(6)	(7)	(8)
	Ore crushed and washed		blick- or legatur		Refined			
			gold		gold[a]		silver	
1754	61,069	0	0	17	0	15	0	2
5	30,428	0	0	9	0	8	0	1
6	5,841	0	0	2	0	2	0	0.2
7	86,436	0	0	29	0	25	0	3
8	55,365	0	1	2	0	36	0	5
9	105,839	0	1	0	0	35	0	5
1760	229,215	0	1	22	1	15	0	7
1	195,859	0	1	23	1	13	0	7
2	296,484	0	1	18	1	10	0	7
3	162,328	0	2	2	1	30	0	11
4	200,534	0	3	18	2	38	0	18
5	213,626	0	3	9	2	33	0	15
6	180,003	0	5	10	4	23	0	24
7	202,800	0	4	25	4	01	0	19
8	198,800	0	4	35	4	10	0	22
9	217,601	0	4	11	3	16	0	20
1770	283,939	0	4	39	4	26	0	8
1	291,026	0	3	27	3	16	0	10
2	284,643	0	2	28	2	21[b]	0	5
3	262,931	0	2	33	2	24	0	9
4	269,834	0	3	5	2	39	0	5
5	326,926	0	4	2	3	32	0	8
6	284,528	0	5	13	5	5	0	7
7	331,393	0	5	23	5	16	0	5
8	420,625	0	5	38	5	33	0	5
9	494,399	0	6	4	5	36	0	5
1780	407,720	0	6	16	6	6	0	8
1	302,624	0	4	5	3	37	0	6
2	287,439	0	3	37	3	31	0	5
3	279,911	0	3	18	3	01	0	15
4	389,557	0	4	11	4	06	0	04
5	318,457	0	3	33	3	28	0	02
6	328,811	0	7	28	7	19	0	05
7	361,388	0	8	05	7	35	0	05
8	401,050	0	7	32	7	16	0	10
9	483,915	0	8	07	7	30[c]	0	10
1790	484,779	0	7	34	7	20	0	09
1	498,728	0	7	36	7	13	0	16
2	428,621	0	9	00	8	16	0	19
3	442,255	0	7	22	6	25	0	34
4	449,089	0	8	10	7	26	0	17
5	494,113	0	10	33	10	13	0	14
6	481,821	0	8	34	8	17	0	11
7	685,306	0	11	06	10	18	0	23
8	788,570	0	14	22	13	29	0	27
9	962,336	0	15	22	13	24	1	22
1800	1,218,849	0	18	4	15	38	1	39
1	1,406,122	0	18	24	17	16	1	01
2	1,069,469	0	14	16	12	26	1	24
3	1,120,578	0	16	14	14	14	1	26
4	1,764,613	0	19	24	17	10	1	37
5	1,722,242	0	16	13	14	03	1	30

	(1) Ore crushed and washed	(2)	(3) blick- or legatur	(4)	(5) Refined	(6)	(7)	(8)
				gold	gold[a]		silver	
6	1,851,670	0	17	15	14	34	1	37
7	1,558,400	0	18	36	15	37[d]	2	12
8	1,499,451	0	22	2	18	19	2	25
1809	1,607,914	0	22	3	22	21	2	14

Notes:
  [a] The five-yearly figures quoted in V.V. Danilevskii, op. cit., pp. 32, 51 incorporated in A. Kahan, op. cit., table 3.6 are derived, as is the table above, from B.F.J. Hermann, *Die Wichtigkeit*..., table 1, and differ from B.F.J. Hermann, 'Sur l' exploitation . . .', which refers to unrefined rather than refined gold.
  [b] W. Coxe, op. cit., p. 284, 'Annual produce (of refined gold) is never more than 200lb and in 1772 it was only 101 lbs.
  [3] H. Storch, op. cit., II, p. 377: 'The mines have furnished annually, 3, 4, 5, 6 and of late 7 puds of gold. From its beginning in 1754 to 1788 it has produced 120 puds worth 1,198,000 roubles.
  [d] B.F.J. Hermann, *Die Wichtigkeit*..., pp. 50–1. Beresov produced 16 pud 35 funt out of a total production during the years 1805–1807 of 20 pud including Miask and a number of brooks with placer gold.

made to separate the gold from the quartz by amalgamation. This involved the carrying of the ore to an open paved yard where it was piled into heaps and water added until a thick slime was produced. Common salt was then mixed in and mercury sprinkled on the amalgam and the whole vigorously agitated, a process which continued over several months. When the gold had been taken up by the mercury the amalgam was removed to washing vats and, being subject to agitation, the heavier particles (containing mercury and gold) sank, leaving an earthy scum which was drawn off and run through settling troughs. The mercury–gold amalgam was poured into canvas bags and compressed to remove the free mercury and to leave a near-solid mass of amalgam which was subsequently heated, the mercury distilled as vapour out of the amalgam to be recovered from the hood of the furnace, and the gold to be left in the hearth. By each of these methods an impure legatur- or blickgold was produced (table A1.5, cols 3–4) still comprising 87 per cent gold, 8.5 per cent silver and 4.5 per cent iron.[38] This was then sent to St Petersburg to the imperial laboratory where the constituent parts were separated into refined gold and silver (table A1.5, cols 5–8).[39]

Copper production

The principal Russian copper mines in the eighteenth century were located in the Altai, Urals, and Olonets. The Urals deposits were by far the most important and by the end of the century occupied great numbers of workshops in the provinces of Perm, Ufa, Vyatka, and Kazan. Considerable quantities were also produced on occasion in the Altai, whilst Olonets and other centres produced annually hundreds rather than thousands of puds of the metal.

Within the industry, unlike in precious-metal production, state-run enterprises were never, even in the 1720s, of primary importance and with the wholesale selling off of state works in the 1750s they made an insignificant contribution to total production.[40] At all times most output was generated by the private sector and historians with only scant collections of private papers, like those of the Demidovs, available to them are forced to view the industry through the eyes of government bureaucrats who enjoyed only limited powers to acquire information from private entrepreneurs. Their initial powers were laid out in Peter's mining law of 1719 which assigned the officials of the newly-created Mining College responsibilities for regulating prospecting activity and licensing works and for collecting the tithe in kind levied on the output of the copper works.[41] In relation to the first of these duties, on discovering mineral deposits prospectors were required to inform the Mining College and despatch thence a sample of the minerals for proving (clause 2 of the mining law). In such circumstances they received a licence to work the mine, which was granted them in hereditary right free from attachment, to build smelting capacity and to acquire such wood and other necessities at a 'reasonable' price for the propagation of their work (clause 5 ibid.). The officials of the Mining College thus enjoyed both licensing and regulatory powers over the private sector. On receipt of a request to develop particular deposits, after testing the ores, they issued an edict (ukaz) authorizing exploitation, keeping a register of such licences detailing:

 (i) the details of the original licence;
(ii) information concerning when the mines were opened and the furnaces blown-in;
(iii) and, on occasion but not consistently, when the works were closed.

They thus had at their disposal data concerning the plants in existence detailing ownership, the number of mines and smelters (including the number of hearths) and (under clause 16 ibid.) the number of workers assigned to the plant, which both contemporaneously and subsequently have been used to construct factory lists and capacity indexes. In relation to the construction of such indexes it should be noted, however, that the simple licensing of the works did not necessarily imply the establishment of a plant at that time but only the establishment of proprietorial rights in the mineral deposits. Thus during the Great Northern War the Demidovs had obtained licences on deposits in the Viso–Kaya and Kungur regions of the Urals but it was not until the late 1720s that they began production here, this only being revealed because in the case of the Viiskii works they chose to obtain a confirmation of their earlier licences which had been issued in 1702–9 before the creation of the Mining College.[42] Unfortunately many modern historians have failed to distinguish between aspects (i) and (ii) above of the licences and their indexes are thus at variance with those presented below. Contemporaries were not, however, so in error and they thus had at their disposal a rough guide to production, obtained by multiplying the number of smelting hearths by an estimate of annual output per hearth. As an index it was only as good, however, as was their knowledge of the metallic content of the ore smelted and of the amount of time the plant was operated, and as events revealed this was highly imperfect.

Theoretically, by clause 9 of the 1719 mining law which endowed the officials of the Mining College with the responsibility for collecting the state's tithe of output, they were afforded more direct information concerning output which has formed the basis of subsequent historians' fiscal production indexes.[43] As will be appreciated from the discussion above, however, officials' ability to check the validity of deliveries of the tithe levied on the private sector was strictly limited, as was to be revealed with the institution of an inquiry into the affairs of Akinfi Demidov in the mid-1730s which had one of its points of investigation defined as requiring an examination of whether he had paid the levy on iron and copper production honestly during the years 1720–33. After two years of assiduously ransacking the Demidovs' factory offices in the Urals and interrogating peasants, the imperial agent in charge of this operation, Guards Captain Kozhukhov, came to the simple conclusion that they had not, the enormous sum of 800,000 roubles' arrears providing some indication of the extent of evasion.[44] The Kozhukhov report of 1735 shattered government confidence in the reliability of the fiscal administration of the mines and highlighted the impossibility of the task imposed upon collegiate officials by clause 9 of the Petrine mining law. Accordingly, whilst the clause was retained in subsequent mining laws and data continued to be collected at least until the late 1740s, under Schoenberg's administration what powers local officials possessed to investigate the books of private entrepreneurs were severely curtailed.[45] Nor, following his fall from grace in 1741, was there anything more than a nominal revival of the old ways which anyway came to an end in 1747/8 as, against a background of declining copper minting and a withdrawal of the state from the industry, government offices already with excessive supplies of the metal on their hands, lost interest entirely in the levy. Only intermittently thereafter, and in the context, as will be shown, of a completely different system of mint supply, did they revive their interest in the fortunes of the industry, and on such occasions *ad hoc* commissions were appointed (in 1767, 1772, 1785, 1790, and 1795) to conduct surveys providing a form of industrial census. Particularly after 1747/8, therefore, information on aggregate production within the industry is rather scanty and it is, accordingly, fortunate that the institution of a new system of mint supply in

1739 affords an opportunity to create a new, if imperfect, index of output.

As part of that reform package embodied in the two-part manifesto which constituted the mining law of 1739 Schoenberg created, as has been shown, a distinct and separate marketing organization for the iron industry.[46] Parallel to this he also totally transformed marketing in the copper industry which, in both private and state sectors had suffered periodic overcapacity crises during the years 1730–4/5 and 1738–9, ensuring that the mint would not only take up the entire output of the state works but also two-thirds of the output of the private sector in the form of compulsory deliveries to be paid for at the prevailing market price.[47] At one stroke the problems of overcapacity within the industry were shifted from the private entrepreneur to the state and the dependence of the industry on the mint, which *de facto* had existed since its inception, was officially recognized. Henceforth, the entire output of the state enterprises and a changing proportion of that generated by the private sector passed to the mint:

1739–52	two-thirds of private output
1752–4	none
1755–6	one-third of private output
1757–68	two-thirds of private output
1769–80	three-quarters of private output
1781–1800	half of private output[48]

Mint consumption of copper was thus, save during the years from about 1747/8–52/4 when the reform of the copper coinage was under way, directly related to output. Unfortunately that relationship was dependent upon two variables: the level of state production which was all delivered to the mint and levels of private production of which only a proportion was delivered. From 1739 to 1747/8 data on state output is fortunately available and thus by substracting this from total mint deliveries and multiplying the residual by 1.5 a check can be provided on the available fiscal production series based on deliveries of the state's tithe. After monetary reforms of 1747/8–52/4 and the 'privatization programme' of 1757, moreover, the state sector was until the 1780s reduced to diminutive levels and its production of copper can be determined by reference to the industrial censuses, the degree of influence it exerted on total mint deliveries thereby being determinable. Accordingly, from about 1757/8–1780 the direct evidence afforded by the production censuses concerning aggregate production can be supplemented by a mint-supply production series which, although an estimate, can be calculated within known tolerances.

Imperfect as the data is, therefore, it is possible to create series of annual production and of industrial capacity for the Russian copper industry during the eighteenth century, an epoch when it rose briefly to a position of total supremacy amongst the world's producers.

Copper production in the Altai: Kolyvan–Voskressensk

As has already been indicated, during the initial phase of mining activity in the Altai under the Demidovs, copper production was of primary importance.[49] Following the discovery of the deposits by peasants traversing southwards from the Ob and the subsequent despatch of ore samples to Neviansk where they came into the hands of the Demidovs, the family had secured a licence to exploit the deposits and in 1727 despatched an expedition to accomplish this task.[50] Thereafter new plant was constructed throughout the period that the works were in their hands, but the objective of plant construction up until 1743 was to increase copper production. Only then was capacity brought on stream to allow the extraction and refining of silver from the argentiferous coppers of the Voskressensk mines, an event which marked the beginning of the end for the family's control over the enterprise.

Table A1.6 Smelting and refining capacity at Kolyvan–Voskressensk, 1726–46

| | Kolyvan Operational hearths | | | Barnaul Operational hearths | |
	Smelting	Refining	Saiger-Prozess	Smelting	Refining
1729	3	—	—	—	—
1730	5	1	—	—	—
1739	5	2	—	2	4
1743	7	2	5	2	4
1746	7	9	5	2	4

Sources: 1729–30, 1739, 1743, B.F.J. Hermann, *Mineralogische Reisen . . .*, I, pp. 305–10; 1739, J.G. Gmelin, *Reise durch Sibirien von dem Jahr 1733 bis 1743* (Gottingen, 4 vols, 1751–2), I, pp. 249–61; 1746, TsGADA, f. Gosarkhiv razr. XI., d. 95 chs 1–2 as quoted in B.B. Kafengauz, *Istoriya khozyaistva Demidovikh v XVIII–XIX vv* (Moscow, 1949), I, pp. 221–2.

Production of copper steadily increased from the blowing in of the first furnaces in 1729, interrupted only by the disastrous fires of 1731/2, but prior to the completion of the first refining hearth in 1730, and in 1733–4 and 1738/9 much of the black (unrefined) copper smelted at the works was transported to the Urals and refined at the Demidovs' Viisk, Nizhnii Tagil, and Neviansk plants, a facet of the family's operational practices which should be borne in mind in considering the data presented in table A1.7 below.[51] Similarly it should be noted with respect to that production data that when the works were seized into the state's hands in 1732 and 1743 current output was simply stockpiled awaiting the outcome of current investigations. On the first occasion, when the enterprise was subject to an investigation by Captain of Engineers Fermor and Assessor Reiser as part of the general inquiry into the Demidovs' affairs, current output was estimated as being worth *c.* 4,000 roubles (*c.* 2,687 puds) an amount which was held by the state against the payment of the family's tax arrears.[52] On the second occasion, when in 1743 the enterprise was subject to investigations by General Beyer, existing stocks of copper derived from that year's production, amounting to 1840 puds of black copper and 4,672 puds of refined and bloom copper, as well as existing stocks of ore – 3,500 puds – and all new production raised between 1743 and 1745 – 42,000 puds – were again stockpiled, only some 13,000 puds being smelted to test their silver content.[53]

Table A1.7 Production of refined copper at Kolyvan–Voskressensk, 1729–46 (in puds)

1729	nil	1735	133	1741	5,993
1730	nil	1736	3,740	1743	822
1731	3,137	1737	4,941	1744	391
1732	nil	1738	nil	1745	53
1733	237	1739	310	1746	979
1734	525	1740	3,235		

Sources and notes: Black copper production 1729–31 2,622 puds p.a. J.P. Falk, *Beitrage zur topographischen Kentniss des russischen Reichs* (St Petersburg, 1785), I, p. 303. 1732 2,687 puds see discussion above. 1733–4 2,920 puds p.a. B.F.J. Hermann, *Mineralogische Reisen . . .*, I, p. 305. Refined-copper production 1731, 1733–4, 1736–7, 1739–41, 1743–6. TsGADA f. Dela Senata po Berg-Kollegii d.8/1510 quoted B.B. Kafengauz, *Istoriya khozyaistva Demidovkh. . ,* I, p. 207. 1735–8. TsGADA f Berg-Kollegii Kn. 653 quoted N.I. Pavlenko, *Razvitie metallurgicheskoi promishlennosti . . .,* p. 290. The figures here for 1736–7 are slightly different from those above: 1736 3,348 pud, 1737 4,461 puds.

During the Demidov epoch from 1726 to 1746, therefore, the primary focus of production within the Kolyvan–Voskressensk works was copper, output of the refined metal rising from *c.* 2,000 puds a year in the early 1730s to *c.* 4,000 puds a year in the latter part of that decade and peaking at *c.* 6,000 puds in the early 1740s. With the seizure of the works by the state in 1746 the situation was totally transformed. Copper sales were suppressed and for almost twenty years the metal produced as a by-product of silver production, which could not be sold locally or used at the works, was stockpiled, so that Coxe, looking back on these years, could note that Kolyvan copper production 'had hitherto been of no use'.[54] Only in 1764 did this epoch, when Kolyvan made no contribution to supplying Russian copper markets, come to an end. In 1761 as a result of the discovery of new argentiferous copper deposits in the Dzunghir mountains by an expedition under the command of one Major Petrulin, the productive potential of the enterprise was considerably expanded.[55] At the same time the state realized the usefulness of exploiting this potential when, against a background of crisis conditions occasioned by dwindling mint supplies from the Urals industry, it was decided to institute an autonomous Siberian copper coinage, releasing Urals supplies to satisfy solely European Russian demand. To this end, accordingly, the construction of a new mint was

Table A1.8 Copper production and minting at Kolyvan–Voskressensk during the years 1767–1809

	Production – minting[a] (puds)	Value of coinage (roubles)
1767	7,145[b]	178,636
8	4,317	107,921
9	5,053	126,331
1770	6,313	157,831
1	6,350	158,747
2	6,344	158,611
3	6,432	160,801
4	6,348	158,696
5	7,612	190,311
6	7,615	190,381
7	7,612	190,291
8	7,425	185,616
9	7,615	190,381
1780	2,694	67,344
1	7,283	182,096
2		
3	29,936[c]	478,975[d]
4		
5	14,404	230,466
6	17,383	278,132
7	11,908	190,537
8	9,079	145,266
9	10,445	167,122
1790	7,207	115,306
1	12,478	199,657
2	12,480	199,673
3	12,479	199,667
4	12,479	199,665
5	12,476	199,622
6	12,481	199,704
7	9,422	150,758
8	12,481	199,709
9	5,079	81,262
1800	5,461	87,371
1	4,260	68,152
2	5,646	90,329
3	5,461	87,373
4		
5		
6	45,736	731,774
7		
8		
9		

Notes: a B.F.J. Hermann, *Die Wichtigkeit des Russischen Bergbaues* (St Petersburg, 1810), table 10 gg (1767–8), ibid hh (1769–1809). b At 25 roubles per pud. c At 16 roubles per pud. d Coxe's £5,000,000 (2,526 million roubles) for mint output seems to relate to total copper coinage in circulation to 1782 *not* annual production.

authorized in 1763 and with its opening in May 1764 at the Susunsk works of the Kolyvan metallurgical complex a new age dawned in copper production at that centre and in the monetary history of Siberia. Henceforth, Siberia enjoyed its own, independent source of money supply as with the conversion of Kolyvan copper into coin a new local, indigenously produced currency was created which by 1781 amounted to some 2.5 million roubles struck at the rate of 25 roubles per pud.[56]

Copper production in the Urals

As will be appreciated from the discussion in the introduction to this section, because of the importance of private entrepreneurs in this sector of the industry and because of the limited powers enjoyed by state officials with respect to obtaining information from them, data concerning copper production is rather scanty.[57] Information concerning production derived from the records of individual enterprises has a very irregular incidence. For state-owned works it is relatively abundant for the years 1720–46 but with the currency reforms of 1747/8–54 and 'privatization' in 1757/8 such data becomes sparser and relates only to the Ekaterinburg works and a number of bankrupt enterprises, which subsequently fell into the state's hands.[58] In the case of the private sector whilst, thanks to the scholarly endeavours of the late Professor B.B. Kafengauz, information is abundantly available on the Demidov enterprises, with regard to the properties of other Urals copper masters it is not.[59] In relation to this numerically dominant group of entrepreneurs there are only the rather unreliable 'fiscal production indexes' derived from information concerning the collection of the state's tithe on output, a levy on the private sector of the industry which, as has been shown, was subject to massive evasion.[60] Direct evidence of *total* copper production must, accordingly, be treated with extreme caution, and where possible cross-checked against indirect evidence of industrial production and capacity, a process best undertaken for reasons given above, in relation to two distinct sub-periods: 1720–46 and 1758–1809.[61]

Urals copper production, 1720–46

Table A1.9 Industrial plant numbers, 1720–46

Owner	1720–4	1725	1726–7	1728–9	1730–4	1735–40	1741–6
State							
Sold	—	—	—	—	—	1	—
Purchased	—	—	—	—	—	—	—
Constructed	2	3	—	—	—	4	—
Closed	—	—	—	—	—	1	1
Cumulative sub-total	2	5	5	5	5	7	6
Private							
Sold	—	—	—	—	—	—	—
Purchased	—	—	—	—	—	1	—
Constructed	—	—	2	4	8	1	4
Closed	—	—	—	—	1	—	2
Cumulative sub-total	—	—	2	6	13	15	17
Cumulative total[a]	2	5	7	11	18	22	23
Estimated productive Capacity (in tons)[b]	112	225	325	500	945	884[c]	1,092

Notes:
a For alternative estimates of the number of plants see N.I. Pavlenko, 'Materiali o razvitii ural'skoi promyshlennosti v 20-40kh godakh XVIII V', *Istoricheskii arkihv*, ix (1953), pp. 165–6 on the state works and the same author's *Istoriya metallurgii v Rossii*. (Moscow, 1962), p. 462 on private ones and R. Portal, *L'Oural au XVIII siecle* (Paris, 1950), pp. 125–8, 157, 162–3. 165, 169–174, 332–5, 350–4 Utilizing this data as well as other information scattered amongst the pages of these works and others, too numerous to mention here, histories of particular works were constructed, primary importance being attached to direct evidence of the original licence and contemporary but retrospective evidence only being used as a last resort in securing information either on whether a plant was in operation or on the number of its operational hearths.
b Obtained by multiplying the number of operational hearths by productive capacity estimated at 14 tons a year to 1730, 16 thereafter.
c See table A1.10 below.

Table A1.10 Urals copper production, 1738: industrial capacity index[62] (map 4.1)

Name	State copper works Smelting hearths	Production (tons)
Operational		
Ekaterinburgskii	9	
Yagoshikhinskii	2	
Pisorskii	6	
Lyalinskii	2	160 (42%)[a]
Visimskii	6	
Nizhnii Yugovskii	2	
Motovilikhinskii[b]	nil	
Closed		
Polevskii[c]	(9)	

Notes: a Figures in parenthesis indicate proportion of capacity in use.
b Refining hearths only in use.
c Converted to iron production in 1736.

over/

Table A1.10 (*cont.*):

Owner and name of plant	Demidov copper works Smelting hearths	Production (tons)
A. Demidov		
Viiskii	3	
Nizhne-Tagilskii	1	NIL[c]
Nevianskii[a]	1	
Suksunskii	6	59 (35%)
Bimovskii[b]	6	
N. Demidov		
Davidovskii	3	NA

Notes: a Eastern Urals. b Western Urals. c Following the reacquisition of the Kolyvan works production was temporarily suspended.

—

Owner and name of plant	Osokin and other copper works Smelting hearths	Production (tons)
P. Osokin		
Irginskii	2	
Biziarskii	3	
G. Stroganov		
Tamanskii	4	
A. Tuchaninov		
Talitskii (Troitskii)	2	
Sisertskii	closed	108 (37%)
S. Inozemstvo		
Taishevski	1	
S. Krasilnikov		
Korinskii	1	
A. Prosorov		
Shurminskii	1	
I. Nebogatov		
Shilvinskii	1	

Notes: On new plants constructed between 1738 and 1746 (map 4.2) see sources listed in table A1.9 and N.I. Pavlenko in *Istoricheskii arkhiv*, IX (1953), pp. 165–6 and *Razvitie metallurgicheskoi promyshlennosti Rossii v pervoi polovine XVIII Veka* (Moscow, 1953), p. 450.

Table A1.11 Urals copper production, 1720–46: 'Fiscal-' and 'money-supply' output indexes

	State works[a]		Private works declared output[a]		undeclared[b]	Estimated total output
1720	5.4		—			5.4
1	nil		—			nil
2	2.8		—			2.8
3	9.3		—			9.3
4	37.6		—			37.6
5	85.7		—			85.7
6	152.3		3.1			368.3
7	161.5	626.9	2.9	48.8	924	389.1
8	143.6		15.4			376.8
9	169.4		27.4			466.4
1730	162.7		50.2			319.4
1	159.9		82.5			363.6
2	137.2	768.2	67.5	372.0	593	307.1
3	131.6		73.3			307.5
4	176.6		98.5			412.6
5	142.3	296.7	112.0	224.2	526	751.1
6	154.4		112.2			295.6
7	162.0		145.5			307.5
8	160.0	725.1	167.7	744.5	185	327.7
9	157.4		203.1			465.4
1740	245.7		228.2			553.7
1	307.2		177.1			483.3
2	283.2	881.2	281.1	688.0	nil[c]	564.4
3	290.8		229.6			520.4
4	337.0	703.7	204.4		nil	541.4
5	366.7		n.a.			n.a.
6	330.0		490.0		nil	820.0

Notes:
 a State works and declared output of private works 'fiscal' production estimate see N.I. Pavlenko, *Istoricheskii archiv*, IX (1953), pp. 178ff, and the same author's *Razvitii metallurgicheskoi promyshlennosti Rossii . . .*, pp. 58, 78, and 288. b Undeclared output calculated from 'money-supply' production series (see p. 334), supplemented by contemporary estimates of total production (see ch. 4 nn. 47 and 49). c From 1739 the incidence of the problems inherent in the existence of excess capacity fell on the state which accumulated unwanted supplies of copper: 1739–40 nil, 1741–3 234 tons p.a. 1744–5 106 tons p.a.

Urals copper production, 1758–1809

Following the monetary reforms of 1747/8–54 and the 'privatization' of state-owned plants, in 1757/8 the state almost completely withdrew from direct participation in this sector of the industry, its activities being confined to the licensing of new works and the institution of periodic surveys of the private sector. It did, however, undertake to purchase a varying proportion of the private sector's output to supply the mint, thereby providing data for the construction of 'money-supply' production series.

Table A1.12 Industrial plant numbers, 1746–1800[63]

Owner	1746–50	1751–5	1756–60	1761–80	1781–90	1791–1800
State						
Sold	—	—	5	—	—	—
Purchased	—	—	—	2	3	4
Constructed	—	—	—	2	—	—
Closed	—	—	—	2	3	—
Sub-total	6	6	1	3	3	7
Private						
Sold	—		—	2	3	4
Purchased	—		5	—	—	—
Constructed	2	12	13	5	—	—
Closed	—	6	—	8	2	—
Sub-total	19	25	43	38	33	29
Total	25	31	44	41[b]	36[c]	36
Estimated[a] productive capacity (tons)	1,236	—	3,500	—	3,100	3,100

Notes: a Calculated by multiplying the number of operational hearths by the yield of smelting 700 tons of ore per hearth per year. b See tables A1.13–14. c See table A1.15.

Table A1.13 Urals copper production, 1762: industrial capacity index (maps 4.3–4.4)

Owner and location of works	Central Urals copper works Hearths	Productive capacity (tons)[a]
State		
Ekaterinburgskii	9	126
Chernishev (ex-state)		
Nizhne Yugovskii	12	
Verkhne Yugovskii	12	504
Anninskii	12	
Vorontsov (ex-state)		
Yagoshikhinskii	8	
Pisorskii	6	
Visimskii	6	448
Motovilkhinskii	12	
Demidov		
Viiskii	3	
Suksunskii	6	
Shakvinskii	2	322
Ashapskii	6	
Bimovskii	6	
Strogonov		
Tamanskii	7	
Yugo-Kamskii	2	
Domrianskii	7	
Kruse-Alexandrovskii	2	420
Pozhevskii	3	
Khoklovskii	3	
Nitvenskii	6	

Central Urals copper works		
Owner and location of works	Hearths	Productive capacity (tons)[a]
Osokin		
Yugovskii	6	
Biziarskii	8	308
Kurashimskii	6	
Irginskii	2	
Glebov		
Sermeitskii	6	84
Pokhodiachin		
Petropavlovskii	17	238
Tuchaninov		
Talitskii	2	28
Sekunov		
Uiiskii	2	28

Southern Urals copper works		
Owner and location of works	Hearths	Productive capacity (tons)
Tverdichev and Miasnikov		
Voskresenskii	7	
Preobrazhenskii	6	
Bogoyavlenskii	8	
Blagoveschchenskii	4	490
Arkhangelskii	4	
Verkhotorskii	3	
Simskii	3	
Osokin		
Nizhne Troitskii	4	
Verkhne Troitskii	4	224
Usen' Ivanovskii	8	
Masalov		
Kano Nikolskii		
Krasilnikov		
Arkhangelskii Sharanskii		
Isheryakovskii		
Luganin		
Zlatustovskii	20	280
Tevkeleev		
Varzino-Alexeevskii		
Yaguzhinskii		
Kurganskii		
Shuvalov	12	168
Pokrovskii		
Von Sivers		
Vosnesenskii		

Note: a   Calculated as in table A1.12 but see ch. 4, pp. 192–3.

Table A1.14 Urals copper production. Effects of Pugachev's Revolt on Industrial Capacity, 1774[64] (map 4.5)

Name of works	Owner	Number of hearths
A. Works put to torch and not reconstructed		
Vosnesenskii	State, ex-von Sivers	
Kurganskii	Yaguzhinskii	12
Prokrovskii	Shuvalov	
B. Works destroyed and reconstructed		
Arkhangelskii		4
Bogoyavlenskii		8
Verkhotorskii		3
Voskresenskii	Tverdichev and Miasnikov	7
Preobrazhenskii		6
Simskii		3
Zlatustovskii		
Satkinskii	Luganin	n.a.
Kano-Nikolskii	Masalov	n.a.
Kaslinskii	Demidov	2
Utkinskii		n.a.
Shermianskii	Yakovlev	n.a.
C. Workshops pillaged and restored		
Anninskii	State, ex Chernisev	12
Ashapskii		6
Bimovskii	Demidov	6
Suksunskii		2
Varzino-Alexeevskii	Tevkeleev	n.a.
Irginskii		2
Verkhne Troitskii		4
Nizhne Troitskii	Osokin	4
Usen Ivanovskii		8
Yugovskii		6
Sisertskii	Tuchaninov	n.a.
Silvinskii	Krasilnikov	n.a.
Yugo-Kamskii	Shakhovskoi, ex Strogonov	2
D. Workshops fortified		
Nytvenskii	Golitzyn, ex Strogonov	6
Polevskii	Tuchaninov	9
Verkhne Yugovskii	State, ex Chernishev	12
Nizhne Yugovskii		12

Table A1.15 Urals copper production. Operational plant in the Kungur region in 1790[65] (map 4.6)

Name of works	Owner	Number of hearths
Nizhne Yugovskii		12
Verkne Yugovskii	State	12
Motovilikhinskii	Vorontsov	12
Suksunskii		2
Ashapskii	Demidov	6
Bimovskii		6
Yugo–Kamskii	Shakovskii	2
Uiiskii	Sekunov	2
Yugovskii		6
Biziarskii	Osokin	8
Kurashimskii		6
Irginskii		2

Table A1.16 Urals copper production: private sector, 1746–95[66] ('000 puds)

1746	30.5	1771	166.1
1751	35.7	1772	164.3
1767	190.4	1785	175.6
1769	173.6	1790	156.2
1770	148.6	1795	89.6

Table A1.17 Estimated Urals copper production, 1757–1800[67] (tons)

| | Private sector | | | State | Total |
	Production	Sales	Mint	mint	output
1757					1,319
1758					2,228
1759					2,030
1760					2,024
1					2,512
2	3,300	1,100	2,200	200	3,500
3					2,740
4					2,540
5					3,088
6					2,044
7	3,061	860	2,201	193	3,254
8					2,586
9	2,791	961	1,830	200	2,991[a]
1770	2,387	nil	2,387	190	2,577
1	2,670	nil	2,670	290	2,931
2	2,641	681	1,045	336	2,977
3	2,093	285	1,807	223	2,316
4					1,093
5					1,819
6					1,950
7					2,224
8					2,784
9					2,323
1780	2,845	491	2,354	209	3,054[b]
1					
2	2,731	1,189	1,542	335	3,066
3					2,612
4					
5	2,820	657	2,163	755	3,575
6					2,600
7					1,598
8					2,559
9					2,391
1790	2,512	1,256	1,256	875	3,387
1					1,905
2					2,087
3					2,256
4					1,689
5	1,440	720	720	1,018	2,458
6					289
7					2,686
8					1,942
9					1,939
1800					2,454[c]

Notes: a From 1757 to 1769 mint supplies of copper were derived from state works, notably at Ekaterinburg, producing about 200 tons a year and the compulsory levy of two-thirds of private output. b During the years 1769–80, apart from the output of three state works, the mint obtained three-quarters of the output of the private sector by compulsory levy and purchased some of the residual, taking after 1770–1 about 86 per cent of the output of the private sector. c After 1780 the state levy was reduced to a half of the private sector's production.

Russia. Monetary stocks, prices, transactions demand, and national income

Monetary stocks and prices

The estimates of Russian monetary stocks in this study (tables 4.2–3, 4.5–6) have been made by two independent methods. The first of these methods is based on data concerning mint output and involves a calculation of the amount of recoinage. The second, based on less comprehensive information concerning the supplies of new metals to the mints, avoids entirely the problem of recoinage.

The first of these methods has also been used by B.N. Mironov in his seminal studies of money and prices in Russia during the eighteenth and nineteenth centuries.[1]

Table A2.1 Russian monetary supply and prices, 1700–1860: estimates of B.N. Mironov

	M^a			P^b	T^c
1701–10	1.12 × 11 = 12.5 =	100		100	100
1711–20	2.46 × 14 = 34.4 =	275		213	129
1761–70	4.75 × 19 = 90.2 =	722		264	273
1791–1800	8.33 × 29 = 240	= 1,920		566	339
1801–10	6.62 × 30 = 195	= 1,560		635	246
1811–20	5.29 × 33 = 179	= 1,432		566	253
1821–30	4.07 × 39 = 161	= 1,288		497	259
1831–40	3.97 × 41 = 164	= 1,312		689	190
1841–50	4.17 × 42 = 175	= 1,400		651	215
1851–60	6.76 × 43 = 290	= 2,320		757	306

Notes: a M = per capita monetary stock (roubles) × population within 1701 boundaries (millions) = total monetary stock (millions roubles). Index 1701–10 = 100. b P = prices with 1701 boundaries. c T = transactions demand for money calculated from M and P.

Unfortunately, although Professor Mironov has published extensively on the methodology employed in his works on Russian grain prices and market structures, his calculations concerning monetary stocks have been but scantily reported.[2] It has accordingly been considered necessary to attempt a reconstitution of his calculations and then to test this against his original data base.

Monetary stock: estimate of B.N. Mironov

The fullest exposition of the methodology employed by Professor Mironov to construct time series of monetary stocks will be found in his '"Revolyutsiya tsen" v Rossii v XVIII veke', *Voprosi Istorii*, no. 11 (November 1971), pp. 54–5 from which it appears that the eighteenth-century series was calculated by backward projection from a base year (1787–9/95) stock estimate. This involved a calculation of the size of the monetary stock in *c*. 1795 when it was considered to comprise:

(1) paper assignats put into circulation since 1769, reduced to their silver equivalent on the basis of their St Petersburg bourse quotation, amounting to 133 million roubles;
(2) silver and gold coins minted since 1764 (70 million roubles) and copper minted since 1763 (20 million roubles);
(3) Siberian copper issues, minted to a standard of 25 roubles per pud during the years 1763–81 (13.5 million roubles) and to a standard of 16 thereafter (3.5 million roubles);

or in total *c*. 240 million roubles, a figure which, though somewhat different in composition, is broadly in line with the near-contemporary estimate made by B.F.J. Hermann.[3] By Hermann's calculation there were in circulation in 1788:

(1) gold and silver coins amounting to 76 million roubles;
(2) copper coins amounting to 54 million roubles;
(3) assignats amounting to 100 million roubles;

or in total *c*. 230 million roubles which, allowing for new issues between 1788 and 1795, again yields a figure of *c*. 240 million roubles in 1795. From this base earlier stocks were then derived by backward projection on the basis of mint issues. Unfortunately Professor Mironov does not reveal how this was accomplished but, in the light of the above calculation, it is clear that he considers the money in circulation in *c*. 1795 as *only* comprising that minted since the monetary reform of 1763/4. Earlier issues, accordingly, must have been recoined subsequent to that date and are thereby included in the mint output of the years 1763/4–95. If this is the case then we are afforded the key to his method of backward projection involving the recoinage of existing monetary stocks after each of the great monetary reforms of 1699, 1718, and 1764. With this in mind, it thus becomes possible to reconstitute his calculations and to estimate stocks from mint output, testing this reconstitution on the basis of the original data to establish how far the results conform to those outlined in table A2.1.[4]

Table A2.2 Reconstituted monetary stock estimates of B.M. Mironov for the years 1699–1796

	Monetary metals (millions 1764 roubles)		
	Ag.Au.Cu		
1699	12.52		
1700–17	21.88		
	Ag.Au	Cu	Total
1718	32.29	2.11 =	34.4
1719–61	41.54	14.26	
1762	73.83	16.37 =	90.2
1763–95	50.96	98.84	
1796	124.79	115.21 =	240

Table A2.2 Supplement Composition of monetary stock in 1795

Type of coin	In circulation	Reserves
Silver and gold	70 million roubles	54.79 million roubles
Copper	37 million roubles	78.23 million roubles
Total	107 million roubles	133.02 million roubles

Total monetary stocks on the eve of each reform in 1699, 1718, and 1763/4 comprised those stocks reminted after the last reform plus issues struck from new monetary metal in the interval. Thus on the occasion of the 1699 reform, as Miliukov suggests, some 6.3 million old roubles were reminted into 9.1 million new roubles (with an intrinsic value of 12.52 million 1764 roubles).[5] Subsequent minting of new monetary metals (amounting to 14.24 million contemporary roubles' worth of gold and silver and 2.63 million copper with an intrinsic value of 21.88 million 1764 roubles) further augmented stocks until on the eve of the 1718 reform some 26 million roubles (34.4 million 1764 roubles) were in circulation. The cycle then began all over again. During subsequent years existing coins were reminted and new ones struck bringing total stocks to 90.2 million (1764) roubles in 1762. With the reforms of 1763/4 the process repeated itself but with the introduction of the assignat in 1769 a proportion of supplies became reserves for these notes.

By such a method stocks were estimated and were shown (tables A2.1–2) to increase from 12.52 to 240 million (1764) roubles during the eighteenth century. Professor Mironov then reduced these to a *per capita* figure in terms of the population within the 1701 imperial boundary.[6]

As is revealed in his *Khlebnie tseni v Rossii za dva stoletiya, XVIII–XIX vv.* (Leningrad, 1985), pp. 35–41, 114–16, subsequent stock estimates were derived for the years 1795–1844 by calculating the silver equivalent of the assignat stock from the contemporary St Petersburg bourse assignat–silver rouble exchange rate and thereafter by a similar calculation utilizing the credit–gold roubles exchange rate.

Thus are Professor Mironov's methods of calculating time series of Russian monetary stocks revealed, the exact fit between his results (table A2.1) and those derived from the reconstitution of his methodology (table A2.2) suggesting the validity of the latter exercise. His series clearly relate only to the roubles-exchange area of the Russian Empire (defined as being encompassed within the 1701 boundaries – of which more below). It takes no cognizance of the circulation of non-rouble coins. In part this is a result of the spatial definition of the roubles-exchange area referred to above, for within that area, from the beginning of the eighteenth century, imported coin was reminted. It also seems to rest on the assumption (ibid., p. 116–17) that silver subsequently imported into the empire (not that sector of the empire encompassed within the 1701 boundaries) was similarly reminted. Yet both contemporary comment and archaeological evidence, as well as his own price data (ibid., pp. 84–8, 90–2, and notes), suggests that this is not the case and that throughout the western provinces of the empire (annexed after 1701) foreign coins provided the main money in use during the eighteenth century.[7] It is also clear from the reconstitution of his methodology that his time series are also based on the assumption that only when the rouble was enhanced (i.e. the price of silver raised in terms of the unit of account), in 1699, 1718, and 1763/4, would coins in circulation be brought to the mint for recoining, thereby giving rise to the characteristic pattern of recycling of monetary stocks noted above. The corollary of this assumption is that no such operation would take place when the coinage was called down (i.e. the price of monetary metals reduced in terms of the unit of account), in 1711, 1731, and 1741/4–55.[8] It is certainly difficult to see why private citizens should bring coins to the mint in these circumstances only to receive a smaller nominal sum in return, but governments intent on pursuing a deflationary policy might be prepared to carry the loss involved, recycling coins received in payment of taxes through the mint over a protracted period. Such a process would be encouraged by tax payers rendering lightweight coins (i.e. of lower intrinsic value) to the government and recipients of state expenditures receiving heavier ones, but the process would be a protracted one. For this reason, as will be suggested below, deflationary policies were much more difficult than inflationary ones to pursue and only Elizabeth's reforms of 1741/4–5 were wholly successful, that of Peter I aborting before it was brought to a successful conclusion, and that of 1731 never getting off the ground. Yet in spite of the limited impact of these policies they will have an effect on any time series of monetary stocks introducing a new phase of recoinage when mint issues are being struck from existing metallic stock rather than from new inputs of monetary metals. For all of these reasons, therefore, an alternative estimate of Russian monetary stocks has been made which forms the basis of the data presented in tables 4.2–3 and 4.5–6 in the text.

Monetary stocks: alternative estimates

In the light of the discussion above any recalculation of Russian monetary stocks in the eighteenth century must incorporate two elements. First, it must involve a re-estimation of rouble stocks utilizing the same data set of mint output employed by Professor Mironov but making new estimates for recoinage.[9] Second, it must provide an estimate of the stock of foreign coins in circulation in the western provinces of the empire.

As will be appreciated from table A2.3 the effect of introducing an analysis of government deflationary policies initiated in 1711, 1731, and 1741/4–55 into the calculations of monetary stocks is quite dramatic. Rates of stock accretion, particularly in the periods 1700–18 and 1718–62, are quite markedly reduced.

After a period of rapid stock growth in the first decade of the century as new supplies of silver were imported, most notably during the years 1704–5 and 1708–9, the decree of 1711 ushered in a new era.[10] Against a

Table A2.3 Alternative estimate of Russian rouble stocks, 1699–1796

	Silver Amount[a]	Silver Mint price[b]	Gold Amount	Gold Mint price	Copper Amount	Copper Mint price	Total Stock Contemporary roubles	Total Stock 1764 roubles
1690–8	5.58	11.37			0.62	12.8	6.2	12
1699	8.07				0.74	15.4		
		16.45			0.96	20		
1699–1710	10.0							
	18.07							
	12.47	5.6			0.96	20		
1711–17	4.24	13[c]			0.96			
1718	15	6.44						
	21.44				1.92	40	23.36	25.4
					6.5[d]			
1718–40	7.53	19.75	1.02	293	0.42[e]	10		
	29.97		1.02		8.42	0.42		
1741	27.68		0.94		8.42	0.42	37.64	39.0
					0.8[f]	10		
					8.42	1.22		
1742–62	5.12	18.24	0.58	270	2.68[g]	8		
					5.36[h]			
					10.28	16		
	32.8		1.52		15.64		50.06	58.6
1763/4	40.9		2.06		15.64			
1764–95	30.0	22.75	14.00	347	64.26	16		
1796	70.9		16.06		79.9		166.86	166.86

Notes: a Amounts in millions of roubles. b Mint price: gold/silver, roubles per funt; copper roubles per pud (= 40 funt). c at 12 roubles for pre-1698 Joachimstalers and roubles. d 1718–34. e 1735–40. f 1741–51. g reminted 1755 at 8 roubles. h reminted 1757.

background of a standstill in the specie trade, reflected in the low level of premiums on the Amsterdam bourse, the institution of a policy to call down silver caused mint activity during the years 1711–17 to be dominated by the recoining of those pre-1698 coins which had escaped the net cast in 1699 and the reminting of the post-1699 coins which slowly percolated into the state's coffers as tax receipts.[11] Prior to the major rise in specie and copper prices in 1718, therefore, net accretions to the stock of monetary metals had largely been concentrated in the first decade of the century when on occasion perhaps as much as 2.5 million contemporary roubles' worth of specie may have been imported in particular years. By far the largest amount of mint activity was concerned with the reminting of existing coins on the occasion of the price adjustments of 1700/2–4 and 1711.

If reminting dominated the years prior to 1718, however, it was to be even more important in the period 1718–40. The massive enhancement in silver prices in 1718 swept up the issues of 1699–1710 which had not been reminted after 1711 (*c.* 13.5 million roubles) as well as the diminutive stock of Petrine 'heavy' roubles (*c.* 4.24 million roubles), converting this heterogeneous collection into 22.64 million new roubles, struck to the weight standard of 1699 but to a much reduced fineness, which remained, in spite of the abortive attempt at reform in 1731, the mainstay of the circulating media until 1741. Nor were these coins augmented much by new issues struck from imported or domestically produced specie. In a period dominated by the effects of the specie crises of 1725–9 and 1734–5 only some 8.5 million roubles' worth of specie was imported into the realm and then mainly in the closing years of the reign of Peter I (viz. 1718–24) when, judging both by the movement of premiums on the Amsterdam bourse and by such direct evidence as is available, perhaps 1 million roubles' worth of specie was on average imported annually, leaving annual imports during non-crisis years in the 1730s at no more than *c.* 200,000 roubles.[12] The once dominant source of new metal supply to the mints, the import trade, was rapidly declining.

Equally significant, however, to the evolution of the currency was a new feature – the minting of copper. On the eve of the 1718 reforms Russian copper stocks, which had been continually recycled during previous years, amounted to a diminutive 960,000 roubles struck at the rate of 20 roubles per pud, which on reminting at the new rate of 40 roubles per pud produced some 1.92 roubles' worth of coin, bringing the total amount of recycled copper in mint output during the years 1699–1725 to 4.25 million roubles, a quantity totally dominating new issues which amounted to a mere 100,000 roubles minted during the years 1723–5.[13] Such was not the case, however, during the subsequent years of acute silver shortages, encompassing the reigns of Catherine I and Peter II (1725–9), when rapidly rising copper prices engendered a massive increase in indigenous production, a host of new plants supplying some 1,600 tons of copper to be converted into almost 4 million roubles' worth of coins. A new force was entering into monetary calculations, a relative abundance of copper only being rendered compatible with its enhanced official price (40 roubles per pud) because of the extraordinarily high silver prices then being offered at the mint to attract supplies of the white metal. As silver prices fell from 1727 to 1730, in response to increased imports, however, the base metal was grotesquely overpriced. Accordingly, with the decision to call down silver to 10 per cent below its 1718 level it was also decided in 1731 to call down copper dramatically – to one-fifth of its former value – establishing a new bi-metallic (silver:copper) ratio of 1:86.[14]

The decision was premature, however, for the inflow of silver was insufficient to warrant such a large abatement in price, and at a price of 8 roubles per pud for copper the infant industry would be obliterated. Accordingly the scheme was abandoned. In 1732 the mint paid 18.75 roubles per pud for silver and left the old market rate for copper unchanged. Yet even this latter measure proved insufficient, and as the new capacity, planned at the end of the boom, came on stream, production at prevailing prices declined and supplies to the mints dwindled, the latter institutions turning out only some 500,000 roubles' worth of copper coins a year on average during the years 1730–4.[15] As the second great specie crisis broke in 1734/5, therefore, some 6.72 million roubles' worth of lightweight copper coins were in circulation. The Russian monetary stock was in the process of being transformed. The previously insignificant copper coinage now, with about 6.72 million roubles' worth of pieces in circulation, was very far from being subordinate to the main gold and silver currency which at that time amounted to *c*. 30 million roubles, and as the copper industry once more responded to crisis conditions in 1734/5, delivering some 750 tons to the mint, the imbalance threatened, at prevailing prices, to become even more acute. The year 1735, accordingly, witnessed the beginning of a reform movement which would dominate the history of mint activity for the next twenty years, the impact of the new copper being reduced as it was minted at a rate of 10 roubles per pud. A deflationary phase in monetary policy was beginning, and whilst, for the moment, the existing stock of lightweight copper coins remained in circulation, new issues were struck at a much reduced rate.

This trend toward the reformation of the coinage, first begun in 1735, reached its apogee under the Empress Elizabeth. Silver was called down in 1741, the new coinage dominating the circulating media until Peter III's reign, and being valuably augmented with some 5 million roubles' worth of coins struck from new silver. The problem of copper was more intractable. Whilst new coins continued to be struck at 10 roubles per pud, however, the existing lightweight coins had their nominal value reduced during the years 1744–6, existing monetary stocks circulating from 1746 to 1755 at rates approximating to 10 and 16 roubles per pud respectively before the whole was reminted at 8 in 1755 creating a new heavyweight copper coinage. By 1755–6, therefore, the whole currency had been reformed, *c*. 30 million roubles' worth of silver coins, struck to a fineness of 75 per cent, circulating alongside less than 3 million roubles' worth of the new copper pieces.

Then in 1757 there was a complete volte-face in government policy as it became committed to minting all available supplies of monetary metals. Annual mint production targets were set at 1 million roubles' worth of coins in that year, 2 million in 1760, 3 million in 1762, and 5 million in 1772. The new programme was not, however, without its problems. Even as it began, a shortfall in the supplies of precious metals forced the government to increase copper prices, leading to a recoinage of existing pieces and bringing forth a new supply of the base metal. This added considerably to the total volume of coins in circulation and provided the basis for a profligate credit operation through the intermediary of the newly created Copper Bank and Bank of the Artillery and Engineering Corps.[16] Mint output levels were, accordingly, maintained, but the copper coinage, with some 15.6 million roubles' worth of coins in circulation, was disproportionately represented in a currency which amounted to no more than 59 million roubles. To correct this bi-metallic (copper-gold/silver) imbalance, therefore, both gold and silver prices were raised, leading to yet another recoinage in 1764, but, as popular demand for copper coins declined the problem remained unresolved – until the introduction of the assignat in 1768/9 which permitted market forces to effect a conversion of surplus copper coins into high-denomination paper notes. Thanks to this innovation, by 1786 the composition of a much enlarged monetary stock accurately reflected popular demand for coins, 15 million roubles' worth of copper coins circulating alongside 80 million roubles' worth of gold and silver coins and 46 million roubles' worth of high-denomination assignats. For some thirty years (1757–86), as monetary stocks rapidly increased, the major problem had been the maintenance of an appropriate compositional balance in accord with popular demand for coins. With the increase in assignat stocks to 100 million roubles in 1787, marginally exceeding the amount of coin in circulation (97 million roubles), a new age dawned in which the stabilization of an unsecured paper currency became the primary concern of those formulating monetary policy.

Considering the period *c*. 1700–86/95 as a whole, therefore, it appears that the incorporation of an analysis of government deflationary policies, in 1711, 1731, and 1741/4–55, significantly affects calculations of monetary stock accumulation, increasing the amount of recoinage in mint output and, particularly during the years to 1762, reducing the importance of new metallic supplies.[7] Indeed, in the case of gold and silver, as new supplies of metal dwindled, the recoinage of existing stocks steadily increased in importance in mint activity, rising to a level of some 83 per cent in Elizabeth's reign. Similarly with regard to copper Elizabeth's reign witnessed an abnormally high level of recoining. In contrast, however, Catherine II's reign saw levels of recoining falling to 50 per cent for silver and gold and 20 per cent for copper. A flood of new metallic supplies ensured a massive absolute increase in monetary stocks.

Table A2.4 Mint output, 1699–1796 (million roubles)

	Silver and gold			Copper		
	Recoinage	New minting	Total	Recoinage	New minting	Total
1699–1718	12.31	10.00	22.31	2.33	nil	2.33
1718–40	21.44	8.55	29.99	1.92	7.00	8.92
1741–62	28.62	5.70	34.32	8.04	7.08 + 4	15.12 + 4
1763–96	42.96	44.00	86.96	15.64	64.62	80.26
Total	105.33	68.25	173.58	27.93	82.70	110.63

Isolation of the amount of coinage which was struck from new supplies of monetary metals, moreover, affords an opportunity to provide a second independent estimate of Russian monetary stocks based on information concerning the amounts of new metal delivered to the mints.[18]

Table A2.5 A second alternative estimate of Russian rouble stocks 1699–1796: monetary metal supplies (million roubles)

	Silver and gold			Copper		
	Minting	Imports	Domestic supply	Minting	Imports	Domestic supply
1699–1718	10.00	10.00	0.00	nil	nil	nil
1718–40	8.55	8.54	0.01	7.00	nil	7.00
1741–62	5.70	nil	5.70	11.08	nil	11.93
1763–96	44.00	nil	46.70	64.26	nil	65.16

As will be seen (table A2.5) monetary metal supplies, whether derived from specie imports or domestically produced metals, which dominated the copper coinages from c. 1720 and silver and gold from c. 1740, were always more than sufficient to satisfy estimated output-levels of new metallic coins.[19] Accordingly, a simple supply-accumulation estimate (table A2.6) provides a striking confirmation of the validity of the original mint output estimate (table A2.3) and yields a new Russian rouble/stock estimate.[20]

The construction of an estimate of the stocks of foreign coins in use within the western provinces of the Empire – the second component element of an estimate of imperial Russian monetary stocks – is beset with difficulties for the source materials relate only to flows and are largely confined to the years after 1757.[21] It is, accordingly, necessary to provide an estimate of monetary stocks to which the flow data can be related. To

Table A2.6 A second alternative, supply accumulation estimate of Russian rouble stocks, 1699–1796 (million roubles)

| | Metallic supply | | Monetary stock |
	Contemporary roubles	1764 roubles	1764 roubles
1699			12
1699–1718	10.00	13.4	
1718			25.4
1718–40	15.55	13.6	
1740			39
1741–62	16.78	19.6	
1762			58.6
1763–96	108.30	108.3	
1796			166.9

this end an estimate of stock levels in 1788 has been made, treating the two monetary zones as autonomous regions, each with its own levels of prices, monetary stocks, and economic activity. Thus given information on prices and economic activity as well as data on roubles stocks it is possible to estimate the stock levels prevailing in the Baltic provinces (table A2.7 supplement).[22]

Table A2.7 Estimated foreign coin stocks: Baltic provinces, 1796–1802 (million roubles)

1762	30.5
1788	60.5
1796	71.8
1802	76.8

Table A2.7 Supplement Monetary stock in the Baltic provinces, 1788

T(million roubles) Muscovy	2,180.0 = 100	Baltics 820.0 = 38
P(kop./pud grain)	25.4 = 100	28.6 = 113
M(million roubles)	141.0 = 100	(60.5) = (43)

Price movements

As has already been suggested, Professor Mironov's price series employed in his monetary analysis ('Revolyutsiya tsen. . .', pp. 54–5; *Khlebnie tseni.* . ., table 20 p. 115, derived from tables 5, pp. 46–7, and 9, pp. 54–5, series A) is based on data relating to the territories encompassed within the national boundaries of 1701 and is accordingly constructed to conform to the same geographical base as his *per capita* monetary stock series.[23] In this study a somewhat different approach has been employed and accordingly an alternative method is required yielding price series for each of the monetary zones for which monetary stock series have been constructed (tables A2.3, 6, and 7). To this end, therefore, each monetary zone has been geographically delineated and a price index constructed, weighting the price data from each intra-zonal region by its share in the total zonal population.[24]

In the case of the Baltic provinces and their hinterlands, where foreign coins constituted the main element in monetary stocks, both archaeological and literary evidence afford the opportunity to delineate the extent of the monetary zone. Archaeological evidence relating to the situation in the early eighteenth century reveals a distribution of foreign silver seemingly closely related to the trading networks of the Baltic ports, encompassing the Baltic provinces (Lifland, Estland, and Kurland), western White Russia (Smolensk, Mogilev, and Vitebsk), the Ukraine (Kharkov, Chernigov, and Poltava) and the south-west (Kiev, Volynia, and Podolia).[25] A century later Oddy, in his description of the trade of the Baltic ports, revealed an exactly similar pattern.[26] Thus the geographical extent of the monetary zone within which foreign silver circulated can be delineated and on the basis of this delineation appropriate regional price data was assembled and used to construct a zonal price series in the manner described above (see table A2.8).[27] An examination of this data reveals, moreover, that whilst prices (money stocks) steadily fell the greater the distance from the port of entry, as a whole prices within this western part of the empire, save during the years 1762–96, rose faster than in Muscovy, and after the Great Northern War were always well above the Muscovite level.

Table A2.8 Monetary Supply and Prices: Baltic provinces 1700–1807

	M^a	P^b	T^c
1700		6.53 = 31	
1718/22		11.52 = 56	
1762	30.5 = 100	20.71 = 100	100
1788	60.5 = 198	28.68 = 138	143
1792/6	71.8 = 235	32.85 = 159	148
1802	76.8 = 252	42.29 = 204	123
1807	69.0 = 226	42.29 = 204	111

Notes: Indices 1762 = 100. a M = monetary stock (million roubles). b P = grain prices (specie kop. per pud). c T = transaction demand calculated from M and P.

A similar procedure was also applied to price data from the rouble-exchange zone comprising the following of Professor Mironov's price regions: north, east, south-east, Volga, central chernozem, central non-chernozem, steppe, Siberia.[28]

Finally, to provide a somewhat longer time-perspective the series has been extended into the period after 1802 when the rouble-exchange area

Table A2.9 Monetary supply and prices: rouble exchange area, 1700–1807

	M		P		T
1700	12	= 20	6.68	= 47	44
1718/22	25.4	= 43	10.46	= 74	59
1762	58.6	= 100	14.21	= 100	100
1788	141	= 241	25.43	= 179	135
1792/6	166.9	= 285	28.35	= 200	143
1802	195	= 332	27.59	= 194	171
1807	268	= 458	27.59	= 194	236

Notes: See table A2.8.

was extended to the whole of the empire, and when, from 1810, the rouble was adapted as the monetary standard. Total monetary stocks (of silver roubles) were estimated utilizing the same method which was employed by Professor Mironov, and this estimate was indexed in relation to the combined monetary stock in 1762 (= 89.1 million roubles).[29] Prices were measured on the basis of data derived from the territories encompassed within the contemporary boundaries of the empire (excluding Poland and Finland), which corresponds to Professor Mironov's series B, and then indexing was undertaken in relation to a weighted 'national' price for 1762.[30]

Table A2.10 Monetary supply and prices: Russian Empire, 1811–1860

	M		P		T
1811–20	178	= 200	31	= 198	101
1821–30	161	= 181	26.3	= 168	108
1831–40	164	= 184	36.3	= 232	79
1841–50	175	= 196	35.3	= 225	87
1851–60	290	= 326	44	= 281	116

Notes: For indexing and definitions of *M*, *P*, and *T* see tables A2.8–9.

Thus on the basis of the data outlined on pp. 339–45 above, series are presented concerning monetary stocks (tables A2.3, 6, 7, and 10) and prices (tables A2.8, 9, and 10) which are embodied in the textual tables (4.2–3, 4.5–6) and provide the basis for a calculation of transactions demand for money during the period 1700–1860.

Transactions demand and national income

Employing the data concerning monetary stocks (tables A2.3, 6, 7, and 10) and prices (tables A2.8–10) transactions demand for money has been calculated utilizing the Fisher equation:

$$MV = PT$$

and the resultant series checked against such contemporary estimates of national income as are available (tables A2.11–13), the resultant high level of correlation between the transactions demand series and that for national income at constant (silver) prices validating the original exercise and allowing absolute levels to be attached to the estimated indices for the

Table A2.11 Transactions demand and national income: Muscovy, 1700–1807

	M	P	T	NI (i)	NI (ii)
1700	20	47	44	397	
1718/22	43	74	59	532	
1762	100	100	100	902	(902)
1788	241	179	135	1,217	2,180
1792/6	285	200	143	1,290	2,586
1802	332	194	171	1,542	(2,991)
1807	458	194	236	2,128	(4,130)

Notes. For indexization and definitions of *M, P,* and *T* see tables A2.8–10. *NI* = National income in millions of 1764 silver roubles (i) at constant 1762 prices derived from *T* and *NI* (ii) contemporary estimates, unparenthesized, in current prices, details of which are given below.

Table A2.12 Transactions demand and national income: 'Baltic Provinces', 1762–1807

	M	P	T	NI (i)	NI (ii)
1762	100	100	100	415	415
1788	198	138	143	594	820
1792/6	235	159	148	614	(976)
1802	252	204	123	510	(1,041)
1807	226	204	111	460	(940)

Notes: See table A2.11.

Table A2.13 Transactions demand and national income: Russian Empire, 1762–1858

	M	P	T	NI (i)	NI (ii)
1762	100	100	100	1,342	(1,342)
1788	225	168	133	1,785	3,000
1792/6	268	195	137	1,839	(3,585)
1802/3	305	194	157	2,107	4,000
1807	378	194	195	2,613	5,070
1811–20	200	198	101	1,355	(2,683)
1821–30	181	168	108	1,449	(2,434)
1831–40	184	232	79	1,060	(2,459)
1841–50	196	225	87	1,168	(2,628)
1851–60	326	281	116	1,557	2,787

Notes: See table A2.11.

years 1788–1858 as well as permitting a tentative backward projection of the national income series into the period 1700–88.

The validity of this exercise rests, however, not only on the original transactions demand series but also upon the veracity of the contemporary estimates of national income. Accordingly, this latter series has been extended into the late nineteenth century and cross-checked against the recent estimates made by Professor Gregory (table A2.14–16).

Contemporary estimates of national income, 1788–1858

1788. The Empire, Muscovy, and the Baltic provinces. B.F.J. Hermann, *Statistische Schilderung von Russland in Rucksicht auf Bevolkerung, Landesbeschaffenheit, Naturproduckte, Landwirtschaft, Bergbau, Manufakturen und Handel.* (St Petersburg and Leipzig, 1790). This study provides an estimate (ibid., pp. 454–7) of the value of market transactions within the empire, comprising two elements:

(i) the author's own estimate of the value of commercial activity within the bounds of the empire as a whole with a population assessed by him as totalling 30 million (and by Kabuzan on the basis of data derived from the IV and V revisions of 1782 and 1795 at 28.4 and 37.4 millions respectively) = 300 million silver roubles;[31]

(ii) Marshall's estimate for the 'Baltics' of £13 million at 6.32 silver roubles per pound sterling for an area with a population of 8 millions = 82 million silver roubles.

The value of market transactions within the Muscovite heartland of the empire thus comprises 218 million silver roubles and in the western provinces 82 million or a total of 300 million which in line with other contemporary estimates is assessed at 10 per cent of gross product – 3,000 million roubles for a population of 30 million.[32]

1792. Muscovy. A.N. Radishchev, 'Pis'mo o kitaiskoi torge', *Polnoe sobranie sochinenii* (Moscow, 3 vols, 1938–52), II, pp. 15–16. A similarly constructed estimate of the total value of market transactions amounting to 300 million (assignat) roubles = 258 million (silver) roubles at the contemporary (1792/3) exchange rate of 86.2 silver kopecks per assignat rouble, but relating only to the Muscovite lands with their population of 26 millions.[33] Converted, in the manner described above, to gross product a figure of 2,586 million silver roubles is arrived at for the Muscovite lands.

1802–7. Russian Empire. I. Golitzuin, *Statisticheskiya tablitsi vserossiiskoi imperii, ili fizicheskoe, politicheskoe, statisticheskoe nachertanie Rossii s XIX stolitiya* (Moscow, 1807). This estimate (ibid. table 6) represents an evaluation of the gross product of the empire during the years 1802–7 when the population was estimated as increasing at half a million a year from 38.5 to 41 millions, annual average product amounting to 4,465 million silver roubles and 4,000 million in 1802/3.

1824. Russian Empire. M.G. Mulhall, *Industries and Wealth of Nations* (London, 1896). A non-contemporary estimate relating to the pre-1861 period and therefore excluded from table A2.13 as it suffers from being calculated as a retrospective figure by the author within his assumptive framework of the early 1890s of which more anon. It was calculated from contemporary data but utilized techniques first employed by Mulhall in the 1890s (ibid., p. 11) giving a gross product of 2,970 million roubles (ibid., p. 169) which was converted into sterling at the 1894 exchange rate of 10 roubles per pound sterling (ibid., p. 13).

1858. Fifty provinces of European Russia. L. Levi, 'On the distribution and productiveness of taxes with reference to the prospective ameliorations in the public revenue of the United Kingdom', *Journal of the Statistical Society of London*, XXIII (1860), p. 42. A similarly calculated estimate from contemporary data giving a gross output of 2,528 million silver roubles, converted into sterling at the prevailing exchange rate of 6.32 roubles per pound sterling, for an area comprising the fifty provinces of European Russia with a population of 60 millions, but excluding those areas of the Caucasus annexed in 1811, Siberia, and Tartary, as well as Poland and Finland. The figure has been enhanced by a factor of 1.1 to incorporate the former territories and render it compatible with earlier data.

The relationship between this figure and that for transactions demand/national income at constant prices is complicated by the growing bimetallic disequilibrium which emerged within the Russian currency after *c.* 1837. Until 1839, as has been noted, both price and monetary-stock data were derived by converting prices in, and stocks of, paper currency into silver/gold at official exchange rates, the relationship between the latter two specie elements remaining constant in conformity with the official bimetallic ratio. Increasing quantities of gold entering circulation from the mid-1830s led, however, to a steady enhancement in the value of silver thereafter, reflected in a fall in silver-commodity prices and in the external value of the rouble. As this was reflected in both monetary stocks (obtained post-1839 in Professor Mironov's studies by deflating paper issues by the *gold* exchange) and prices (quoted by him in gold roubles) the trends indicated in tables A2.10 and A2.13 remain unaffected even if in the 1850s silver rouble prices were at a 36-per-cent discount on gold prices and monetary stocks amounted to 185 million silver roubles.[34] The current gold/credit rouble valuation of national income of 4,375 million must, however, be adjusted to take account of the fall in silver prices, yielding a figure of 2,787 million silver roubles.

As will be appreciated, from the data presented in tables A2.13, a high correlation seems to exist between the estimates of transactions demand for money (tables A2.8–10) and those made by contemporaries of national income when the latter are reduced to constant (1762) prices (tables

A2.11–13). The latter authors, whilst benefiting from an intimate knowledge of the contemporary economy when formulating the assumptive framework of their calculations, did not conform, however, to modern national-income accounting practices in their estimates. To check the validity of their work, therefore, the series of contemporary estimates has been extended into the late nineteenth and early twentieth centuries and a comparison made with the most recent estimates of Professor Gregory.

Table A2.14 Contemporary estimates of Russian national income, 1858–1914

	NI current prices (million roubles)	Population (million)	NI pop (roubles per capita)
1858	£400 × 6.32 = 2,528	60	42.0
1868	£513 × 6.32 = 3,242	70	46.3
1870	£566 × 6.32 = 3,577	74	48.3
1880	£632 × 6.32 = 3,944	80	49.3
1882	£760 × 6.32 = 4,803	84	57.2
1887	£975 × 6.32 = 6,162	98	62.9
1894	£1004 × 6.32 = 6,345	100	63.5
1900	10,111	118	85.7
1914	18,476	162	114.3

Contemporary estimates of national income, 1858–1914

1868. A. de Buschen, *L'Empire de Russie* (Paris, 1867), adapted R. Dudley Baxter, *National Debts* (London, 1871) pp. 69–70.

1870–80 M.G. Mulhall, *Balance Sheet of the World for Ten Years, 1870–1880* (London, 1881), table 20, p. 33.

1882. M.G. Mulhall, *Dictionary of Statistics* (London, 1884), pp. 245, 356.

1887. M.G. Mulhall, *Dictionary of Statistics* (London, 1892), pp. 320, 447.

1894, M.G. Mulhall, *Industries and Wealth of Nations* (London, 1896) pp. 169–70, and V.I. Pokrovsky, *Sbornik' svedenii po istorii i statistik' vneshei torgovli Rossii* (St Petersburg, 1902).

1900, 1914. Prokopovich–Falkus–Gregory. The original estimates of S.N. Prokopovich as modified by M. Falkus, 'Russian national income, 1913: a revaluation', *Economica*, XXXV, 137 (1968), analysed and extended in P.R. Gregory, *Russian National Income 1885–1913* (Cambridge, 1982), pp. 64–77 with particular reference to tables 3.4–6.

An examination of the techniques employed in the construction of these estimated reveals that whilst their authors certainly did not conform to modern national-income accounting procedures in their calculations they did work within a series of consistent assumptive frameworks rooted in their perceptions of the workings of the Russian economy and its structural forms, which might change over time but which remained anchored in the realities of Russian economic life. Prior to *c.* 1880, moreover, authors shared a common assumptive framework with their predecessors, the economic arithmeticians of the late eighteenth and early nineteenth centuries. These latter writers had divided the economy into four sectors:

(a) Agrarian production, which had been divided into two parts:
   (i) foodstuffs, valued at retail prices thereby including processing (milling, butchering, etc.) and distribution costs, whether these were confined to intra-village transactions or extended to inter-regional trade;
   (ii) non-foodstuff raw materials, including grain for spirit production, flax, hemp, tobacco, hides, wood valued at farm-gate prices as well as forest products, timber, game, furs, fisheries similarly valued, all of which were regarded as basic inputs of manufactory.
(b) Manufacturing, perceived as an additional element to the products generated by sector a(ii), the value-added being 36 per cent of the raw-material costs and, initially at least, including both processing and transport/distribution outlays.
(c) Mining and metallurgy which was treated as a separate, and implicitly integrated, entity in which the value of mining activity, processing, and fabrication were all encapsulated in a valuation of the final product.

National product was simply an aggregation of these elements (Hermann, Golitzuin). Nor did this perception of the economy change in the immediate pre- or post-emancipation periods. Until *c.* 1880 foodstuff production continued to be treated as a separate element of agrarian output, but as the output of grain steadily increased there was a corresponding decline in the other categories of tillage and animal husbandry until they were eliminated from the calculations of the statisticians of the 1850s and 1860s (Levi, de Buschen). Even more marked was the decline in agrarian raw-material production, contemporaries perceiving the elimination of field-crop production and a marked attenuation of the output of forest products. The rural economy was rapidly assuming the character of a grain monoculture and industry was becoming divorced from its agrarian roots. To accommodate these perceived changes information on industry, other than mining, was derived directly from contemporary industrial censuses, output being valued at factory-workshop sales prices and thereby including both raw materials and processing within the valuation of final output. In such a manner raw materials from the agrarian sector (wool, flax, hemp, tallow, hides, and timber other than construction lumber and firewood), excluded from their calculations of agrarian output, were encompassed within the production of the industrial sector. Similarly, imported raw materials were engrossed within the valuation of production emanating from the 'modern' industrial sector, embracing cotton manufactory and metal fabricating. Indeed, within ferrous metallurgy and fabrication a clear distinction was maintained between the 'new' mechanical-engineering industry, which during the years 1856–85 drew on foreign supplies of raw materials, and the 'old' mining and metallurgical complex which continued to be treated in the same manner as it had been in the late eighteenth and early nineteenth centuries. Having eliminated the value-added concepts of an earlier age in the measurement of industrial output, however, authors were now left with the problem of measuring the output of the tertiary sector – commercial, transport, and professional services – previously engrossed within industrial output. Initially (e.g. de Buschen, 1867) they were simply omitted but an English author (Baxter, 1870), utilizing the Frenchman's work, felt compelled to incorporate into his calculations a new element, 'professions and other employments', adding 30 per cent to the value of primary and secondary production, to encompass the tertiary sector of the economy. Only imperfectly did the conceptual framework, first formulated in the late eighteenth century, now represent the fundamental structural changes taking place within the economy.

As these structural changes became more pronounced during the 1880s, accordingly, authors were forced to seek a new conceptualization, ushering in a new statistical methodology which would finally be realized in the 1890s. Nowhere was the necessity for a changed approach more apparent than with regard to agriculture. As the 'crisis' of the 1880s engendered fundamental structural changes within Russian agriculture it was no longer possible to regard that sector as being simply a grain monoculture. Indeed, for authors (e.g. Mulhall, 1894) examining the official statistics relating to 1887 it was clear that grain was being subordinated to other products.

Table A2.15A Russian agricultural production, 1887 (million £ sterling)

All grain (wheat, rye, barley, oats, buckwheat)	181.1
Other (straw, hay, potatoes, flax, etc.)	109.4
Sundries (wine, beetroot, vegetables, tobacco)	43.0
Timber	40.0
Total agricultural	373.5
Total animal husbandry	190.0

This was not a new revelation. As early as 1881, having abandoned his predecessors' assumptions concerning the central importance of grain production in the light of the new information available to him, Mulhall was already aware that 'the people [of Russia] are by no means so ill-fed as is commonly supposed, but rather better than in many countries' (Mulhall, 1881). To encapsulate this newly perceived reality within his statistics, accordingly, a new technique had to be evolved embracing all aspects of the agrarian regime.[35]

Table A2.15B Russian agricultural production, 1870–94
(million £ sterling)

	1870	1880	1882	1887	1894
Agriculture					
Grain	215⎤		266⎤	181⎤	258⎤
Other crops	⎬350	370	72⎬483	152⎬523	112⎬540
Animal husbandry	135⎦		145⎦	190⎦	170⎦
Timber	24		27	40	40
Gross output	374		510	563	580
Net output	374		510	507	364

The new perception, moreover, extended not only to the increasing diversity of production but also to the increasing dependence of the agricultural sector on inputs from the non-agricultural sector of the economy which caused Mulhall in the 1890s (1892, 1896) to deduct inputs at a rate, at first, of 10 and then 40 per cent from gross output. This changed perception of the agrarian regime and an increasing awareness of intersectoral dependency within the economy also inevitably affected the conceptualization of manufactory and service activity. Gross output continued as before to be estimated by providing a valuation of final output on the basis of either direct evidence or proxies but allowance was made now for intra- and intersectoral inputs. Initially (Mulhall, 1881, 1884) this involved a simple 40-per-cent deduction from estimated gross output but gradually this crude technique was replaced by a more specific series of estimations to calculate sectoral value-added (Mulhall, 1892, 1896).

By 1894 Mulhall had accomplished his reconceptualization of the economy embodying his perception of increased diversification and interdependency within his statistical framework (Mulhall, 1896). His technique was not subsequently seriously challenged by other investigators until the recalculations of recent years. Yet the question remains as to how well his and other contemporary estimates hold up in the light of the most recent research. To afford an answer to this question Professor Gregory's latest figures have been recast in table A2.16 in a form which permits the reader to make direct comparisons with table A2.14.

As will be appreciated from a comparison between tables A2.14 and A2.16 the estimates of contemporary statisticians hold up remarkably well. There is a differing geographical basis for the two series, occasioned by Professor Gregory's inclusion of Poland and Finland in his data set and contemporaries sometimes varying their sets to include the empire without Poland and Finland and sometimes to include only the fifty provinces of European Russia (which puts them at variance with both tables A2.13 and A2.16). But when reduced to a *per capita* basis the contemporary series falls within a variance of only ± 8 per cent of the modern one. It may also be observed that this degree of variance does not significantly alter over time

Table A2.16 Russian national income: estimates of P.R. Gregory, 1860–1913

	NI 1913 prices (million roubles)	Current prices 1913–100 (index)	NI current prices (million roubles)	Population (millions)	NI/pop (roubles per capita)
1861	5,269	52.9	2,787	74	37.7
1887	9,258	71.7	6,637	113	58.4
1894	11,463	72.5	8,301	122	68.0
1900	13,940	79.0	11,013	133	79.0
1913	20,266	100.0	20,266	171	118.5

Sources: P.R. Gregory, *Russian National Income 1885–1913* (Cambridge, 1982), pp. 56–7, 155, converted to current (silver-rouble) prices from S.G. Strumilin, 'Oplata truda v Rossii', *Planovoe Khozyaistvo*, no. 4 (1930), reprinted in *Ocherki ekonomicheskoi istorii Rossii i SSSR* (Moscow, 1963), p. 82.

in spite of contemporaries' three separate attempts at reconceptualization of their statistical methods – a striking confirmation of their perceptiveness in observing and accommodating structural changes within the economy within their methodology.

Bearing these observations in mind it may be possible, accordingly, to assemble the data presented in this appendix to construct a long-term national-income series and to set the results of Professor Gregory's investigations in the context of these findings (table A2.17). First it may be

Table A2.17 Russian economic growth, 1700–1913 (in constant 1762 prices)

	Muscovy		Baltic provinces		Empire	
	NI[a]	NI/POP[b]	NI	NI/POP	NI	NI/POP
1700	397	34				
1718/22	532	42				
1762	902	54	415	77	1,342	58
1788	1,217	59	594	74	1,785	63
1792/6	1,290	50	614	73	1,839	53
1802/3	1,542	53	510	57	2,107	55
			Russian Empire			
1807/9			2,613	68		
1811–20			1,355	32		
1831–40			1,060	20		
1851–60			1,557	24		
1868/70			2,851	40		
1880/94			4,778	48		
1900			6,878	58		
1913			9,775	60		

Notes: a NI = national incomes in millions of (1764) silver roubles. b POP = population in millions.

noted that the period prior to the Emancipation Act (1700–1861) formed a discrete entity, but one very far removed from the stable-state traditional economy often pictured in the literature. Russia experienced rapid economic growth in the mid-eighteenth century which, though slackening during the years 1788–1807, only came to an end in the acute crises of the 1810s and 1830. As a result of these crises the populace experienced abysmally low levels of *per capita* income in the 1830s. Second, it may be observed that recovery began during the years 1840–51/60 and that the economy experienced a phase of extremely rapid growth from that time until the 1870s which only came to an end in the 1880s. The 1880s, the base years of Professor Gregory's series, thus marked a phase of retardation in an economy which was already firmly established on the path of rapid economic growth. Only in the 1890s was growth resumed, as he reveals, continuing thereafter to 1913 when *per capita* national income (in 1762 prices) attained *c.* 60 roubles – or still slightly below the levels of 1807.

Notes and references

1 The South and Central American mining 'crisis'

1 On conditions prevailing in European metal markets prior to the events of the 1570s see my 'England and the international bullion crisis of the 1550s', in H. Kellenbenz (ed.), *Precious Metals in the Age of Expansion* (Nuremberg: Beitrage zur Wirtschaftsgeschichte, Bd 2 1981), pp. 87–118.

2 Apart from small deposits of native silver found as far apart as Scotland and Austria the only major workings of European argentite were at Kongsberg in Norway, on which see p. 53 below.

3 Because of the absence of 'rebellious' elements in the ore treated, the amalgamation process was first applied to auriferous quartz, the technology being transferred from central Asia to the western Islamic lands in the twelfth century where it continued to be employed throughout the remainder of the Middle Ages. Known in Europe first from the works of Avicenna, the earliest documented use of these techniques dates from the fourteenth century when they were applied to gold-bearing ores in Bohemia and silver-bearing ones in Austria. I should like to thank Karl-Heinz Ludwig for bringing this latter reference to my attention.

4 The Fugger Asiento of 1563–72 and the attempt at monopolization of mercury supplies during these years is discussed in A. Matilla Tascon, *Historia de las Minas de Almaden*, 1 (Madrid, 1958), pp. 87–102 and P.J. Bakewell, *Silver Mining and Society in Colonial Mexico, Zacatecas 1546–1700* (Cambridge, 1971), pp. 171–2 which provides valuable information on the Crown's pricing policies during these years. For the impact of these policies on silver production in New Spain see A. Szasdi, 'Preliminary estimate of gold and silver production in America 1501–1610', in Kellenbenz (ed.), op. cit., p. 166.

5 Figure 1.1: mercury production: Huancavelica. G. Lohmann Villena, *Las Minas de Huancavelica en los Siglos XVII* (Seville, 1949) and A.P. Whitaker, *The Huancavelica Mercury Mine* (Cambridge, Mass., 1941) as well as M.E. de Rivero y Ustariz, *Coleccion de memorias científicas, agricolas e industriales* (Brussels, 1857), pp. 11, 86–176 from which the statistics presented are drawn.

Figure 1.5: price of mercury. At Mexico City during the years 1559–1700, Bakewell, op. cit., pp. 171–3 and D.A. Brading, *Merchants and Miners in Bourbon Mexico* (Cambridge, 1971), pp. 10–11 who continues the story until the late 1810s, pp. 140–2, 159–60. For the period 1825–48 R.W. Randall, *Real del Monte. A British Mining Venture in Mexico* (Austin, Texas, 1972), p. 168. The data used for the construction of mercury price series at Almaden and Huancavelica will be found scattered about the pages of the classic studies of Matilla Tascon and Lohmann Villena.

6 Figure 1.6. At mercury prices prevailing during the years 1570–1820 the lowest exploitable level of metal content in ores seems to have been 7.5 oz per cwt. See A. von Humboldt, *Political Essay on the Kingdom of New Spain* (London, 3 vols, 1822), edited by J. Ortego y Medina in a critical edition published in Mexico City in 1966, p. 381, at a time, namely the 1790s, when average Mexican yields were said to be about 20 oz per cwt: M.P. Laur, 'De la metallurgie de l'argent au Mexique', *Annales des Mines*, 6th series, xx (1871), p. 165. By the early nineteenth century average metallic content of ores had fallen to the 10 oz level, Brading, op. cit., pp. 152–6, or roughly on parity with European ores. This marked the end of a long decline in yields from in excess of 200 oz a ton in the 1580s to 70 oz between 1620 and 1740 and, as has been shown, 40 oz in the 1770s. See on Zacatecas, Bakewell, op. cit., pp. 129–131; Guanajuato, Brading, op. cit., p. 287. The amount of silver left behind in the residue amounted before *c.* 1780 to some 30 per cent but, thereafter, was reduced to 12.5–25 per cent: see D.A. Brading and H.E. Cross, 'Colonial silver mining: Mexico and Peru', *Hispanic American Historical Review*, LII (1972), p. 556. On silver-extraction rates by the cupelation process see my 'The British silver–lead industry and its relations with the Continent 1470–1570', in W. Kroker and E. Westermann (eds), *Montanwirtschaft Mitteleuropas vom 12 bis 17 Jahrhundert. Stand, Wege und Aufgaben der Forschung* (Bochum: Der Anschnitt. Beiheft 2. 1984). pp. 180–2 and R.F. Tylecote, *Metallurgy in Archaeology* (London, 1982) p. 82.

7 The literature on the distribution of American silver to Europe and beyond is enormous but for the period 1570–1670 see A. Attman, *The Russian and Polish Markets and International Trade, 1550–1650* (Göteborg, 1973) and the same author's *Dutch Enterprise in the World Bullion Trade, 1550–1800* (Göteborg, 1983).

8 Figures 1.1.–1.4: mercury production. On Huancavelica see sources listed in note 5. Concerning Almaden the carefully recorded data contained in Matilla Tascon for the years to 1645 has been used to check the longer series in M.H. Kuss 'Memoire sur les mines et usines d'Almaden', *Annales des Mines*, 7th series, XIII (1878), pp. 149–50. Finally on production at Idria see I. Mohoric, *Rudnik zivego srebra v Idrizi* (Idrizi, 1960).

9 Lohmann Villena, op. cit., chs 2, 4, 10.

10 On Peruvian schemes to import Chinese mercury in 1604–15 and their subsequent plans in the 1640s, 1660s, and 1690s see Bakewell, op. cit., pp. 152–3, whilst eighteenth century ventures, including an ingenious one of 1786 to exchange Californian sea-otter skins for Chinese mercury, are considered in F. Fonseca and C. de Urrutia, *Historia general de real hacienda. . .* (Mexico City, 6 vols, 1845–53), I, pp. 320, 373–7.

11 Figure 1.2, inset: Matilla Tascon, op. cit., chs 8–9.

12 Szaszdi, op. cit., p. 166.

13 See note 8.

14 See note 8 and for further details on the Idrian assientists see Mohoric, op. cit., pp. 74–5 which may be supplemented by reference to Bakewell, op. cit., pp. 157–160 and A. Matilla Tascon, op. cit., pp. 224f.

15 Kuss, op. cit., p. 148.

16 Lohmann Villena, op. cit., pp. 199–205.

17 ibid., ch. 13.

18 Kuss, op. cit., p. 148.

19 See note 14 above.

20 See note 5 above.

21 A. Nöggarath, 'Mittheilungen über die Quecksilberbergwerke zu Almadén und Almadénejos in Spanien nebst einem Ueberblik der Vorkommnisse von

Quecksilber in Allgemeinen', *Zeitschrift für das Berg-, Hütten- und Salinenwesen in dem Preussischen Staate*, Bd X, (1862), p. 364.

22 Lohmann Villena, op. cit., pp. 301–4, 311–13, 337.

23 Kuss, op. cit., pp. 148–9.

24 See figures 1.1–1.4.

25 See figure 1.2.

26 Nöggarath, op. cit., p. 364.

27 F. Bernaldez and R. Rua Figueroa, *Memoria Sobre los Minas de Almaden y Almadenejos* (Madrid, 1861) and J.M. Hoppensack, *Ueber den Bergbau in Spanien überhaupt und den Quecksilbergbau zu Almaden insbesondere* (Weimar, 1796) provide excellent surveys on mining developments at Almaden, 1750–1850.

28 See works cited in note 27 above.

29 Nöggarath, op. cit., pp. 381–4: J. Reeves, *The Rothschilds, the Financial Rulers of Nations* (London, 1887), pp. 180–1 and F. Morton, *The Rothschilds, a Family Portrait* (New York, 1963), p. 79.

30 See figures 1.4–1.5.

31 On silver production at Potosi, A. von Humboldt, *Political Essay. . .* III, ch. 2 and L. de Sierra, 'Reales quintos paquados a S.M. desde 1 de enero de 1556 hesta 31 diciembre de 1783', *Coleccion de documentos ineditos para la historia de Espana* (Madrid, 1844), V, pp. 173–84.

32 M.O. de Mendizabal, 'Los minerales de Pachuca y Real del Monte en la epoca colonial', *El tramestre economico*, VIII/2 (1941), pp. 273, 275.

33 Figure 1.6. The *Correspondencia*, or ratio of silver produced to mercury consumed, was reckoned in most manuals written towards the end of the mining boom, like those of J. Moreno y Castro, *Arte o nuevo modo de beneficiar los metales de ora y plato* (Mexico, 1758), or F. Sonnenschmid, *Tratado de amalgamacion de Nuevo España* (Mexico and Paris, 1825) at about 1 lb of mercury per mark of silver (2 lb hg/l lb ag), a figure which has been indiscriminately utilized by authors like Brading and Cross, op. cit., p. 579 as representing the technological parameters of the process throughout the whole of the colonial period whereas in reality it varied with the tractability of the ores and altered spatially and through time. A treatise compiled shortly after the introduction of the amalgamation process suggests that in New Spain and New Galicia between 1556 and 1559 the prevailing rate was only 1.25 lbs Hg/l lb Ag and calculations of the relationship at Potosi suggests the incredibly low rate of 0.86 lb Hg/1 lb Ag at least until *c.* 1595 when it began to deteriorate. These figures, presented below, are calculated by dividing mercury production at Huancavelica, less 2,000 quintals a year which were shipped until *c.* 1600 to New Spain, by silver production at the mine. After 1600 supplies imported to Peru are added to the quantities produced there. On mercury production see sources listed in note 5, on silver production those listed in note 31 above, and on the early treatise and mercury imports see Matilla Tascon, op. cit., pp. 211–13 and 234–5.

Correspondencia at Potosi, 1585–1645 (lbs Hg/lb Ag)

1585	0.86	1595	0.86	1610	1.0
1615–20	1.25	1625–30	1.5	1635–40	2.1
1645	2.32	1650	1.5		

Similar calculations have been made for Zacatecas by Bakewell, op. cit., pp. 188–3 suggesting the following figures for the *Correspondencia*:

1595	1.34	1610–30	1.5	1640–5	2.15
1650	1.5	1670–5	1.6		

A century later only Potosi and Guanajuato conformed to these levels and at Zacatecas, Bolanos, Real del Monte, and San Luis Potosi some 2.42 lb of mercury were required to produce 1 lb of silver: Brading, op. cit., pp. 153–6; Fonseca and de Urrutia, op. cit., I, 383; P. Canete y Dominquez, *Guia historica. . . de la provincia de Potosi* (Potosi, 1967), p. 70.

34 On Zacatecas see Bakewell's study whilst on the Parral mining there is R.C. West, *The Mining Community in Northern New Spain: The Parral Mining District* (Berkeley and Los Angeles, 1949). For production at Zacatecas which reached its maximum level in 1623, when it amounted to 2.2 million pesos or about half of total Mexican production, see Bakewell, op. cit., table 4, pp. 242–5. All statistics of silver production quoted in this study are in pesos de oro comun at 65 reales (i.e. 8 pesos = 1 real) to the mark of 8 oz.

35 See note 33 above.

36 Matilla Tascon, op. cit., pp. 234–5.

37 On exports from Europe to New Spain and Peru see Bakewell, op. cit. table 10a, pp. 254–5.

39 See sources cited in notes 5 and 33 above.

40 The main source of the lead used in the Americas from 1580–1620 was England. An anonymous report of 1591 preserved in state papers (*CSPD. Elizabeth I 1591–4*, CXXL, no. 133) relates how once more, after an interval of twenty years, the king of Spain had become dependent upon this product for refining his gold and silver and thereafter there are numerous references to the trade in English lead to San Lucar (ibid., CCXLIII, no. 57 and Public Record Office, London E 190/313/8, 314/14). On production by amalgamation see note 43 below. From about 1625 to 1640 output of amalgamated silver from the main production centres fell below exports and accordingly a *minimum* production of smelted lead of 0.8 million pesos a year between 1625–1635 seems not unreasonable, falling to negligible proportions thereafter. Such a quantity, amounting to 21–34 per cent of total Mexican output, would certainly not be incompatible with production at prevailing *correspondencia* at Zacatecas and would confirm Dr Bakewell's belief that during these years much rescate silver was being 'covered' to avoid paying the quinto. Bakewell, op. cit., pp. 242–53.

41 See notes 40 above and 43 below.

42 See pp. 44–5 below.

43 On production and exports see sources listed in note 47. It may perhaps be noted that there was only a marginal shift towards the retention of silver within the colonies in the years after 1630. In the main exports followed production trends in the principal centres and the residual production of the lesser centres does not seem to have increased its share of total output. The colonies thus conformed to the 'classic' pattern of major metal producers consistently running a major adverse balance of trade.

Estimated precious-metal production and export 1610–1644 in millions of pesos is as follows:

| | Amalgamation Process | | Smelting | Total | Exports |
	Potosi	Mexico	Mexico		
1610–4	6	3.3	Nil	9.3	9.2
1615–9	6.2	3.3	Nil	9.2	8.6
1620–4	5.4	4.4	Nil	9.8	9.0
1625–9	5.3	4.1	1	10.4	9.0
1630–4	4.9	3.6	1.8	10.3	6.8
1635–9	5.9	1.0	1	7.9	5.4
1640–4	4.6	2.9	Nil	7.5	4.8

44 Zacatecas prior to *c*. 1650 drew upon the silver–lead mines of Izmilquilpan for lithage, subsequently drawing upon New Leon (Bakewell, op. cit., p. 147) for supplies. On the provisioning of Parral see West, op. cit., pp. 29–30.

45 See note 47.

46 Based on a simple projection involving a calculation of production on the basis of contemporary levels of mercury production (see note 8), the prevailing rate of *correspondencia* (see note 33) and the proportion of mercury distributed to the Americas prior to the 1660s (see notes 48–9).

47 Figure 1.7. For the production of silver at Potosi see the sources listed in note 31. The problem of estimating Mexican production is more intractable during the years prior to 1690. The figures presented are derived by multiplying mercury imports (Matilla Tascon, op. cit., pp. 234–5 and Brading and Cross, op. cit., p. 572 which is based on Bakewell, op. cit., table 10a, pp. 254–5) by the relevant *correspondencia* recorded in note 33 to give the amount of silver produced by the amalgation process. As has been suggested in note 43, however, from about 1625 the Mexican industy began to undergo a process of structural change as more and more ores were smelted, the *minimum* output derived from these estimations increasing from 21 per cent of the total in 1625–30 to 34 per cent in 1635 and 50 per cent in 1640, figures broadly compatible with smelted output at Zacatecas. Subsequently at Zacatecas/Sombrerete the figure rose to 60 per cent during the 1670s at which level it remained until the 1720s. Bakewell, op. cit., pp. 243–4, 249. Applying this figure to the industry as a whole one thus obtains an average annual output of:

1660–9	4.0 million pesos	1670–9	5.8 million pesos
1680–9	4.5 million pesos	1690–9	4.2 million pesos

The latter figure, derived by this method of estimation, conforms exactly to the official record for that decade: see von Humboldt, op. cit., p. 333.

48 The residual quantities of mercury distributed to markets other than those provisioning the South and Central American silver mines were simply calculated by deducting shipments to the Americas from total production. By 1660 Idria had completely divorced itself from supplying the Spanish American mines and its output of about 2,000 quintals a year was distributed 60 per cent to Amsterdam and 40 per cent to Venice. On production and distribution at this time see Mohoric, op. cit., pp. 81, 96.

49 Figure 1.8. On exports to New Spain and Peru, Bakewell, op. cit., table 10a, pp. 254–255; the residual is calculated by deducting this total from aggregate Old World production derived from the sources listed in note 8.

50 Brading, op. cit., p. 130.

51 von Humboldt, op. cit., pp. 353, 356; J. Burkhart, 'Memoradia de la explotacion de las minas de Pachuca y Real del Monte', *Anales de la Mineria* (1861), p. 44; H.G. Ward, *Mexico in 1827* (London, 1828), II, 440.

52 Randall, op. cit., pp. 10ff.; Brading and Cross, op. cit., pp. 550–1.

53 Brading, op. cit., pp. 134–5.

54 Randall, op. cit., pp. 101ff.

55 Brading, op. cit., pp. 139–140.

56 See p. 21 above.

57 The increasing importance of Amsterdam, already apparent in the 1660s (note 48), had become totally dominant before the decade was out; see Mohoric, op. cit., p. 96.

58 Calculated by dividing output – 1630/40 10,000 lbs – by mercury exported thence – 280 quintals. Matilla Tascon, op. cit., pp. 234–5 and A. Soetbeer,

Edelmetal-Produktion. . . (Gotha, 1878), p. 60. It should be noted that this figure approximates with contemporary *correspondencia* in the silver industry.

59 Bakewell, op. cit., pp. 254–5.

60 On gold production in New Granada, Soetbeer, op. cit., p. 60, cross-checked by estimates of output derived by multiplying mercury imports (note 59) by the appropriate *correspondencia* (note 58).

61 The story of the gold trade during this period remains to be told but on conditions prevailing in the lands about the Niger Bend see the anonymous *Tedzkiret en-Nisian* and al-Zayani, *Le Maroc de 1631 à 1812*, translated by O. Houdas and published in Paris in 1899 and 1886 respectively.

62 Estimated on the basis of calculations of the size of the trans-Saharan trade contained in Jackson's *Account of the Empire of Morocco* (London, 1811) and on contemporary estimates of the coast trade cited in note 64.

63 On production in New Granada, Soetbeer, op. cit., p. 670 and on Brazil, ibid., p. 92 whose figures hold up well in the light of the study of M. Morineau, *Incroyables gazettes et fabuleux metal* (London and Paris, 1985).

64 W. Bosman, *Nauwkeurige beschryving van de Guinese Gould-Tanden Slavekust* (Amsterdam, 1737).

65 Soetbeer, op. cit., p. 62. 2.73 million pesos of which 98 per cent was delivered to Cartagena and 2 per cent to Caracas.

66 A.J.R. Russell-Wood 'Technology and society: the impact of gold mining on the institution of slavery in Portuguese America', *Journal of Economic History*, XXXVII, 1 (1977), pp. 59–60.

67 ibid., pp. 63–4.

68 ibid., p. 65.

69 On production see W.L. von Eschwege, *Pluto Brasiliensis* (Berlin, 1833), and V. Magalhaes Godhino, 'Le Portugal, les flottes du sucre et les flottes de l'or (1660–1760)', *Annales, économies, sociétés, civilisations* (1950), no. 5, pp. 190–7.

70 Figure 1.10. See sources cited in notes 31, 47, 60, 62, 69.

71 See pp. 24–5 above.

72 F.P. Braudel and F. Spooner, 'Prices in Europe from 1450–1750', *Cambridge Economic History of Europe*, IV (Cambridge, 1967), pp. 458–9.

73 On these changes, which have attracted a considerable literature, see the brief summary in K. Glamann, 'The changing patterns of trade', *Cambridge Economic History of Europe*, V (Cambridge, 1977), pp. 263–4.

2   The European producers' response

1 See L. Suhling, *Der Seigerhuttenprozess. Die Technologie des Kupferseigerns nach dem fruhen metallurgischen Schriftum* (Stuttgart, 1976).

2 I. Blanchard, 'English lead and the international bullion crisis of the 1550s', in D.C. Coleman and A.H. John (eds), *Trade, Government and Economy in Pre-Industrial England* (London, 1976), p. 36.

3 E. Westermann, 'Die Bedeutung des Thuringer Saigerhändels fur den mitteleuropaischen Handel an der Wende vom 15 zum 16 Jahrhundert', *Jahrbuch für Geschichte Mittel-und Ostdeutschlands*, XXI (1972), pp. 67–91, and 'Das "Leipziger Monopolprojekt" als Symptom der mitteleuropaischen Wirtschaftskrise um 1527/8', *Vierteljahrschrift für Sozial- und Wirtschaftsgeschichte*, LVIII (1971), pp. 1–23. I. Blanchard, 'The British silver–lead industry and its relations with the Continent, 1470–1570', in W. Kroker and E. Westermann (eds), *Montanwirtschaft Mitteleuropas vom 12 bis 17*

Jahrhundert (Bochum: Der Anschnitt. Beiheft 2, 1984), pp. 183–5 and 'Commercial crisis and change: trade and the industrial economy of the north-east, 1509–1535', *Northern History*, VIII (1973).

4 Westermann, 'Das "Leipziger Monopolprojekt" . . . ', p. 17.

5 E. Westermann, 'Zur Silber-und Kupferproduktion Mitteleuropas vom 15 bis zum fruhen 17 Jahrhundert', *Der Anschnitt*, 83, 5/6 (1986), pp. 187–211; R. Hildebrandt, 'Augsburger und Nürnberger Kupferhandel, 1500–1619. Produktion, Marktanteile und Finanzierung im vergleich zweier Städte und ihrer wirtschaftlichen Fürhrungsschichten', *Zeitschrift für Wirtschafts-und Sozialwissenschaften*, CXII (1972), 1–31.

6 *Stora Kopparbergs Berglag Aktiebolog* (Falun, 1949), p. 9; B. Boethius, *Koppar-Bergslagen fram till 1570-talets genombrott. Uppkomst, medeltid, tidig vasatid* (Uppsala, 1965), pp. 267 ff.

7 A.S. Lindroth, *Gruvbrythyng och Kopparantering vid Stora Kopparberget intill 1800-talets borjan* (Uppsala, 2 vols, 1955), I, pp. 50–1.

8 G. Hammersley, 'Technique or economy? The rise and decline of the early English copper Industry, *c.* 1550–1650', in H. Kellenbenz (ed.) *Schwerpunkte der Kupferproduktion und der Kupferhandel in Europa 1500–1650* (Cologne and Vienna: Kolner Kolloquien zur internationalen Sozial-und Wirtschaftsgeschichte, Bd 3. 1977).

9 On the fortunes of the Swedish industry see Lindroth, op. cit., I, pp. 189–248, 389–471; II, pp. 245–266.

10 K. Glamann, *Dutch-Asiatic Trade 1620–1740* (Copenhagen and The Hague, 1958), p. 176.

11 When Japanese imports were at their peak in 1672–5 they amounted to between a third and a half of Swedish exports. Their importance is attested by their inclusion alongside Norwegian copper, from 1669 to 1688, in the Amsterdam price currents; see K. Glamann, 'The Dutch East India Company's trade in Japanese copper, 1645–1736', *Scandanavian Economic History Review*, I (1953).

12 On the English tariff and its impact see p. 37 and the Russian one p. 38 below.

13 See ch. 4, p. 186.

14 See ch. 3, p. 92 below.

15 See ch. 4, p. 180.

16 On the value of imports before and after the tariff of 1697 see S.E. Astrom, *From Stockholm to St Petersburg. Commercial Factors in the Political Relations between England and Sweden 1675–1700*, (Helsinki: Finnish Historical Society. Studia Historica 2. 1962) appendix 1, pp. 120–2, table 4., p. 144 and on prices E.B. Schumpeter, *English Overseas Trade Statistics, 1697–1808* (Oxford, 1960), table 47, p. 71.

17 English domestic production of copper increased from *c.* 160 tons a year in 1697 to *c.* 375 tons in 1721. See contemporary estimates of Houghton, *A Collection for Improvement of Husbandry and Trade*, nos 256–7 and the anonymous, *The State of the Copper and Brass Manufactures in Great Britain*, 1721.

18 Bengal, however, seems to have been a point of transhipment. Japanese copper imported by the Dutch, for whom it seems to have been the largest single Asiatic market, was thereafter reshipped by the English Company to the domestic market.

19 H. Hamilton, *The English Brass and Copper Industries to 1800* (London, 1926), pp. 107–8.

20 See Astrom, op. cit., p. 145 (1725–6); House of Commons Committee Reports, X, appendix 37, p. 727 (1720–1730).

21 Hamilton, op. cit., pp. 105ff.

22 D.B. Barton, *A History of Copper Mining in Cornwall and Devon* (Truro, 1961), pp. 20–1.
23 On exports from England to the Indies see H. Hamilton, op. cit., p. 280 and on Dutch schemes J.F. Kuiper, *Japan en de buitenwereld in de achttiende eeuw*, ('s-Gravenhage, 1921), p. 152. The English trade was only maintained, however, by producers' involvement in a discriminating monopoly, English consumers subsidizing the Company's trade.
24 J.R. Harris, *The Copper King. A Biography of Thomas Williams of Llanidan* (Liverpool, 1964), ch. 2.
25 Hamilton, op. cit., pp. 180ff.
26 C.J. Schmitz, *World Non-ferrous Metal Production and Prices, 1700–1976* (London, 1979), tables 4/7, 33; 17/6.
27 See ch. 4, pp. 199–200.
28 European production, 1700–1750. (tons):

	Britain	Sweden	Norway	Mansfeld	Harz	Erzgebirge	Total
1700	160	1,275	340	110	1,633	232	3,750
1725	375	908	370	239	2,648	709	5,249
1750	900	782	570	485	2,466		5,203
1775	4,000	856	447	448	1,795		7,746

29 On exports of Russian copper see p. 186 below and on English exports Harris, op. cit., p. 12n.
30 See p. 46 below.
31 On Norwegian production see K. Sprauten, 'Die Norwegischen Kupferwerke als Investionsobjekte, 1720–1760', E. Westermann (hrsgb.) *Quantifizierungsprobleme bei der Erforschung der europaischen Montanswirtschaft des 15 bis 18 Jahrhunderts* (St Katharinen, 1988), pp. 154–69, and on Mansfeld, Herrn Schrader, 'Der Mansfeldsche Kupferschiefer-Bergbau', *Zeitschrift für Berg-, Hütten-, und Salinenwesen in dem Preussischen Staate*, Bd 17 (1869), pp. 251–303.
32 Schmitz, op. cit., table 4/7, 33; 17/6.
33 I. Blanchard, 'English lead and the international bullion crisis of the 1550s', in D.C. Coleman and A.H. John (eds), *Trade, Government and Economy in Pre-Industrial England* (London, 1976), pp. 21–44.
34 I. Blanchard, 'Lead mining and smelting in medieval England and Wales', in D.W. Crossley (ed.), *Medieval Industry* (London: CBA Research Report, no. 40, 1981), pp. 79–80; D.T. Kiernan, 'Technological, economic and social change in the Derbyshire lead industry 1540–1600' (unpublished Ph.D. thesis, 2 vols, University of Sheffield, (1985), I, pp. 216–91.
35 Somerset Record Office, Taunton, DD/SAS/BA 3–5; DD/SHY 9; DD/WG 16/1–15; J.W. Gough, *The Mines of Mendip* (Oxford, 1930), pp. 152–3; P.M. Hembry, *The Bishops of Bath and Wells, 1540–1640* (Oxford, 1967), appendix D.
36 I. Blanchard, 'Derbyshire lead production, 1195–1505', *Derbyshire Archaeological Journal*, XCI (1971), p. 130n; Kiernan, op. cit., I, pp. 148–9.
37 Blanchard, 'English lead. . .', p. 43n and Kiernan, op. cit., II, p. 368.
38 H.-J. Kraschewski, 'Der Bergbau des Harzes im 16 und zu Beginn des 17 Jahrhundert. Stand und Aufgaben der Forschung', in W. Kroker and E. Westermann (eds), *Montanwirtschaft Mitteleuropas vom 12 bis 17 Jahrhundert* (Bochum: Der Anschnitt, Beiheft 2, 1984), pp. 137–40; E. Henschke, *Landherrschaft und Bergbauwirtschaft. Zur Wirtschafts-und Verwaltungsgeschichte des Oberharzer Bergbaugebietes im 16 und 17 Jahrhunderts* (Berlin, 1974), pp. 199ff.; D. Molenda, *Kopalnie rud olowiu na*

terenie zlóz ślasko-/krakowskich w XVI–XVII w. (Wroclaw, Warsaw,-Kraków and Gdansk, 1972), pp. 311ff.

39 Kiernan, op. cit., II, pp. 368, 421 on Derbyshire production and sources listed in note 35 above on Mendip.

40 Public Record Office, London, SP 16/41/30 as quoted in Kiernan, op. cit., II, p. 421.

41 See pp. 19–20 above.

42 On the fortunes of the Polish industry see Molenda, op. cit., pp. 311ff. and 362ff. The history of the Scottish industry has yet to be related.

43 I. Blanchard, 'La loi minière anglaise 1150–1850. Une étude de la loi et de son impact sur le developpement économique. 11ᵉ partie. Mythe, 1550–1850', unpublished paper, presented at EHESS, Paris, 1985.

44 ibid.

45 ibid.

46 R. Burt, 'Lead production in England and Wales, 1700–1770', *Economic History Review*, 2nd series, XXII, 2 (1967), p. 251.

47 I. Blanchard, 'Resource depletion in European mining and metallurgical industries, 1400–1800', in A. Maczak and W.N. Parker (eds), *Natural Resources in European History* (Washington; RFF Research Papers, R-13, 1979), p. 105.

48 Gough, op. cit., pp. 157ff.

49 I. Blanchard, 'Labour force organization and labour productivity in the English mining industries, 1400–700', *XVIIIa Settiman di Studio, Miniere e Metallurgia. Instituto Internazionale di Storia Economica "Francesco Datini", Prato 11–15 April 1986* (in press).

50 Burt, op. cit., p. 251.

51 Blanchard, 'La loi minière. . .'

52 Burt, op. cit., p. 259.

53 ibid., pp. 253–4.

54 R. Burt, *The British Lead Mining Industry* (Redruth, 1984), pp. 234ff.

55 Calculated from R. Burt, 'Lead production. . .', table 2, p. 259.

56 See figure 2.6.

57 See pp. 34, 46 above.

58 On the technological parameters of the cupelation process see Blanchard, 'The British silver–lead industry. . .' pp. 179–82

59 The history of the Serbian and Bosnian mines in the seventeenth century has yet to be written but some information may be gleaned from R. Anhegger, *Beitrage zur Geschichte des Bergbaues im osmanischen Reich*, (Istanbul, 3 vols, 1943–5).

60 On Middleton's activities at Cwmsymlog see W.G. Lewis, 'The Cymsymlog lead mine', *Ceredigion*, 11 (1952), pp. 27ff, whilst de Beausoleil's activities are chronicled by R. Gandilhon, 'Une mine d'argent dans la Baronie de Brugny découverte par le Baron de Beausoleil (1632)', *Histoire des Enterprises*, XII (1963).

61 See figure 2.6.

62 K. Sternberg, *Umrisse einer Geschichte der Böhmischen Bergwerke*, 1 vol in 2 parts (Prague, 1836–7).

63 A. Jager, *Beitrag zur Tirolisch – Salzburgischen Bergwerkgeschichte* (Vienna: Archivs fur Osterreichische Geschichte, Bd 53, 1875), pp. 337–436; J. von Sanger, *Beitrage zur Geschichte des Bergbaues in Tirol*, (Innsbruck: Sammler für Geschichte und Statistik in Tirol, Bd 1, 1807).

64 J. Vlachovič *Slovenská Med v16 a 17 Storočí* (Bratislava, 1964), pp. 218–46.

65 Bjorn Berg, 'Produktion, Belegschaft und Produktivitat beim Kongsberger Silberbergwerk 1623–1805', in E. Westermann (hrsgb.)

Quantifizierungsprobleme bei der Erforschung europaischen der Montanwirtschaft des 15 bis 18 Jahrhunderts (St Katharinen, 1988), pp. 127–53 deals comprehensively with the Norwegian silver mine providing much more complete production data than is available in the brief *Kongsberg, Foredrag i tilknything til ekskursjon 11 September 1976* (Oslo, 1977) published by Det Nordsk Viderskap-Akademi.

66 On Harz production at this time see C. Bose, *Generale Haushalts-Principia vom Berg-, Hutten-, Salz- und Forstwesen in Specie vom Harz* (Berlin, 1877), the latter work providing invaluable information on the value of each mine Kux (share) which highlights the importance of the crisis of 1725, and C. Gatterer, *Anleitung den Harz und andere Bergwerke mit Nuzen zu bereisen*, vols 1–3 (Göttingen, 1785–1790), vols 4–5 (Nürnberg, 1792–1793).

67 Information on the output of the Freiberger Revier is taken from G.E. Beneseler, *Geschichte Freibergs und seines Bergbau* (Freiberg, 1843), and J.F. Gmelin, *Beitrage zur Geschichte des teutschen Bergbau* (Halle, 1783), the latter work also providing information on silver output at Johanngeorgenstadt, where production commenced in 1662 and thereafter increased until the early eighteenth century when some 4,149 marks a year were produced before a relentless decline set in. The information on this mine is unfortunately scanty before 1762 when C.G.A. Weissenbach's *Sachsens Bergbau, National-ökonomisch betrachtet* (Freiberg, 1883) provides comprehensive data on Saxon production.

68 J. Ferber, *Physikalisch-metallurgische Abhandlungen über der Gebirge und Bergwerke Ungarns* (Berlin, 1780) and M. von Schwartner, *Statistik des Königreichs Ungarn*, in 2 parts (Ofen, 1809 and 1811).

69 See note 65.

70 See note 65.

71 See note 65 and 67.

72 See ch. 3. p. 59f. below.

73 See ch. 5, p. 215f. below.

74 On barrel-amalgamation, its utilization in Europe after 1780, and its introduction into the Americas see the papers in *200 výročie zavendenie nepriamej amalgamácie a založenia I. medizinárodnej vedeckej spoločnosti na svete-Zbornik*. (Donovaly-Sklené Teplice-Banská Stiavnica, 1986) and Randall, op. cit., pp. 118–25.

3 Russia. Precious-metal production

1 This chapter is based largely upon four main groups of sources. First, there are the accounts of mining operations by contemporary mining officials – H.M. Renovantz, *Oberbergmeister* of the Kolyvan mining region and teacher in mining sciences at the Imperial Mining School of St Petersburg, and B.F.J. Hermann, who held the same office from 1785 to 1796 before being moved to Ekaterinburg and promoted to *nachal'nik* – whose studies were based on documents, many now lost, kept in the offices of the local mine administration and in the archives of the central mine administration at St Petersburg:

H.M. Renovantz, *Mineralogische – geographische und andere vermischte Nachrichten von den Altaischen Gerburgen Russische Kanserlichen Anteil* (Reval, 1788) and the Russian translation, *Mineralogischeskie, geographicheskie i drugie smeshanie izvestiya o Altaiskikh gorakh* (St Petersburg, 1792).

B.F.J. Hermann, *Versuch einer mineralogischen Beschreibung der uralischen Erzgeburges* (Berlin and Stettin, 1789), published together with *Statistiche Schilderung von Russland* (St Petersburg and Leipzig, 1790) after his second tour of duty, 1785–9. He subsequently made a further tour of duty 1790–6 and on returning published 'Sur l'exploitation des mines de l'empire de Russie', *Nova Acta Academiae Scientiarum Imperialis Petropolitanae*, 11 (St Petersburg, 1798); *Mineralogische Reisen in Sibirien vom Jahr 1783 bis 1796* (St Petersburg, 1797–8) and *Sochineniya o sibirskikh rudnikakh' i zavodakh'* (St Petersburg, 1797–1801), incorporating materials updating his earlier data published in 1790. After his appointment at Ekaterinburg he published, *Die Wichtigkeit des russischen Bergbaues* (St Petersburg, 1810a); *Istoricheskoe nachertanie gornogo proizvodstva v rossiiskoi imperii* (Ekaterinburg, 1810b), of which only one volume was published relating to the history of the industry to 1740, and, 1808, *Opisanie zavodov pod vedomostvom Ekaterinburgskogo gornogo nachal'stva sostoyavshikh* (Ekaterinburg, 1808).

Henceforth, each of these works will be cited by reference to the author's surname and the date of publication.

Second, there are the accounts of mining operations by individuals who either had the opportunity to discuss conditions with mining officials or who had access to the archives of the central mine administration at St Petersburg.

H. Storch, *Tableau historique et statistique de l'Empire de Russie à la fin du dix-huitième siècle*. (Basle and Paris, 2 vols 1801) or the German translation of this work (Leipzig, 1803).

W. Coxe, *Travels into Poland, Russia, Sweden and Denmark* (London, 1784).

Again these henceforth will be cited by reference to the author's surname and the date of publication.

There are, third, the accounts of contemporary visitors to the mines who also gained access to the records of the chancery of the local mining administration. The most important of these are: Gmelin, who visited the Kolyvan works as part of the Great Northern Expedition of 1733–43, and Pallas and Falk, who visited the works as part of the so-called 'Orenburg Expedition' of 1768–74.

J.G. Gmelin, *Reise durch Sibirien von dem Jahr 1733 bis 1743* (Göttingen, 1751–2).

J.P. Falk, *Beitrage zur topographischen Kentniss der russischen Reichs* (St Petersburg, 1785).

P.S. Pallas, *Reise durch verschiedene Provinzen des russischen Reichs in einem ausfuhrlichen Auszuge* (Frankfurt and Leipzig, 1776–8), translated into French, *Voyages de M.P.S. Pallas en differentes provinces de l'empire de Russie et dans l'Asie septentrionale* (Paris, 1788–93).

Henceforth these works will also be cited by reference to the author's surname and the date of publication.

Finally, there are modern Soviet studies, based on surviving archival materials, which have been culled for their source materials, and modern editions of documents relating to mining matters.

N.I. Pavlenko, *Razvitie metallurgicheskoi promyshlennosti Rossii v pervoi polovine XVIII veka. Promyshlennaya politika i upravlenie* (Moscow, 1953a) as well as the same author's 'Materiali o razvitii ural'skoi promyshlennosti v 20–40 kh godakh XVIII v.', *Istoricheskii Arkhiv*, ix, (1953b) and *Istoriya metallurgii v Rossii XVIII veka. Zavodi i zavodovladel'tsi* (Moscow, 1962).

B.B. Kafengauz, *Istoriya khozyaistva Demidovikh v XVIII–XIXv. Opit issledovaniya po istorii Ural'skoi metallurgii* (Moscow, 1949).

V.V. Danilevskii, *Russkoe zoloto, Istoriya otkritiya i dobichi do seredini XIXv* (Moscow, 1959).

Z.P. Karpenko, *Gornaya i metallurgicheskaya promyshlennost' zapadnoi Sibiri v 1700–1860 godakh* (Novosibirsk, 1963).

Again these works will be henceforth cited by the author's surname and the date of publication. In order to afford the reader an impression of the surviving source materials in Soviet archives and to relate these materials to those cited by contemporaries, however, Russian archival references are also cited. The reader wishing to set these in context is referred to appendix 1 pp. 307–8 where these sources are discussed and the abbreviations used here are explained.

Of a more general character are the imperial edicts collected in *Polnoe sobranie zakanov rossiskoi imperii*, henceforth abbreviated as PSZ.

2 See p. 55 above.

3 See chapter 1, n. 6 above.

4 On Norway see A. Soetbeer, *Edelmetall-produktion und Werthverhältniss zwischen Gold und Silber seit der Entdeckung Amerika's bis zur Gegenwart* (Gotha, 1879), p. 35. On silver production in the Rhineland D. Haberle, 'Die bergbaulichen Verhaltnisse der Pfalz', *Pfalzer Heimatkunde*, 15 (1919), pp. 72ff.

5 M. Levy, *Der Silber – und Blei-Bergbau zu Przibram* (Vienna, 1875).

6 See p. 55 above.

7 Soetbeer, op. cit., pp. 60–4, 89.

8 See p. 55 above.

9 See note 7 above.

10 Soetbeer, op. cit., p. 111.

11 ibid.

12 Danilevskii, 1959.

13 T. Armstrong, *Russian Settlement in the North* (Cambridge, 1965), p. 23.

14 ibid., pp. 53–4.

15 ibid., p. 24.

16 V.I. Shunkov, 'Geograficheskoie razmeshchenie Sibirskogo zemledeliya v XVII veka', *Voprosi geografi*, 20 (1950), pp. 203–38.

17 J.R. Gibson, *Feeding the Russian Fur Trade. Provisionment of the Okhotsk Seaboard and the Kamchatka Peninsula, 1639–1856* (Madison, Wisc., 1969), p. 9.

18 C.M. Foust, *Muscovite and Mandarin. Russia's trade with China and its Setting, 1727–1805* (Chapel Hill, 1969), pp. 11–12, and B.G. Kurts, *Gosudarstvennaya monopoliya Rossii s Kitaem v pervoi polovine XVIIIs.* (Kiev, 1928), pp. 1–27.

19 Between 1676 and 1710 the population of the lower Yenisei increased from 480 to 1,360 Russians: V.A. Aleksandrov, 'Russkoe naselenie Mangazeysko-Turukhanskogo kraya v XVIII- pervoi polovine XVIII veka, *Institut Etnografii imeni N.N. Miklukho-Maklaya. Kratkie soobshcheniya*, 35 (1960), pp. 14–24, and M. Klochkov, *Nasleniya Rossii pri Petre Velikom po perepisyam togo vremeni*, I (St Petersburg, 1922), pp. 61–9. Between 1696/7 and 1710 the population of the Lena basin and the lands beyond similarly increased from 1,222 to 4,161, Klochkov, op. cit., 61–9.

20 Foust, op. cit., p. 13.

21 S.P. Krasheninnkov, *Opisanie zemli Kamchatki* (Moscow, 1949), p. 476; V.K. Andrievich, *Sibir v XIX veka*, (St Petersburg, 1889), II, pp. 370–1.

22 Gibson, op. cit., p. 11.
23 Ibid., p. 12.
24 Foust, op. cit., p. 13.
25 ibid.
26 On the trade in the early 1720s see L. Lange 'Journal of the Residence of Mr. de Lange, Agent of his Imperial Majesty of All the Russians, Peter the First, at the Court of Peking during the Years 1721 and 1722', in J. Pinkerton, *A General Collection of the Best and Most Interesting Voyages and Travels in All Parts of the World*, VII (London, 1811), p. 456 and PSZ, 1st series, t.VII, no. 4992. The subsequent history of the trade will be found in Foust, op. cit., chs 4, 6, and 8.
27 Foust, op. cit., pp. 332–3.
28 See this chapter, pp. 70f.
29 Amongst a large literature on this subject see the work of Gibson referred to in note 17 and S.B. Okun, *The Russian–American Company* (Cambridge Mass., 1951).
30 Yu. V. Kozhukhov, *Russkie krest'yane vostochnoi Sibiri v pervoi polvine XIX veka. 1800–1861* (Leningrad, 1967), pp. 59–60.
31 On the history of the Nerchinsk mines from 1704 to 1745 see Pavlenko, 1953b – a special report compiled by officials of the Mining College in 1746/7 for the Senate and preserved in the latter's archive: TsGADA, Senate archive, dok 1510. On the years 1704–9 see ibid., pp. 176–7 and on the Troitsk mine, Hermann, 1797–8, II, pp. 279ff.
32 See pp. 83–4 below.
33 Pavlenko, 1953b, pp. 176–9.
34 See p. 171 below.
35 Pavlenko, 1953b, pp. 182–3, 194–5, 200–1, 212–213, 218–9, 228, 230–1, 236–7.
36 See p. 84 below.
37 Table 3.1. Renovantz, 1788, pp. 263–6; Hermann, 1797–8, II, pp. 289–96.
38 Table 3.2. Sources as table 3.1.
39 Table 3.3 Free workers and convicts 1704–36, Pavlenko, 1953a, pp. 236–7. Peasants 1708–36, I Bogoslovskii, *Istoriko-statisticheskii ocherk proizvoditel'nosti Nerchinskogo gornogo okruga s 1703 do 1871* (St Petersburg, 1873), p. 3, 1772. Coxe, 1784, II, p. 286; 1782. Storch, 1801, II, p. 380, Hermann, 1797–8, II, pp. 11–12, Hermann, 1790, pp. 323–5. 1794. Hermann, 1797–8, II, pp. 266–7.
40 Pavlenko, 1953a, pp. 236–7.
41 ibid.
42 Pavlenko, 1953a, pp. 112, 143–4. Ukaz dated 19 August 1720 founding the Nerchinsk *nachal'stva*, recorded in register of edicts issued by the Mining College – TsGADA, archive of Mining College (fond 271), bk 2, fo. 89.
43 On the size of the obrok in grain see V.N. Sherstoboyev, 'Zemledelie severno Predbaikalya v XVII–XVIII vv', in B.D. Grekov (ed.), *Materialia po istorii zemledeliya SSSR* (Moscow and Leningrad, 1952), I, pp. 281–6.
44 Just as the group of ascribed peasants engaged in grain cultivation can be estimated as numbering 238 by dividing the consumption requirements of the full-time workers (i.e. number 95 × per capita consumption 125 kg, see p. 239 below) by the *obrok* obligations of the peasantry (3 puds = 108 lb, see note 43 above), leaving a residual 244 peasants to engage in other activities, the obligations of this latter group may be estimated by dividing the charcoal consumption of the works (ore smelted 105,754 puds divided by consumption 15.5 puds per koroba = 6,822 koroba or the produce, at 70 koroba per kucha,

of 98 charcoal beds) by the number of teams of five employed i.e. 49, giving a team obligation of cutting wood for, and making, two beds. On technological parameters of charcoal-making see p. 78 below.

45 Coxe, 1784, II, p. 286.
46 Calculated in the manner described in note 44. 5,000 peasants delivering an obrok in kind of 108 lb of grain render grain to feed 1,800 full-time workers viz 24,500 kg. Similarly 6,000 peasants, organized into teams of five each making two charcoal beds yielding at this time 100 korobi, produce 120,000 korobi or enough to smelt 1.86 million puds of ore.
47 Dividing average output per man into total mine output per year. Renovantz, 1788, pp. 263–6; Hermann, 1797–8, pp. 289–96.
48 See p. 85 below.
49 Storch, 1801, II, pp. 399–400.
50 That is 1,963 ex-colonists organized into 982 teams making 1,963 beds at 63 korobi each = 123,669 korobi or enough to smelt 1.92 million puds of ore and requiring 39,260 cubic sazhen of wood, produced by 6,000 peasants at a rate of 6.5 cubic sazhen per man/year or 4 man days per cubic sazhen.
51 1.8 million puds output in 1794 divided by workforce of 2,000.
52 Hermann, 1797–8, pp. 278–89.
53 Pavlenko, 1953b, pp. 175ff.
54 Figure 3.3 See appendix 1, table A1.
55 Pavlenko, 1953b, p. 228.
56 ibid., pp. 200–1, 212–13.
57 See p. 100 below.
58 Figure 3.4. Hermann, 1797–8, pp. 286–8.
59 Pavlenko, 1953b, pp. 262–7.
60 Coxe, 1784, II, 286; Hermann, 1790, p. 325.
61 Hermann, 1797–8, pp. 286–8.
62 ibid.
63 See p. 64ff. above.
64 For convenience bounded in the east by the lower Yenisei.
65 On the raids into the Tiumen district in 1634–6 see P.N. Butsinskii, 'Sibirskie arkhiepiskopi. Makarii, Nektarii, i Geramsin, 1625–1650', *Vera i razum*, 10 (1891), pp. 233–4; on the incursions along the Ishim and Tobol in 1647, P.A. Slotsov, *Istoricheskoe obozrenie Sibiri* (St Petersburg, 1886), I, pp. 65–6; whilst attacks on Kuznets in 1630 and Tiumen and Tara in 1634 are noted in Slotsov, op. cit., I, p. 41 and *Russkaya istoricheskaya biblioteka*, II, pp. 587–8.
66 N.S. Orlova, *Otkritiya russkikh zemleprokhodtsev i polyamykh morekhodov XVII veka na severo-vostoke Azii: sbornik documentov* (Moscow, 1951), pp. 88–9.
67 ibid., pp. 80–2.
68 On population in 1660, Slotsov, op. cit., I, pp. 84–5.
69 M.I. Belov, *Istoriya otkritiya i osvoieniya severnogo morskogo puti* (Moscow, 1956), I, 290.
70 S.V. Bakhrushin, *Nauchnie trudi* (Moscow, 1955), III–1, pp. 297, 334.
71 *Akti istoricheskie* (St Petersburg, 1841–2), II, pp. 313–14.
72 V.M. Kabuzan, *Narodonaselenie Rossii v XVIII – pervoi polovine XIX v* (Moscow, 1963), p. 161, and A. Schultz, *Sibirien: eine Landeskunde* (Breslau, 1923), p. 167.
73 Falk, 1785, I, pp. 284–9.
74 Schultz, op. cit., p. 167.
75 Hermann, 1797–8, I, p. 305; Renovantz, 1788, p. 230.
76 Renovantz, 1788, p. 230.

77 See p. 169 below.
78 Kafengauz, 1949, p. 186. Reports of Mining College compiled from registers of grants to establish works. TsGADA, archive of Mining College (fond 271), dok. 978.
79 Hermann, 1797–8, I, p. 305; Falk, 1785, I, p. 295.
80 Hermann, 1797–8, I, p. 305.
81 Appendix 1, table A1.7.
82 Hermann, 1797–8, I, p. 305.
83 Appendix 1, table A1.7.
84 Falk, 1785, I, p. 303.
85 Appendix 1, table A1.6; Renovantz, 1788, p. 231.
86 Kafengauz, 1949, p. 199. TsGADA, archive of Mining College (fond 271), dok. 980.
87 Hermann, 1797–8, I, p. 305.
88 Gmelin, 1751–2, I, pp. 249–61.
89 Hermann, 1797–8, I, pp. 308–9.
90 ibid.
91 Kafengauz, 1949, p. 222: 'General essay' on the estate of Akinfi Demidov *c.* 1745–7. TsGADA Gosarkhiv, razr. XI, dok. 95/2, comprising a separate essay on the works in Kuznets *uezd*.
92 Appendix 1, table A1.7.
93 Falk, 1785, I, p. 303; Kafengauz, 1949, p. 192; Gmelin, 1751–2, IV, pp. 400–10.
94 Kafengauz, 1949, p. 207: Beyer Commission report, compiled from data at Mining College for the Senate, TsGADA. Dela Senata po Berg Kollegii, dok. 8/1510.
95 Kafengauz, 1949, p. 222: 'General essay'.
96 B Rozhov, 'A.N. Demidov na svoikh Kolivano-Voskresenskikh zavodakh', *Gornii zhurnal* (1891), t. 3, pp. 343–4.
97 Kafengauz, 1949, p. 179.
98 Gmelin, 1751–2, I, pp. 249–50.
99 See pp. 301–2 below.
100 See pp. 323–4 below.
101 See pp. 323–4 below.
102 Hermann, 1797–8, I, p. 314.
103 Appendix 1, table A1.7.
104 Falk, 1785, I, p. 304.
105 PSZ, 1st series, t. XII, no. 9403.
106 E.g. Coxe, 1784, II, p. 284; Storch, 1801, II, p. 369.
107 Hermann, 1797–8, I, pp. 312–14.
108 ibid., I, pp. 312–14.
109 ibid., I, pp. 317–18: Report 13 December 1750 of General Beyer.
110 Renovantz, 1788, pp. 239–41, 252.
111 ibid., pp. 10–12.
112 Appendix 1, table A1.2.
113 Hermann, 1797–1801, II, p. 286.
114 Renovantz, 1788, pp. 44–47; Falk, 1785, I, p. 319.
115 Falk, 1785, I, pp. 320–1; Renovantz, 1788, pp. 910, 10–12.
116 Appendix 1, tables A1.1–2.
117 Renovantz, 1788, pp. 84–218.
118 Table 3.5, 1747–50. Hermann, 1797–8, I, p. 318. 1751–8, derived by subtracting output for 1745–50 and 1759–64 from total 1745–64 in Renovantz, 1788 description of Schlangenberg (pp. 89–180). 1759–71. Falk, 1785, I,

p. 318. 1787. Hermann, 1790, pp. 318–319. 1791. Hermann, 1797–8, III, p. 280. 1772–83, derived by deducting 1745–71 from total 1745–83, Storch 1801, II, p. 379.

119 Falk, 1785, I, pp. 314–15.
120 ibid., I, pp. 311–13.
121 ibid., I, p. 316; Hermann, 1797–8, I, p. 327.,
122 Falk, 1785, I, pp. 328–9.
123 ibid., I, pp. 330–1.
124 Hermann, 1797–8, I, pp. 324, 327; Falk, 1785, I, pp. 330–1.
125 Falk, 1785, I, p. 331.
126 See pp. 196–7 below.
127 Falk, 1785, I, p. 307.
128 Table 3.6. Renovantz, 1788, pp. 204–5; Falk, 1785, I, p. 320.
129 Falk, 1785, I, pp. 331–2.
130 ibid.
131 Appendix 1, table A1.8.
132 Falk, 1785, I, pp. 332–3.
133 Falk, 1785, I, p. 318: Hermann, 1790, pp. 318–22; Hermann, 1797–8, III, p. 280.
134 Table 3.5 above and appendix 1, table A1.2.
135 Falk, 1785, I, p. 318; Hermann, 1790, pp. 318–22.
136 Table 3.7. 1763–71, Falk, 1785, I, pp. 318–19; 1763–83, Renovantz, 1788, pp. 206–17; 1791, Hermann, 1797–8, III, p. 280.
137 On the Schlangenberg crisis of 1762–5 and the subsequent decline in yields see Falk, 1785, p. 318, and on the washery constructed on the lower Korbolicha in 1769, Falk, 1785, I, p. 316, Hermann, 1797–8, I, p. 327.
138 Renovantz, 1788, pp. 10–12.
139 ibid., pp. 12ff.
140 ibid., pp. 27–31.
142 ibid., p. 189; Hermann, 1797–1801, II, pp. 81–7; Hermann, 1797–8, III, p. 280.
143 Hermann, 1790, pp. 318–22; Hermann, 1797–8, III, p. 280.
144 Hermann, 1790, pp. 318–22.
145 See figure 3.1.
146 Karpenko, 1963, pp. 102–10. Table 3.8. Hermann, 1797–1801, III, pp. 51–3, 95–104, 120–5.
147 On production from 1788 to 1795 see table 3.8. 1840, Karpenko, 1963, p. 142. On the size of the labour force 1791 see table 3.9, 1795, Hermann, 1797–1801, I, p. 58. 1809 and 1850, Karpenko, 1963, p. 142 from workers' lists in Barnaul mining office GAAK, Chancery archive (fond 1) op. 2 sv 616 dok. 1078 and mining section archive (fond 2) op. 5 dok. 480 on which see discussion in note 163 below.
148 Table 3.9. Hermann, 1797–1801, III, pp. 60–2 and Hermann, 1810a, pp. 66–8.
149 Table 3.9. Hermann, 1797–1801, III, pp. 60–2 and ibid., II, pp. 184–266.
150 Table 3.10. Hermann, 1797–1801, II, pp. 275–7.
151 See p. 102 above.
152 Appendix 1, table A1.8.
153 Renovantz, 1788, pp. 44–57; Hermann, 1797–1801, III, p. 280.
154 Appendix 1, table A1.8.
155 Hermann, 1797–1801, I, pp. 259–67. In 1798 the main works employed 2,323 men: Barnaul 969, Pavlovsk 595, Susunsk 759. Karpenko, 1963 from GAAK, Chancery archive (fond 1) op. 2 sv 50 dok. 89.
156 Hermann, 1797–1801, I, pp. 259–67.

157 Hermann, 1797–8, I, pp. 330; Renovantz, 1788, pp. 57–60.
158 ibid.
159 Table 3.11. Hermann, 1797–8, III, p. 280.
160 Table 3.12 Hermann 1797–1801, I, pp. 259–67.
161 Kolyvan, 1746, appendix 1 table A1.6. 1771, Renovantz, 1788, pp. 221–72,
     Falk, 1785, I, pp. 327–8. 1795, Hermann, 1797–1801, I, pp. 260–1. Barnaul,
     1746, appendix 1 table A1.6.1749–1750, Falk, 1785, I, p. 305. 1769, Hermann,
     1797–8, I, p. 327. 1771, Falk, 1785, I, pp. 328–9. 1795, Hermann, 1797–8, II,
     pp. 6–8. Pavlovsk 1763/4, Hermann, 1797–8, I, p. 324. 1769, ibid., I, p. 327.
     1771, Falk, 1785, I, pp. 329–330. 1794, Hermann, 1797–8, II, pp. 6–8.
     Susunsk 1764/5, Hermann, 1797–8, I, pp. 325–326, Falk, 1785, I, p. 307. 1771,
     ibid., pp. 330–2. 1788, Hermann, 1797–8, I, p. 350. 1795, ibid., II, pp. 6–
     8.Aleisk 1775, Hermann, 1797–8, I, p. 327, Renovantz, 1788, pp. 27–31. 1795,
     Hermann, 1797–8, II, pp. 6–8. Loktevsk 1781/2, ibid., I, p. 330, Renovantz,
     1788, pp. 57–60. 1795, Hermann, 1797–8, II, pp. 6–8. Gavrilovsk 1795, ibid.
162 Appendix 1 tables A1.2 and A1.8.
163 Table 3.14. The data contained herein is derived entirely from materials
     collected and stored in the Barnaul chancery (or copies at St Petersburg) and
     as such it resembles the serf lists found on estates elsewhere (see for example
     S.L. Hoch, *Serfdom and Social Control in Russia. Petrovskoe, a Village in
     Tambov* (Chicago, 1986)) rather than the records of the 'Revizions'. It seems
     to have been based on two main groups of documents both of which were
     consulted by Falk at the time of his visit to the mining camps (Falk, 1785, I, pp.
     311–19) and works (ibid., I, pp. 328–32). First, there seems to have been at
     each settlement at that time a series of annual registers containing information
     on the numbers of *all* inhabitants, men and women, children and adults, as
     well as information on their religious affiliations, none of which, judging by the
     work of Soviet historians, have survived. Second, there were the files of
     workers' passports. These still survived in the records of the Kolyvan mining
     nachalstva, bureau 35, in the late nineteenth century when they were consulted
     by D.N. Belikov, *Pervie russkie krest'yane-nasel'niki Tomskogo kraya*
     (Tomsk, 1898), many dating back to the earliest days of the enterprise. These
     have since been lost, only one set for the eighteenth century, for the Salairsk
     mine in 1795 (Karpenko, 1963, p. 218 GAKO fond 6 op. 3 dok. 171 containing
     841 passes), surviving, containing the following data: 1 Name of worker. 2
     Age. 3 Social origin (eg. master worker's son or peasant). 4 Career. 5 Whether
     literate or not. 6 Details of any misdemeanours during career. 7 Marital status.
     8 Number of children.
        The data available to both contemporaries and modern historians therefore
     relates to *all* persons, not just taxable males contained in the Revizion returns.
     Contemporaries, moreover, in referring to 'revisional souls' (male *and* female)
     spoke only of those ascribed to the works, a category encompassing both
     master workers and their dependents and the ascribed peasantry and
     excluding those *urochniki* recruited after 1783 who were not ascribed to the
     works and who have thus not been included in table 3.14. These numbered
     2,400 in 1795–1796/7. With these facts in mind the available source material
     may be considered. 1747, Kafengauz, 1949, p. 228: 'General Essay' 736 master
     workers; ibid., p. 207 source not given, 1758, masters with dependents, Falk,
     1785, I, p. 305, 3,121 peasants of whom 2,004 were said to be at work. Of these
     736 workers, 316 were factory workers, 166 of peasant origin (482 in all,
     recruited before 1738), and the remaining 254 were incomers of unknown
     origin. Of the total ascribed peasants numbering 3,121 only 7 were from the
     period prior to 1738, the remainder being incomers: 'General essay'. 1763–5,

on the total see Falk, 1785, I, p. 307; Hermann, 1797–1798, I, pp. 319–322.
40,008. On master workers and their dependents, Karpenko, 1963, p. 174,
quoting Beimarn's contemporary manuscript chronicle of the works. 1771:
Falk, 1785, I, p. 335, detailed list of master workers. 1782–6 peasants 54,000,
master workers 5573: Hermann, 1790, pp. 318–22; peasants 54,750, master
workers and dependents 13,151 (broken down by place of residence) = 67,901:
Hermann, 1797–1801, I, pp. 286–7. 1795–7, total 69,612 and by fifth revision
70,000 Hermann 1797–8, II, pp. 8–11. Peasants 55,306 (after emancipation of
Tomsk peasantry) Karpenko, 1963, p. 74, quoting GAAK, Chancery archiv
(fond 1). op. 2 sv 32 dok. 58, list of peasants broken down by district of
residence. 14,306 master workers and dependents of whom 7,453 are at work
Hermann, 1797–8, II, pp. 8–11, which provides a detailed breakdown of this
group. The larger figures of 9,311 quoted in N. Zobin, 'Masterovie altaiskikh
gornikh zavodov do osvobozhdeniya', *Sibirskii Sbornik*, II (1892), p. 11, and
10,000 quoted in Hermann, 1810a, pp. 66–8 and dated in A.G. Rashin,
Formirovanie promishlennogo proletariata v Rossii (Moscow, 1940), p. 46 to
1802–6 include also 1,858 and 2,547 *urochniki* respectively.

164 Table 3.14 and Schultz, op. cit., p. 167.
165 Hermann, 1797–8, I, p. 305.
166 Belikov, op. cit., p. 24, and Hermann, 1797–8, I, p. 305.
167 V.M. Kabuzan and N.M. Shepukova, 'Tabel' pervoi revizii narodonaseleniya
    Rossii (1718–1727gg)', *Istoricheskii arkhiv*, II (1959), p. 163.
168 Karpenko, 1963, p. 50, quoting Beimarn's chronicle.
169 Belikov, op. cit., pp. 28, 30, 33, 119; Hermann, 1797–8 I, p. 305; Gmelin, I,
    pp. 249–61.
170 Gmelin, 1751–2, I, pp. 249–61.
171 ibid.
172 Falk, 1785, I, p. 302; Hermann, 1797–8, I, p. 305.
173 Kafengauz, 1949, p. 228: 'General essay'.
174 Kafengauz, 1949, p. 222: 'General essay'.
175 Falk, 1785, I, p. 305.
176 PSZ, 1st series, t. IX, no. 6559.
177 Karpenko, 1963, p. 50, quoting response of management to a question posed
    on the occasion of the Beyer commission and preserved in the archive of its
    papers TsGIA-L fond 468 op. 315/476 dok. 47.
178 The 'actual' cubic sazhen diverged from that cut by a factor of 2.5. On time
    expenditure on coaling and woodcutting see the 'response' note 177 and p. 139f.
    below.
179 See work referred to in note 177.
180 Table 3.15. See work referred to in note 177.
181 Table 3.16. Storch, 1801, pieces justificatives, pp. 43–4.
182 Appendix 1, pp. 227ff.
183 Hermann, 1797–8, I, pp. 311–18.
184 ibid.; Falk, 1785 I, p. 334; Hermann, 1790, pp. 318–22; Hermann 1797–8, II,
    pp. 8–11.
185 Hermann, 1797–8 pp. 8–11.
186 Karpenko, 1963, pp. 91–5.
187 Hermann, 1797–8, I, pp. 319–22; II, pp. 8–11; Falk, 1785, I, p. 334.
188 Hermann, 1810b, I, pp. 49–70; PSZ, 1st series, t.XVIII. no. 133.03; PSZ, 1st
    series, t.XX, no. 14878. Renovantz, 1788, p. 179; Storch, 1801, II, p. 395.
189 Table 3.18. Falk, 1785, I, p. 335.
190 Falk, 1785, I, pp. 311–19.
191 ibid., I, pp. 328–33.

192 Table 3.19. ibid., I, pp. 328–9.
193 ibid., I, p. 327.
194 ibid., I, p. 385; Hermann, 1790, pp. 318–22.
195 Storch, 1801, II, p. 396; Renovantz, 1788, p. 179.
196 Falk, 1785, I, p. 355.
197 Renovantz, 1788, pp. 89–180.
198 Hermann, 1797–8, II, pp. 8–11.
199 Karpenko, 1963, pp. 83, 86–8 (Salairsk, 1795), pp. 155–7 (1820/31, 1840) based upon the files of passports referred to in note 43 above.
200 Falk, 1785, I, p. 335; Hermann, 1797–8, II, pp. 8–11.
201 On the ordinance of ascription PSZ, 1st series, t. XII no. 9403 and the mining ordinances of 1747, Hermann, 1797–8, I, p. 311. Karpenko, 1963, p. 73 for data in table 3.20.
202 Falk, 1785, I, p. 305.
203 Karpenko, 1963, p. 73; Falk, 1785, I, p. 305.
204 Table 3.21. Hermann, 1797–1801, I, pp. 286–7; Karpenko, 1963, p. 74. By an edict of 3 March 1797 (PSZ, 1st series, t. XXIV, no. 17862) 7,787 peasants in Tomsk oblast were freed from industrial work.
205 Table 3.22: Urals. PSZ, 1st series, t. VII, no. 4425; PSZ, 1st series, t. XVIII, no. 13303; PSZ, 1st series, t. XX, no. 14878. V.I. Semevskii, *Krestiane v tsarstvovanie imperiatritsi Ekaterini II* (St Petersburg, 1901), pp. 514ff. Altai, Karpenko, 1963, p. 53; 'General Essay'; Renovantz, 1788, pp. 334–5; Hermann, 1790, pp. 320–2.
206 On Altai wages for non-obligatory work see table 3.15 above and for an exactly contemporaneous description of 'free'-market wages at the Bimovskii works in the Urals see A.S. Savich, *Ocherki istorii krestianskikh volnenii na Urale v XVII–XVIII v.* (Moscow, 1931), pp. 15–16, 25.

Category	Altai	Urals
	(k. per day)	
Peasant without horse		
winter	7	5
summer	6–7	7
Peasant with horse		
winter	10	9
summer	12	10

207 On the shift to payments by the summer tariff, Renovantz, 1788, pp. 334–5.
208 Semevskii, op. cit., pp. 450, 508.
209 W. de Hennin, *Opisanie ural'skikh i sibirskikh zavodov 1735*, ed. M.A. Pavlov (Moscow 1937), pp. 356ff; Semevskii, op. cit., pp. 450–1.
210 ibid.
211 Table 3.23. Karpenko, 1963, p. 53: 'General essay'; Renovantz, 1788, pp. 334–5; Hermann, 1790, pp. 320–2, 414–15. On changes in the size of the cubic sazhen see note 212 below.
212 The sazhen was steadily increased in size as an accounting unit in the period to 1778: Wood cutting and coaling. The 'cubic sazhen' and koroba.

1720 6-chetvert measure cut in lengths of 3.5 arshin = 1.16 sazhen. Cubic measure accordingly (1.16 × 1.16 × 1.16 sazhen) 1.56 cubic sazhen. The official charcoal bed or *kucha* of 20 sazhen is thus actually 32, which at variable conversion ratios produces charcoal measured in koroba proportionate to size of wood input unit viz. 16 puds.
1737 The 9.5-chetvert unit cut in lengths of 5.5 arshin = 1.83 sazhen

accordingly produces a larger ($1.16 \times 1.16 \times 1.83$) 2.5-cubic-sazhen load, a larger 50-cubic-sazhen bed and a larger koroba of 26 puds.

1767 The 11-chetvert measure similarly, being cut in lengths of 6.4 arshin = 2.13 sazhen produces a larger wood measure (2.9 cubic sazhen), charcoal bed (58 cubic sazhen) and koroba 30 puds, see S.G. Strumilin, *Istoriya chernoi metallurgii V SSSR. T.1. Feodal'nii period* (Moscow, 1954), pp. 62–3.

Subsequently, following the reforms of 1779–83 the 6-chetvert measure was employed, which in real rather than accounting terms measured 1.75 arshin wide × 6 arshin long and 0.5 arshin wide, 7–15 such measures making a charcoal bed of 20 cubic sazhen which produced 51–75 koroba (according to type of wood) each of 15 puds: Hermann, 1790, p. 415, and Hermann, 1797–8, II, p. 13.

213 Table 3.24. Karpenko, 1963, p. 53; 'General Essay'; Renovantz, 1788, p. 334–5, 60; Hermann, 1790, pp. 320–2.
214 Karpenko, 1963, p. 53, Renovantz, 1788, pp. 217–18.
215 See p. 131 above.
216 Storch, 1801, II, p. 399.
217 ibid., II, p. 400.
218 Table 3.25. Falk, 1785, I, pp. 334–5; Hermann, 1797–8, I, pp. 333–4; Hermann, 1797–1801, I, pp. 259–67.
219 Tables 3.14 and 3.25 above. Hermann, 1797–1801, I, p. 287.
220 Table 3.26 Calculated from data on wage outlays in Karpenko 1963, table 5, p. 80 and wage rates pp. 79–80.
221 PSZ, 1st series, t. XXII, no. 16,312.
222 Karpenko, 1963, pp. 79–80.
223 A.N. Radishev, 'Zapiski puteshestviya po sibiri', *Polnoe sobranie sochinenii* (Moscow, 3 vols, 1932–52), III, pp. 275–6.
224 On the total employed in mining activity and at the Salairsk mine, N. Zobin, 'Pripisnie krest'yane na Altae', *Altaiskii Sbornik*, I, (1894), p. 45, and on their employment at the Schlangenberg see my 'Il "Potosi russo"; la miniera di Schlangenberg o Zmeinogorsk nel Settocento', *Quaderni storici* (in press).
225 Hermann, 1797–1801, I, pp. 227–81.
226 ibid., I, pp. 277–81.
227 ibid., I, pp. 277–81.
228 Falk, 1785, I, p. 307; Hermann, 1797–98, I, pp. 311–18.
229 S.G. Karpenko, 'Promyshlennie goroda i poselki Zapadnoi Sibiri v XVIII stoletii', *Materiali po istorii Sibiri. Sibir' perioda feodalizma*, Vip. 1: *Sibir' XVII–XVIII vv.* (Novosibirsk, 1962).
230 Falk, 1785, I, pp. 327–32; Renovantz, 1788, pp. 24–5, 27–31, 60, 178–9; Storch, 1801, II, pp. 396–7.
231 Karpenko, 1963, p. 77; Falk, 1785, I, pp. 328–9 and for comparison of these prices, B.N. Mironov, *Klebnie tseni v Rossii za dva stoletiya (XVIII–XIX vv.)* (Leningrad, 1985), appendix table 3.
232 Table 3.27. (1733–1734) Hermann, 1797–8, I, , p. 305; (1746) ibid., I, pp. 311–18; (1783) ibid., I, pp. 333–4; (1747–1765, 1766–1776, and 1799) Karpenko, 1963, table 23, p. 197.
233 Appendix I, table A1.2.
234 ibid.
235 Hermann, 1797–8, I, pp. 333–4. Table 3.28.
236 On state budgetary income see A. Kahan, *The Plow, the Hammer and the Knout* (Chicago, 1985), pp. 339–43.

237 Figure 3.8. appendix 1, tables A1.1–2, 4–5.
238 On the nineteenth-century industry see Danilevskii, 1959.
239 Danilevskii, 1959, pp. 35–40; Coxe, 1784, II, p. 283.
240 Danilevskii, 1959, pp. 75–8.
241 ibid., pp. 99–103.
242 A. Ermann, *Travels in Siberia, including excursions northward down the Obi to the Polar circle and southward to the Chinese frontier*, transl. W.D. Cooley (London, 2 vols, 1848), II, pp. 267–71.
243 ibid., II, pp. 267–71; Coxe, 1784, II, p. 283.
244 Table 3.30. Danilevskii, 1959, table 10, p. 105.
245 Calculated by deducting total for the years 1748–1800 (appendix 1, table A1.5) from total 1748–1860 (*Gornii zhurnal*, t. 4/2. 1862 as quoted Danilevskii, 1959, p. 104).
246 Hermann, op. cit., pp. 267–71.
247 Hermann, 1810a, p. 45.
248 Appendix 1, table A1.5.
249 Hermann, 1810a, table 6a.
250 Hermann, 1797–8, IV, pp. 101–2.
251 Ermann, op. cit., II, pp. 267–71.
252 Hermann, 1797–8, IV, pp. 96–8.
253 Hermann, 1808, pp. 50–8; Hermann, 1810a, pp. 50–1, table 6a, and p. 45.
254 Hermann, 1810a, pp. 45ff.
255 Ermann, op. cit., pp. 267–71.
256 Hermann, 1810a, p. 45.
257 Hermann, 1810a, pp. 50–1, appendix 1, tables A1.1–2, 5.

4   Russia. Money supply

1 This section could not have been written without the generous assistance afforded me by Dr Jennifer Newman who graciously allowed me to utilize the material on Russian overseas trade contained in her study, *Russian Foreign Trade, 1680–1780: The British Contribution* (University of Edinburgh, Ph.D thesis. 1985) which is shortly to be published. All subsequent citations of trade, shipping, exchange, and price statistics relating to activity in those ports servicing the Russia trade, unless otherwise stated, are derived from the appendices of this study. All errors in interpretation of this data are, needless to say, mine.
2 By far the most useful survey of Russian trade in the period from *c.* 1570–1670 will be found in Artur Attman, *The Russian and Polish Markets in International Trade, 1500–1650* (Göteborg; Meddelanden fran Ekonomisk-historiska institutionem vid Goteborgs universitet, 26, 1973).
3 B.N. Florya, 'Torgovlya Rossii so stranami zapadnoi Evropi v Arkhangel'ske. konets XVI – nachalo XVII v.', *Srednie veka*, XXXVI (1973), p. 144; Attman, op. cit., pp. 88–93; A. Izyumov, 'Razmeri russkoi torgovli XVII v. cherez Arkhangel'sk v svyazi s neobsledovannimi arkhivnimi istochinkami', *Izv. Arkhang. ob-va izucheniya Russkogo Severa*, no. 6 (1912), pp. 250–3.
4 G. Jensch, 'Der Handel Rigas im 17 Jahrhundert', *Mitteilungen aus der livlandischen Geschichte*, XXIV, 2 (1936), 87; Kh. A. Piirimae, 'Tendenciya razvitiya i ob'em torgovli pribaltiiskikh gorodov v period shvedskogo gospodstva v XVII veke', *Skandinavskii Sbornik*, VIII (1964), p. 111.
5 Piirimae, op. cit., p. 111.
6 Jensch, op. cit., pp. 109, 111; K.G. Mityaev, 'Oboroti i torgovie svyazi

Smolenskogo rinka v 70- godakh XVII veka', *Istoricheskie Zapiski*, XXXII (1942), p. 59; *Knigi Moskovskoi bolshoi tamozhni, 1693–1694; Novgorodskaya, Astrakhanskaya, Malorossiiskaya* (Moscow: Trudi gosudarstvennogo istoricheskogo muzeia, t. XXXVIII, 1961).

7 See pp. 20–1 above.

8 E. Harder, 'Seehandel zwischen Lubeck und Russland im 17/18 Jahrhundert' *Zeitschrift des Vereins fur lubeckische Geschichte und Altertumskunde*, XLI (1961), pp. 43–118, XLII (1963), pp. 5–45; E. Harder-Gersdorff, 'Handelskonjunkturen und Warenbilanzen im lubeckischen seeverkehr im 18 Jahrhunderts', *Vierteljahrschift fur Sozial- und Wirtschaftsgeschichte*, LVII (1970), pp. 1–45.

9 E.I. Kamentseva and N.V. Ustyugov, *Russkaya metrologiya* (Moscow, 1965), pp. 143–60; B.N. Mironov ' "Revolyutsiya tsen" v Rossii v XVIII veke', *Voprosi Istorii*, 11 (November 1971), table 6. p. 55, and the same author's *Khlebnie tseni v Rossii za dva stoletiya, XVIII–XIX vv.* (Leningrad, 1985), table 18, p. 112; table 5, pp. 46–7.

10 Given the greater purchasing power of silver in Russia than in western Europe it was worthwhile for merchants to buy riksdalers at a rate which was above the official exchange which was fixed in terms of the relative national valuations of silver. Sellers of riksdalers were thus afforded a premium, over and above the official exchange, which was related to the enhanced purchasing power of silver in Russia when compared with western Europe: 1651–1700 1:6; 1701–1740 1:3.5–4 (Mironov, ' "Revolyutsiya tsen". . .', p. 54). In estimating the size of the premium that he was willing to pay, however, the merchant also had to take into account other transactions costs which even in 'normal' circumstances might amount to as much as 50 per cent of the sales price of the wares he bought in Russia and in wartime could rise astronomically above this level. Premiums, accordingly, were established within a much narrower range – normally within a 1:2 ratio – than was implied by the relative purchasing power of the coin.

11 A. Semenov, *Izuchenie istoricheskikh svedenii o rossiiskoi vneshei torgovle i promyshlennosti s polovini XVII-go stoletiya po 1858 god* (St Petersburg, 3 vols, 1859), III, 23ff.

12 See pp. 20–1 above.

13 On silver commodity prices in the southern part of the Danish peninsula see E. Waschinskii, *Wahrung, Preisentwicklung und Kaufkraft des Geldes in Schleswig-Holstein, 1226–1864* (Neumunster, 2 vols, 1952–9). On Russian price movements see sources in note 14.

14 On prices in the Baltic provinces, Mironov, *Khlebnie tseni.* . ., pp. 68–9, 84–6. Index 1690–9 = 100, 1721–30 = 59, 1731–40 = 48, 1741–50 = 66, 1751–60 = 58. The base-year price statistics involve an average excluding data from the three years of the 'great famine' 1695–7.

15 See pp. 20–1 above.

16 1758–68, 1773–7, H. Storch, *Tableau historique et statistique de l'Empire de Russie à la fin du dix-huitieme siècle* (Basle and Paris, 2 vols, 1801), notes et pieces justificatives p. 49, note 65 to p. 402 derived indirectly from M.D. Chulkov, *Istoricheskoe opisanie rossiiskoi kommertsii pri vsekh portakh i granitsakh ot drevnikh vremen do nine nastoiashchego i vsekh preimuschestvennykh ukazonenii po onoi gosudaria imperatora Petra Velikago i nine blagopoluchno tsarstvuiushchei gosudarini imperatritsy Ekaterini Velikiia* (St Petersburg, 21 vols in 7, 1781–8), VII, 1, appendix. 1783–967, *Supplementband* to the German edition of Storch's work (Leipzig, 1800–3). Further information on specie imports through Riga may be obtained from

J.J. Oddy, *European Commerce* (London, 1805), pp. 143ff. and W. Ch. Friebe, *Physische-okonomisch und statistische Bermerkungen von Lief-und Ehstland oder von den beiden Statthalterschaften Riga und Reval* (Riga, 1794) who provides useful data on the types of specie imported through the port in 1792:

Ducats (Au) 119,400 = 0.48 million roubles
Albertsthalers (Ag) 1,135,819 = 2.36 million roubles
Funferthalers (Ag) 20,038 = 0.04 million roubles

17 The archaeological evidence of the distribution of Albertsthalers through White Russia (Vitebsk, Mogilev, and Smolensk) and the Ukraine (Kharkov, Chernigov, and Poltava) in the early eighteenth century presented by V.N. Riabtsevich, 'Monetnie kladi XVII i pervoi chetverti XVIII v na territorii Chernigov – severskoi zemli i vostochnoi Belorossii', *Numizmatika i Sfragistika*, 1 (January 1963), is echoed in J.J. Oddy's description of their distribution during the opening years of the nineteenth century – *European Commerce* (London, 1805), p. 145. In the light of the evidence from Ukrainian hoards presented in A. Mikotajczyk's important study – 'W kwestii znalezisk nowozytnych monet niderlandzich na ziemiach koronnych', *Zapiski historyczne*, XLIX (1984) – it appears, however, that gold played an increasingly important role in a trade which was far smaller than it had been in the seventeenth century.
18 Oddy, op. cit., p. 145.
19 ibid.
20 Semenov, op. cit., III, pp. 23f.
21 Figure 4.1. Baltic provinces, Mironov, *Khlebnie tseni* . . . , pp. 68–9, 84–6. Russian heartland, ibid., pp. 69–84. On construction of series see appendix 2, pp. 344f.
22 See pp. 21–2, 52 above.
23 Figure 4.2: Silver. A(market price) calculated 1720–45 from N.I. Pavlenko, 'Materiali o razvitii ural'skoi promyshlennosti v 20–40kh godakh XVIII v.', *Istoricheskii Arkhiv*, IX (1953), pp. 176ff. From 1745, when mint supplies were derived exclusively from domestic producers the latter's output, whether sold to the mint or to private customers, was valued at mint price: B.F.J. Hermann, *Die Wichtigkeit des russischen Bergbau* (St Petersburg, 1810), table 10gg. B(Mint price): Kamentseva and Ustyugov, op. cit., pp. 143–60, 213–16; I. I. Kaufman, *Serebryanii rubl' v Rossii ot ego vozniknoveniya do kontsa XIX veka*, (St Petersburg, 1910), p. 218 and appendix 2, note 9. Copper A (Market price): Pavlenko, op. cit., pp. 176ff. It should be noted that during the years 1720–45, save for the period 1735–40, the data represents production costs rather than sales prices. For subsequent price charge see notes 77, 80, 82, 116 below to which reference should be made concerning data on mint purchase prices.
24 See p. 339 below.
25 See Jennifer Newman, 'Anglo-Dutch commercial co-operation and the Russia trade in the eighteenth century', *The Interactions of Amsterdam and Antwerp with the Baltic Region, 1400–1800* (Leiden: Werken uitgegeven door de vereeninging het Nederlandsch Economisch-Historisch Archief gevestigd te Amsterdam, 16, 1983).
26 During the period 1760–90 there was a gradual assimilation of price levels (measured in terms of silver) between the two regions (Mironov, *Khlebnie tseni*, table 13, pp. 68–9) leading to an integration of rates on intra- and inter-national bill transactions (Oddy, op. cit., p. 145).

27 See p. 151 above.
28 On specie imports to 1796 see table 4.1. Subsequently the trade fluctuated violently as supplies were diverted between Riga and St Petersburg; see table 4.4 derived from Oddy, op. cit., pp. 117, 136; Storch, *Supplementband*, p. 7; and F.G. Wurst, *Bemerkungen uber einige Gegenstande der russischen Staatswirthschaft* (1806), tables 10–11. On the impact of these inflows on the bourse quotations for assignat-coin exchange see M. Kashkarov, *Denezhnoe obraschchenie v' Rossii* (St Petersburg, 2 vols, 1898), I, pp. 24–6, containing two series, one compiled by the author, the other, contemporaneously with the events, by Sperenskii. Both are slightly at variance with those quoted by P.A. Shtorkh, 'Materiali dlya istorii gosudarstvennikh denezhnikh znakov v Rossii s 1653 po 1840 god' *Zhurnal ministerstvo narodnogo prosveshcheniya*, CXXXVII (1868), pp. 822–3.
29 Kashkarov, op. cit., I, pp. 24–6.
30 Kaufman, op. cit., p. 190.
31 On monetary policy under the 'double standard' see H. Keller, *Die Geld und Kreditpolitik des russischen Reiches in der Zeit der Assignaten, 1768–1839/43* (Wiesbaden, 1983), and W. McKenzie Pinter, *Russian Economic Policy under Nicholas I* (Ithaca, NY, 1967) pp. 184–98.
32 On the reforms of 1839/43 and the introduction of the credit rouble, which will be discussed more fully below, see, apart from the works referred to above, A.D. Druian, *Ocherki po istorii denezhnogo obrashcheniia Rossii v XIX veke* (Moscow, 1941).
33 See figure 4.2
34 See pp. 310–11 above.
35 B. Neumann, *Die Metalle, Geschichte, Vorkommen under Gewinnung nebst ausfuhrlicher Produktions- und Preis-Statistik* (Halle, 1904), p. 115.
36 On the history of the state works in the period 1720–45 see 'Vedomost' uchinennaya v Kantselyarii glavnogo pravleniya sibirskikh i kazanskikh zavodov . . .' TsADA, f. Senata 1748 g., d. 1510 11 370–99 transcribed and edited in N.I. Pavlenko, 'Materiali o razvitii ural'skoi promyshlennosti v 20-40-kh godakh XVIII v.', *Istoricheskii Arkhiv*, IX (1953), pp. 175–283.
37 The Moscow mint had been established in 1700 and that at St Petersburg in 1726 yet whilst Hennin was instructed by the Senate, by an ukaz dated 14 June 1725 (*Polnoe sobranie zakonov rossiiskoi imperii* = PSZ, 1st series, t. VII, no. 4736), to construct a mint at Ekaterinburg it was not until 1735 that, under the direction of Tatischev, it was operational.
38 Pavlenko, op. cit., pp. 165–6, 232–7, and V. Hennin, *Opisanie ural'skikh i sibirskikh zavodov 1735* (Moscow, 1937).
39 The Demidovs, for instance, had secured copper leases in 1702/9, 1705, and 1720 for the exploitation of deposits in the Viso-Kaya and Kungur regions but had failed to bring them into operation at that time: B.B. Kafengauz, *Istoriya Khozyaistva Demidovikh v XVIII–XIX vv.* (Moscow and Leningrad, 1949), I, p. 186.
40 Pavlenko, op. cit., pp. 175–283.
41 ibid.
42 ibid.
43 Kafengauz, op. cit., pp. 186 and pp. 91–2 above.
44 R. Portal, *L'Oural au XVIIIᵉ-siècle* (Paris, 1950), pp. 169–71.
45 Pavlenko, op. cit., pp. 175–283.
46 ibid.
47 During the years 1734–6 the state works, with some 24 hearths, each capable of producing 16 tons of metal, had a productive capacity of some 384 tons and

an output of 195 tons: Portal, op. cit., p. 396ff., and Pavlenko, op. cit., pp. 175–283, which also provides data on marketing. The private sector with 27 hearths and a productive capacity of 432 tons produced 118 tons: B.F.J. Hermann, *Istoricheskoe nachertanie gornogo proizvodstva v Rossiiskoi imperii* (Ekaterinburg, 1810), I, p. 157, supplemented from B.B. Kafengauz, op. cit., p. 206. The latter author, ibid., p. 190, also provides an estimate of total production: 313 tons.

48 With the closure of the Polevskii works in 1735 capacity was reduced from 816 tons to 750.

49 On total production in 1735 see N.I. Pavlenko, *Razvitie metallurgicheskoe promyshlennosti Rossii v pervoi polovine XVIII veka. Promyshlennaya politika i upravlenie* (Moscow, 1953), p. 290.

50 See pp. 172–3 above.

51 See Appendix 1.

52 See pp. 92–3 above.

53 Kafengauz, op. cit., p. 186.

54 On the Biziarskii works, see Portal, op. cit., p. 168, and P.G. Liubomirov, *Ocherki po istorii rosskoi promyshlennosti XVII, XVIII i nachale XIX v.* (Moscow, 1947), pp. 353ff., whilst on the Kushvinsk works the same works may be supplemented by reference to Pavlenko, 'Materiali . . .', p. 166 and Storch, *Tableau historique . . .*, notes et pieces justificatives p. 38, note 50 to p. 375.

55 See p. 171 above and figure 4.2.

56 Prices calculated from Pavlenko, 'Materiali . . .', pp. 175–283.

57 Ukaz 3 March 1739 required the delivery of two-thirds of output for which the entrepreneur was paid the current market price; see Storch, *Tableau historique . . .*, notes et pieces justificatives p. 37, note 50 to p. 374; on 3 July, however, the price was fixed at three roubles per pud: Hermann, *Istoricheskoe nachertanie . . .*, p. 214.

58 Calculated from Pavlenko, 'Materiali . . .', pp. 175–283 and *Razvitie . . .*, p. 462.

59 Kafengauz, op. cit., ch. 7/2, pp. 188f.

60 Portal, op. cit., p. 396f.

61 See pp. 93–4 above.

62 Pavlenko, 'Materiali . . .', pp. 175–283.

63 On total production in the private sector see Pavlenko, *Razvitie . . .*, p. 290, and the same author's *Istoriya metallurgii v Rossi v XVIII veka. Zavodi i zavodovladel'tsi* (Moscow, 1962), p. 462. Sales to private customers are derived by deducting state deliveries to the mint abstracted from Pavlenko, 'Materiali . . .', 175–283 from total mint production as presented in B.F.J. Hermann, *Die Wichtigkeit des russischen Bergbaues* (St Petersburg, 1810), table 10gg., leaving a residual private sale to the mint which was subsequently deducted from total production to give sales to private customers.

64 See pp. 95f. above.

65 Pavlenko, 'Materiali . . .', pp. 175–283.

66 See for instance the implementation of the Demidovs' programme described in Kafengauz, op. cit., pp. 188ff.

67 G.M. Romanov, *Moneti tsarstvovaniya imperatritsi Elizabeti I im imperatora Petra III* (St Petersburg, 1896), I, pp. 272–7, and Hermann, *Die Wichtigkeit . . .*, tables 1, 8aa/bb, 9cc/dd, 10gg.

68 Portal, op. cit., pp. 134, 141ff.

69 ibid.

70 Public Record Office, Chancery Lane, London. Heath papers, Chancery Masters Exhibits. C 104/143 (1), 145. I should like to express my gratitude to Dr Newman for bringing these references to my attention.

71 Hermann, *Die Wichtigkeit* . . . , table 10gg.

72 See pp. 82, 151 above.

73 *PSZ*, 1st series, t.XIV nos 19370, 10374 and 10624.

74 Portal, op. cit., pp. 138ff.

75 See pp. 179–80.

76 Portal, op. cit., pp. 160–1.

77 V.N. Vitevski, *I.I. Nepliuev i Orenburgski kray v prezhnem ego sostave do 1758 g.* (Kazan, 1897), pp. 656ff., suggests that sales prices at the plant varied between 1.25–2.25 roubles per pud and averaged 1.70 roubles in the mid-1750s; ibid., p. 646, he suggests a contemporary production cost of between 1–1.5 roubles per pud in the southern Urals.

78 Calculated from data in Hermann, *Die Wichtigkeit* . . . , tables 1–10 and Romanov op. cit., I, pp. 272–7.

79 See pp. 311, 313–14.

80 Vitevski, op. cit., pp. 656f., and Hermann, op. cit., table 10gg.

81 Portal, op. cit., p. 134.

82 ibid., p. 160.

82 See pp. 171–2 below.

84 By far the most detailed survey of the southern Urals fields at this time is provided by T.I. Rychkov, *Topografia Orenburgskoi gubernii* (Orenburg, 1762).

85 See p. 185 above.

86 Portal, op. cit., pp. 131ff.

87 H. Hermann, op. cit., tables 1–10.

88 Neuman, op. cit., p. 115. A complete discussion of the materials embodied in maps 4.3–4.4 and the following two paragraphs will be found in Appendix 1.

89 The Nytvenskii plant built by the Golitsins in 1756 had passed by 1762 to the Strogonovs.

90 Liubomirov, op. cit., pp. 468–71.

91 See figure 4.1.

92 Mironov, *Khlebnie tseni* . . . , pp. 68–9.

93 The present data on mint output and all subsequent references in this section are derived from G.M. Romanov, *Moneti tsarstvovannya imperatritsi Ekaterini II* (St Petersburg, 1894), I, 335–41 and Hermann, op. cit., tables 1–10.

94 See pp. 202–3 below.

95 See pp. 343–4.

96 See p. 192 above.

97 I hope to discuss the effects of the labour reforms of 1779–83 more fully in a future article, 'Urals–Siberian industrial labour markets and the reforms of 1779–1783'.

98 Portal, op. cit., pp. 304–5.

99 ibid., pp. 309–10.

100 Kafengauz, op. cit., I, table 34, pp. 248–9, 260; Portal, op. cit., p. 305.

101 Portal, op. cit., pp. 304–5, 310.

102 ibid., pp. 165, 350–1.

103 ibid., pp. 352–3. Inaccuracies in this latter table necessitated the consultation of the work upon which it was based, N.B. Baklanov, *Tekhnika, metallurgicheskogo proizvodstva XVIII veka na Urale* (Moscow, 1935).

104 Data on output of the Pokhodiachin works, before they were bought by the

Bank of Assignation in 1792 for 2 million roubles, will be found scattered amongst the pages of a number of works; Storch, *Tableau historique* . . ., notes et pieces justificatives pp. 44–5, note 60, p.386; P.S. Pallas, *Voyages de M.P.S. Pallas en differentes provinces de l'Empire de Russie et de l'Asie septentrionale* (Paris, 8 vols, 1788–93), II, p. 341–4; Portal, op. cit., p. 355n.

Output of the Pokhodiachin works
(tons)

1766	220
1770	642
1777	736
1779	820
1782	888

105 On plant closures see p. 194 above. These years also saw the replacement of Luginin's Zlatoustskii works by a new plant (Miasskii) on the upper reaches of the Mias river and the rebuilding of the Simskii works, destroyed in Pugachev's rebellion, on a new site where it became known as the Nizhne–Simskii factory. Other works, like those of the Tverdichev–Miasnikov concern, increased production during the years 1769/73–80:

Plant	Number of hearths	Output (tons) 1769/73	1777	1780	1783–96
Preobrazhenskii	6	102	—	137	—
Bogoyavlenskii	8	132	172	188	139
Arkhangelskii	4	71	23	101	97
Verkhotorskii	3	62	78	72	108
Voskresenskii	7	171	—	171	141

(Portal, op. cit., pp. 356–7)

106 W. Coxe, *Travels in Poland, Russia, Sweden and Denmark Interspersed with Historical Relations and Political Enquiries* (London, 2 vols, 1784), II, p. 285.

107 On the establishment of the Susunsk mint see B.F.J. Hermann, *Mineralogische Reisen in Sibirien vom Jahr 1783 bis 1796* (St Petersburg, 1797), pp. 325–6 whilst in the same author's *Die Wichtigkeit* . . ., table 10gg/hh, there is complete data on this mint's output. Information concerning the amount of coin in circulation in 1781, when Siberian mint prices were brought into line with those at Ekaterinburg, is provided by Mironov, ' "Revolyutsiya tsen" . . .', p. 55.

108 Coxe, op. cit., II, pp. 287–8.

109 Storch, op. cit., II, pp. 389–90; Coxe, op. cit., II, p. 288.

110 In 1773 the Urals industry, comprising 37 copper smelting works and 12 copper–iron works belonging to private entrepreneurs and 7 state plants, produced 144,037 puds = 2,315 tons of copper of which 126, 315 puds = 2,030 tons were delivered to the mint:

13,868 puds (at 16r per pud = 221,888 roubles) from state works
97,627 puds (at 16r per pud = 1,562,032 roubles) of compulsory deliveries from the private sector, being 75 per cent of 130,169 puds
14,820 puds (at 16r per pud = 237,120 roubles) bought from the private sector.
126,315 puds (16r per pud = 2,021,040 roubles: Total

285 tons were sold to private customers (Coxe, op. cit., II, pp. 287–8).

111 On the size of the coinage in circulation in 1788–95 see Mironov, ' "Revolyutsiya tsen" . . .', p. 55, adjusted by the methods described in Appendix 2 to conform to the situation in 1786. Cf. the contemporary estimates for 1788 in B.F.J. Hermann, *Statistiche Schilderung von Russland in*

Rucksicht auf Bevolkerung, Landesbeschaffenheit, Naturproduckte, Landwirtschaft, Bergbau, Manufakturen und Handel (St Petersburg and Leipzig, 1790), p. 457:

Gold and silver	70 +	6 Siberian issues deposited against paper assignats
copper	20 +	<u>34</u> copper coins deposited against paper assignats
paper	100 + 40	
Total	190 + 40	= 230 million roubles

The paper currency had already begun to exceed its coin backing in 1784 and with the large issue of assignats in 1787 which brought the total in circulation to 100 million roubles the amount of unsecured paper exceeded the amount of specie in circulation.

112 On the commission appointed by Catherine II in 1786 to inquire into grain price movements and the external and internal trade in the product see Mironov, *Khlebnie tseni . . .*, p. 4. This commission gave rise to an extensive literature on grain balances and prices in the 1780s which was again resumed in the first decade of the nineteenth century and which is briefly surveyed ibid.

113 See pp. 59ff.

114 Coxe, op. cit., II, p. 288 and note quoting *Journal of St Petersburg* 1780, p. 53; Storch, op. cit., ., II, p. 390.

115 On total output within the industry in 1780 and 1782 viz. 200,000 puds and 190,000 puds respectively see Liubomirov, op. cit., p. 222, and B.F.J. Hermann, *Versuch eine mineralogischen Beschreibung des Uralischen Erzgebirges* (Berlin and Stettin, 1789), p. 327. On minting, Hermann, *Die Wichtigkeit . . .*, table 10gg/hh.

116 Storch, op. cit., II, pp. 391–2.

117 Portal, op. cit., pp. 304–5.

118 ibid., pp. 356–7.

119 Pavlenko, *Istoriya metallurgii . . .*, p. 462.

120

	Private sector		State works	Mint deliveries	Private
	Production (tons)	Mint deliveries (tons)	Mint deliveries (tons)	total (tons)	sales (tons)
1785	2,823	1,411 +	752 = 2,163 (2.15 mill. roubles)		1,411
1790	2,511	1,256 +	876 = 2,132 (2.12 mill. roubles)		1,256
1795	1,440	720 +	1,018 = 1,738 (1.73 mill. roubles)		720

(Sources: Hermann, *Die Wichtigkeit . . .*, table 10; Pavlenko, *Istoriya . . .*, p. 462.)

121 Kashkarov, op. cit., I, pp. 24–6.

122 Hermann, *Die Wichtigkeit . . .*, tables 1–10.

123 See p. 167 above.

124 Storch, *Supplementband . . .*, p. 7; Oddy, op. cit., pp. 117, 136; Wurst, op. cit., tables 10–11.

125 G.M. Romanov, *Moneti tsartvovaniya imperatora Pavla I i imperatora Aleksandra I* (St Petersburg, 1891), pp. 21–3, 78–9; Kashkarov, op. cit., II, prilozhenie, pp. 13–15.

126 The phrase is that of M.W. Bernatskii, minister of finance in the provisional government of 1917.

127 Russia, Ministerstvo finansov, *Ministerstva finansov, 1801–1902* (St Petersburg, 1902) I, p. 66.

128 Kaufman, op. cit., p. 190.

129 Kashkarov, op. cit., I, pp. 24–6.
130 Kashkarov, op. cit., II, prilozhenie, pp. 13–15.
131 ibid.
132 Pinter, op. cit., p. 194.
133 *Ministerstva finansov . . .* I, p. 59.
134 Pinter, op. cit., p. 203.
135 The materials for the discussion of the 'popular exchange rates' contained in the following three paragraphs are all drawn from Pinter, op. cit., ch. 5, pp. 184–220 and appendix, pp. 254–63. As anyone consulting this work will realize, however, the analysis of this data below is somewhat different from that of Professor Pinter.
136 From about 1823, because of the steady rise in the price of copper coins of this metal disappeared from circulation, A. Schmit, 'Das russische Geldwesen wahrend der Finanzverwaltung des Grafen Cancrin (1823–1843)', *Russische Revue*, VII (1875), pp. 37–41. In 1846 platinum was demonetized: V.V. Danilevskii *Russkoe zoloto* (Moscow, 1959), pp. 283–4.
137 See on monetary conditions in this period Kashkarov, op. cit., I, pp. 72–3; II, prilozhenie, pp. 13–15; and Kaufman, op. cit., p. 218.
138 Figure 4.3 based on data contained in tables 4.2–3, 4.5, and appendix 2, tables A2.3, 5–7.

5   Russia. Economic Growth

1 See ch. 2, pp. 49–55.
2 See ch. 3, pp. 59–162.
3 See ch. 4, pp. 174–93.
4 ibid, pp. 162–73.
5 For a brief description of these changes taking place within the Russian economy in as far as they affected its foreign-trade sector see J. Newman, *Russian Foreign Trade, 1680–1780: The British Contribution . . .*, ch. 9. As I understand that Dr Newman intends to develop these arguments further in a future study, in the discussion of Russian economic growth in the eighteenth century that follows I have only attempted to delineate its magnitude, on the basis of the independent analysis outlined in appendix 2 below, and to describe some of its salient features. Even this latter task has been made considerably easier, however, by her generosity in allowing me to consult the work cited, and the following of her works prior to their publication: 'Anglo-Dutch commercial co-operation and the Russia trade of the eighteenth century', *Interactions of Amsterdam and Antwerp with the Baltic Region, 1400–1800* (Leiden: Werken uitgegeven door de vereeniging het Nederlandsch Ekonomisch-Historisch Archief gevestigd te Amsterdam, 16, 1983); 'The Russian grain trade 1700–1779', in W. Minchinton (ed.), *The Baltic Grain Trade. Five Essays* (Exeter, 1985); as well as 'The English contribution to the economic revolution in Russia in the eighteenth century', currently in press, and working papers on the Russian fruit trade 1760–1840, henceforth 'Fruit trade', and Moscow *c*. 1800 henceforth 'Moscow'. My interpretation of these works in the following pages is not necessarily that of their author and any errors of interpretation are, of course, my own.
6 Newman, *Russian Foreign Trade . . .*, ch. 5, pp. 123–30.
7 See pp. 346f. below.
8 ibid.

9 V.M. Kabuzan, *Narodonaselenie Rossii v XVIII – pervoi polovine XIX v. po materialam revizii* (Moscow, 1963), table 17, pp. 159–65.

10 P.J. Strahlenberg, *Das Nord und ostlich Theil von Europa und Asia* (Stockholm, 1730), p. 186; P.H. Bruce, *Memoirs of Peter Henry Bruce, Esquire* (London, 1782), pp. 238–43.

11 J. Perry, *The State of Russia under the Present Czar* (London, 1716), pp. 87–8.

12 Bruce, op. cit., p. 228.

13 ibid.

14 Perry, op. cit., pp. 118–19.

15 Bruce, op. cit., p. 110; B.B. Kafengauz, 'Ekonomicheskie svyazi Ukraini i Rossii v kontse XVII-nachale XVIII stoletiya', *Vossoedinenie Ukraini s Rossiei 1654–1954. Sbornik statei* (Moscow, 1954), pp. 421–39.

16 For international comparisons see pp. 281–3 below.

17 F.C. Weber, *Das veranderte Russland* (Frankfurt, 1721), p. 68.

18 Perry, op. cit., pp. 262–5 on the situation prior to 1716 and on the impact of the diversion of grain supplies described in Newman, 'Russian grain trade . . .', pp. 51–2 see table 5.1 below.

19 C. Whitworth, 'An account of Russia as it was in the year 1710', *Fugitive Pieces* (London, 1758), II, p. 180.

20 On the early history of the trade see my 'The continental European cattle trades, 1400–1600', *Economic History Review*, 2nd series, XXXIX, 3 (1986), pp. 427–60; Gwagini, *Sarmatiae Europeae Descriptio* (1581), p. 40 as quoted in R. Rybarski, *Handel i Polityka Handlowa Polski w XVI stuleciu* (Warsaw, reprint 1958), I, p. 69, freely translated from the Latin by the present author.

21 J. Baszanowski, *Z dziejow handly polskiego w XVI–XVIII wieku. Handel wolami* (Gdansk, 1977), pp. 130f.

22 Kafengauz, op. cit., pp. 421–39.

23 ibid.

24 Whitworth, op. cit., p. 180.

25 See appendix 2, table A2.17.

26 ibid.

27 Newman, 'Russian grain trade . . .', pp. 55–8.

28 Perry, op. cit., pp. 40–1 quoted in Newman, 'Russian grain trade . . .', p. 55 and W. Coxe, *Travels into Poland, Russia, Sweden and Denmark* (London, 1784), II, pp. 293–4.

29 Newman, 'Russian grain trade . . .', pp. 56–7, on the imposition of a seigneurial regime on these lands.

30 On the economy of this region 1720–62 see Whitworth, op. cit., pp. 178–80; Perry, op. cit., p. 83; C. le Brun, *A new and more correct translation of Mr. Cornelius le Brun's Travels* (London, 1758), pp. 175–6.

31 I.D. Koval'chenko, *Krest'yane i krepostnoe khozyaistvo Ryazanskoi i Tambovskoi gubernii v pervoi polovine XIX veke* (Moscow, 1959), p. 83.

32 ibid., p. 69; E.I. Indova, *Krepostnoe khozyaistvo v nachale XIX veka* (Moscow, 1955), p. 22; I.Y. Bulygin, *Polozhenie krest'yan i tovarnoe proizvodstvo v Rossii, vtoraya polovina XVIII veka* (Moscow, 1966), pp. 103–4. pp. 103–4.

33 R.E.F. Smith and D. Christian, *Bread and Salt: A Social and Economic History of Food and Drink in Russia* (Cambridge, 1984), pp. 208–10.

34 I.D. Koval'chenko, *Russkoe krepostnoe krest'yanstvo v pervoi polovine XIX veka* (Moscow 1967), appendix, table 4, pp. 392–4.

35 On yields in Moscow and Vladimir see E.I. Indova, 'Urozhai v tsentral'noi Rossii za 150 let (vtoraya polovina, XVII–XVIII v.)', *Ezhegodnik po agrarnoi istorii vostochnoi Europi za 1965* (Moscow, 1970), pp. 146–51, as quoted

A. Kahan, *The Plow, the Hammer and the Knout* (Chicago, 1985), p. 50 which also contains comparative materials suggesting a differential of 20–5 per cent. Recent studies by E. Melton, 'Two estates in nineteenth century Russia', *Past and Present*, (May 1987), and S.L. Hoch, *Serfdom and Social Control in Russia. Petrovskoe, a village in Tambov* (Chicago, 1986), by studying data in contemporary accounts and modern monographic estate studies, suggest that Indova's data, based on governors' accounts, overestimates non-black-earth yields and underestimates black-earth yields.

36 I.D. Koval'chenko and L.V. Milov, *Vserossiiski agrarnii rinok XVIII – nachalo XIX veka* (Moscow, 1974), pp. 211–14 and Newman, 'Russian grain trade . . .', pp. 47–59, upon which the following paragraphs are based.

37 Newman, 'Russian grain trade . . .', pp. 47–59.

38 ibid., the data contained in table 2 therein has been extended to incorporate information on Moscow prices, utilizing data in B.N. Mironov, *Klebnie tseni v Rossii za dva stoletiya XVIII–XIX vv* (Leningrad, 1985), appendix, table 2, and employing the methodology outlined in Newman, *Russian Foreign Trade*, pp. 352–5.

39 Yu. Tikhonov, *Pomeschich'i krest'yane v Rossii: Feodal'naya renta v XVII i nachale XVIII v.* (Moscow, 1974), pp. 218–91; V.M. Kabuzan, *Izmeniya v razmeschenii naseleniya Rossii v pervoi polovine XVIII V.* (Moscow, 1971), pp. 96–7.

40 Newman, 'Moscow'.

41 R.K. Porter, *Travelling Sketches in Russia and Sweden, 1805–1808* (London, 1809), I, p. 206.

42 Y.A. Fedorov, *Pomeschich'i krest'yane tsentral'no promishlennogo raiona kontsa XVIII- pervoi polovini XIX veke* (Moscow, 1974), pp. 56–7.

43 I.V. Meshalin, *Tekstil'naya promishlennost' krest'yan moskovskoi gubernii v XVIII i pervoi polovine XIX veke* (Moscow, 1950); W. Tooke, *View of the Russian Empire during the Reign of Catherine II and to the close of the present century* (London, 1799), III, pp. 512–13.

44 M. and C. Wilmot, *The Russian Journals of Martha and Catherine Wilmot*, ed. Lady Londonderry and H.M. Hyde (London, 1934), p. 180.

45 See pp. 228–9 above.

46 See Fedorov, op. cit., pp. 95–7.

47 ibid., pp. 133–6 and Tooke, op. cit., III, pp. 512–13.

48 The distinction between kustari and craft used here involves a consideration of the degree of the workers's dependence upon the grain market for their sustenance.

49 See pp. 223–4 above.

50 H. Storch, *Tableau historique et statistique de l'empire de Russie à la fin du XVIIIe siècle* (Paris and Basle, 1801), I, pp. 157, 232.

51 M. Holderness, *New Russia* (London, 1823), pp. 90–1.

52 Coxe, op. cit., I, p. 504.

53 Kafengauz, op. cit., pp. 421–39.

54 Newman, *Russian Foreign Trade*, pp. 65ff.

55 See pp. 232–3.

56 See pp. 224–5.

57 Storch, op. cit., II, pp. 159, 168–9, 235; J.F. Geogel, *Voyage à St Petersbourg en 1799–1800* (Paris, 1818), pp. 250–1.

58 B.F.J. Hermann, *Statistiche Schilderung von Russland in Rucksicht auf Bevolkerung, Landesbeschaffenheit, Naturproduckte, Landwirtschaft, Bergbau, Manufakturen und Handel* (St Petersburg and Leipzig, 1790), pp. 454–7, the nature of whose estimates is discussed in appendix 2, p. 348

below and Al'b Vainstein, *Narodnii dokhod Rossii i SSR* (Moscow, 1969), pp. 32–7.

59 On the total volume of trade see table 5.2. This calculation excludes grain which Hermann considered as supplied only through household provisioning systems. Even if commercially marketed grain (see note 79 below) was included in the total it would barely alter this figure. On foreign-trade turnover see N.L. Rubinstein, 'Vneshnaya torgovlya Rossii i russkoe kupechestvo vo vtoroi polovine XVIII v.', *Istoricheskie zapiski*, LIV (1955), pp. 343–61. Internal trade is the residual derived by subtracting foreign from total trade.

60 Information of the total value and volume of 'feed' and 'drink' grain, for conversion into vodka, is provided by Hermann. The 576 million puds of 'feed' grain has been converted to volumetric measure at the rate of 8.74 puds rye per chetvert = 66 million chetverts. The 99-million eimar/vedro of vodka at conversion rates of 2.25–2.5 eimar/vedro per chetvert, given in Kahan, op. cit., p. 56 = 44 million chetverts or a total net harvest of 110 million chetverts. This gives a figure which is very close to an independently calculated (32.6 million desyatini × 4.78 chetvert per desyatina gross or 3.42 net = 156 million chetverts gross, 112. million net at a seed/yield ratio of 1:3.5) harvest estimate.

61 Calculated on the basis of the regional distribution of population, in Kabuzan, *Nardodonaselenie . . .*, table 17, pp. 159–65, landholding and yields, in N.L. Rubinstein, *Sel'skoe khozyaistvo Rossii vo vtoroi polovine XVIII v. Istoriko-ekonomicheskii ocherk* (Moscow, 1957), pp. 232, 355–6, and Kahan, op. cit., pp. 49–50, quoting Indova. By breaking down patterns of population and landholding in the black-earth regions between the lands of the Volga and eastern Don and the rest (including the rich lands south of Moscow) and applying Indova's yields only to the former, a residual figure is obtained for the latter region, broadly in line with the higher yields quoted by Melton and Hoch (see note 35). Thus is the circle squared between Indova and Rubinstein, employing governors' reports and army surveys, and their critics, employing *individual* estate studies or individual responses of estate owners to questionnaires of the Free Economic Society.

62 Hermann, as suggested in note 60, distinguishes between 'feed' and 'drink' grain.

63 Such consumption patterns had not been known in western Europe since the Middle Ages, the peasantry there, if not the denizens of the great urban metropoli, undergoing a process of 'depecoration' in the period 1500–1700; see for example my 'Consumption and hierarchy in English peasant society 1400–1600', *Chicago Economic History Workshop. Papers* No. 20 (1980) and 'English peasant consumption. The end of an epoch 1580–1680', *Kwartalnik historii kultury materialnij*, XXX, 1 (1982).

64 Coxe, op. cit., I, p. 437.

65 The peasant family, according to Hermann, numbered five in all, or exactly the same figure indicated by Hoch, op. cit., p. 41 for the period 1810–56, who were endowed, accordingly with about 6 desyatini of land and, if the same production relations held true as later, 1 desyatina of garden and 1.86 desyatina of hay field or in total 6 and 13.5 million desyatini respectively, producing in the former case, according to Hermann, wares worth at farm-gate prices *c*. 250 million roubles, together with 245 million derived from off-farm sales and value added in manufactory and in the latter instance 243 million. Average returns per tenth of an acre of garden ground thus equalled 2.6 roubles which at prevailing production levels noted in Hoch, op. cit., p. 49, could be yielded by putting the ground under cabbages, hemp or flax, tobacco or fruit. On hemp and flax production see note 67 below, for vegetables and

tobacco data from the nineteenth century appendix 2, pp. 352–3, has been extrapolated, leaving a residual for orchards and giving the following values; flax and hemp (at farm-gate prices) 12.5 million roubles, 26.5 million for tobacco, and 156 million each for vegetables and fruit.

66 According to Hermann the production of 'feed' grain was worth 144 million roubles, iron 7 million.

67 Calculated on the basis of data in G.S. Isaev, *Rol' tekstil'noi promishlennosti v genezise i razvitii kapitalizma v Rossii 1760–1860* (Leningrad, 1970), pp. 52–3.

68 Tooke, op. cit., III, pp. 357–8; Isaev, op. cit., p. 118f.

69 Isaev, op. cit., pp. 52–3.

70 On total land area (485.5 million ha) see M.A. Tsvetkov, *Izmenenie lesistosti evropeiskoi Rossii s kontsa XVII stoletiya po 1914 god* (Moscow, 1957), pp. 110–14, of which 42.5 million ha were encompassed in fields and garden grounds.

71 The extent of woodland clearance was calculated from Tsvetkov, op. cit., pp. 110–14 whilst the average stand of timber is estimated at 63.5 cords per hectare (on the basis of contemporary English data, e.g. British Library, (Lansdowne MS 166 fo 348^r) = 863 cubic sazhen, giving a total product of 532 million cubic sazhen worth according to Hermann 397 million roubles or 74 kopecks per cubic sazhen, a price confirmed elsewhere in his works (Hermann, 1790, pp. 414–15, and Hermann, 1797–1801, II, p. 13).

72 See table 5.1.

73 Storch, op. cit., II, p. 247; Tooke, op. cit., III, pp. 101–7.

74 I.e. 297 million roubles divided by 60 kopecks a pud, the price of fish in S.G. Strumilin, 'Oplata truda v Rossii', *Ocherki ekonomicheskoe istorii Rossii i SSSR* (Moscow, 1966), p. 60.

75 PSZ, 1st series, t.XXII, no. 16736.

76 See pp. 226ff.

77 The problem of calculating production of animal husbandry is one of the most intractable ones. Hermann affords data on commercial sales, and information on the numbers and value of cattle exported exists in *Gosudarstvennaya torgovla v raznikh eya vidakh 1802–7 gg* and Baszanowski, op. cit., pp. 130ff. Assuming the ratio of foreign to domestic sales is as given in note 59 above and deducting the 3.5 million shipped from the southern steppe to St Petersburg then one is left with a residual of 30 million cattle or 5 per family, a figure very similar to that of peasant farms in western Europe in an earlier age and suggestive of a family herd of *c*. 10 animals, i.e. some 60 million cattle in all and some 13 million sheep as indicated in Isaev, op. cit., p. 55. A killing-out rate for the steers of 44 per cent may be assumed and an allowance of 2.4 per cent made for hides and tallow. Of the remaining cows a yield of *c*. 3.5 vedros per animal is probable.

78 Kahan, op. cit., p. 58.

79 Koval'chenko, *Russkoe krepostnoe krest'yanstvo* . . ., pp. 392–4, projecting data on western and eastern regions backwards from *c*. 1800 on the basis of central region trends.

80 H. Storch, *The Picture of St Petersburg* (London, 1801), pp. 85–8. Supplies from the western region at this time are estimated by deducting exports through the Baltic ports (Newman, *Russian Foreign Trade* . . ., p. 286) and Black Sea (B.N. Mironov, 'Eksport russkogo khleba vo vtoroi polovine XVIII – nachale XIX veka', *Istoricheskie zapiski*, XCIII (1979), pp. 164–5) from total in Koval'chenko, *Russkoe krepostnoe krest'yanstvo* . . ., pp. 392–4.

81 Calculated by deducting western supplies from the consumption requirements of the capital and then deducting this figure from grain passing down the Volga

beyond Liskovo, leaving 590,00 chetverts for the population of the region or 1.8 chetverts per head.

82 Kabuzan, op. cit., pp. 159–65.

83 On rate of commercial returns see works referred to in appendix 2, p. 351, yielding an income flow of 39.5 million roubles for some 395,000 persons.

84 A. Swinton, *Travels into Norway, Denmark and Russia in 1788, 1789, 1790 and 1791* (London, 1792), p. 229.

85 Storch, *Tableau historique* . . . , II, pp. 278–80.

86 Hermann, *Statistische Schilderung* . . . , pp. 378–9, 454–57 (copper and iron), Storch, *Tableau historique* . . . , II, pp. 408–9 (salt), Isaev, op. cit., pp. 175, 119, 135, 146, 156, and Strumilin, *Istoriya chernoi metallurgii* . . . , pp. 348, 360, 370 (textiles, metal working, and others). The element of artisanal-handicraft production is calculated as a residual by deducting the above from product and employment totals in Hermann.

87 Storch, *Picture* . . . , p. 274.

88 Isaev, op. cit., pp. 52–3, 135.

89 See pp. 246–7 above.

90 See pp. 246–7.

91 See p. 237 above.

92 L.N. Semenonova, *Rabochie Peterburga v pervoi polovine XVIII veka* (Leningrad, 1974); K.A. Pazhitnov, *Ocherki istorii tekstil'noi promishlennosti dorevolytusionnoi Rossii. Khlopchato-bumazhnaya, l'no-pen'kovaya i shelkovaya promishlennost'* (Moscow, 1958), pp. 177–8.

93 Isaev, op. cit., pp. 122f.

94 See pp. 198–9 above.

95 Strahlenberg, op. cit., p. 351; L.G. Beskrovnii, *Russkaya armiya i flot v XVIII veke. Ocherki* (Moscow, 1958;), pp. 76–7, 86–7, 94–5; E.D. Clarke, *Travels in Various Countries*, (London, 1800), I, p. 183.

96 Kahan, op. cit., pp. 99–101.

97 ibid. pp. 101–5.

98 Calculated by deducting non-peasant consumption of these items from totals in Hermann and dividing by peasant population and by the numbers of peasants attending fairs, on which see B.N. Mironov, *Vnutrenni rinok Rossii vo vtoroi polovine XVIII-pervoi polovine XIX V.* (Leningrad, 1981), p. 160.

99 Kafengauz, *Istoriya khozyaistva Demidovikh* . . . , pp. 398ff.

100 Newman, *Russian Foreign Trade* . . . , p. 144.

101 Newman, 'Fruit trade'.

102 ibid.

103 Mironov, op. cit., pp. 190–2.

104 ibid., pp. 74–5, 190–2.

105 Newman, *Russian Foreign Trade* . . . , p. 12f.

106 Kahan, op. cit., p. 271.

107 Clarke, op. cit., I, p. 206.

108 Mironov, op. cit., pp. 74–5.

109 ibid., pp. 161–3.

110 See note 83 above.

111 I. Golitzuin *Statisticheskiya tablitsi vserossiiskoi imperii* (Moscow, 1807), table 6, the nature of whose estimates is discussed in appendix 2, p. 348, and Vainstein, op. cit., pp. 41–3.

112 M.A. Tsvetkov, op. cit., pp. 110–14.

113 PSZ, 1st series, t. XXIII, no. 17222 and ibid., t. XIXV, no. 18278.

114 A. Kahan, 'Natural calamities and their effect upon the food supply in Russia', *Jahrbucher fur Geschichte Osteuropas*, XVI (1968), pp. 355–77.

115 ibid.
116 See p. 223 above.
117 Koval'chenko, op. cit., p. 386.
118 See my *Consumption and Hierarchy*.
119 The question of the use of consumption-allocatory systems for the purpose of risk alleviation is considered ibid. whilst D. McCloskey in various studies, of which see for instance *Research in Economic History*, I (1796); *Explorations in Economic History*, XIV 4 (1977); *Journal of European Economic History*, VIII, 1 (1979) and IX, 1 (1980), considers production reorganization, and particularly the 'scattering' of landholdings, for the purpose of risk alleviation.
120 Calculated on the basis of estimates in note 122 below and Tsvetkov, op. cit., pp. 110–14.
121 For a useful brief survey of existing agricultural systems in *c*. 1807 see J. Blum, *Lord and Peasant in Russia* (New York, 1974), pp. 336–7. As will be indicated from the discussion in the text, neither the temporal perspective of this static picture nor the assumptions about relative efficiencies of existing systems is accepted here.
122 Regional distribution of land-holding is calculated by dividing the quantities of grain sown by contemporary sowing rates. Yields are derived directly from the sources listed below as are regional output patterns. On the situation in 1788 see sources in note 61 above whilst subsequent yields to 1802 are taken from Kahan, op. cit., pp. 49–50, adjusted in the manner described therein. To break down the averages for the whole period 1802–11 into sub-periods 1802–4, 1805–11 I have followed the indications of Koval'chenko, *Russkoe krepostnoe krest'yanstvo* . . ., p. 386 and used the data in his 'Dinamika urovnya zemledel'cheskogo proizvodstva Rossii v pervoi polovine XIX v.', *Istoriya SSSR*, 1 (1959), appendix, in conjunction with that in E. Zyablovskii *Zemleopisanie rossiiskoi imperii* (St Petersburg, 6 vols, 1810).
123 Clarke, op. cit., I, p. 189.
124 See note 122 above.
125 Calculated from data in table 5.7.
126 The following discussion on the grain trade is based on data in Koval'chenko, *Russkoe krepostnoe krest'yanstvo* . . ., pp. 96, 392–393.
127 Newman, 'Russian grain trade . . .', p. 58, and Mironov, 'Eksport russkogo khleba . . .', pp. 149–88.
128 A.G. Rashin, *Naseleniie Rossii za 100 let, 1811–1913* (Moscow 1956), pp. 93–6; Swinton, op. cit., p. 229.
129 As calculated in note 122.
130 On changes in manufactory see p. 251 above.
131 See pp. 247ff. above.
132 Calculated by taking total grain supply and subtracting requirements of that proportion (50 per cent) of the population of Vladimir and Moscow dependent upon grain imports, on which see Fedorov, op. cit., pp. 56–7. The residual was then divided between the population of the remaining non-black-earth provinces yielding a figure of 10-per-cent dependence.
133 See sources listed in note 86 above.
134 Isaev, op. cit., pp. 61ff.
135 See pp. 255ff.
136 Tsvetkov, op. cit., pp. 110–14.
137 E.D. Clarke, *Travels in Various Countries* (London, 1800), I, pp. 199ff)
   J. Reuilly, *Voyages en Crimée et sur les bords de la Mer Noire* (Paris, 1806);M. Guthrie, *A Tour performed in the Years 1795–96 through the Tauride or Crimea* (London, 1802); M. Holderness, *New Russia. A Journey from Riga*

to the Crimea (London, 1823). These works provide an excellent picture, all apart from Holderness providing eye-witness accounts of conditions prior to 1805 when 'russification' began to destroy the old order, creating a new, described in E.I. Druzhinina, *Yuzhnaya Ukraina 1800–1825 gg* (Moscow, 1970).

138 Reuilly, op. cit., p. 185; Guthrie, op. cit., pp. 207–25; Holderness, op. cit., pp. 135–6, 166.

139 Reuilly, op. cit., p. 58; Guthrie, op. cit., pp. 128–30.

140 Reuilly, op. cit., pp. 129, 140, 262, 280; Oddy, op. cit., pp. 168ff.

141 Holderness, op. cit., p. 138.

142 Reuilly, op. cit., pp. 129, 280, 140.

143 See sources listed in note 126.

144 Clarke, op. cit., I, p. 199.

145 ibid., I, pp. 275–99.

146 ibid., I, pp. 304–12.

147 ibid., I, pp. 195, 259.

148 Reuilly, op. cit., p. 200.

149 See sources in note 126 and Mironov, 'Eksport russkogo khleba . . .', pp. 164–5.

150 See pp. 258ff.

151 On the later fruit trade see Newman, 'Fruit trade' and on the trade in fruit brandies, a woefully neglected aspect of Russian alcohol production, Clarke, op. cit., I, p. 195.

152 Estimated in Storch, *Tableau historique* . . ., II, pp. 408–12 as 3 million puds a year.

153 On agricultural impact see pp. 258–9 above.

154 Storch, *Tableau historique* . . ., II, pp. 157–9.

155 See pp. 242–3 above.

156 Storch, *Tableau historique* . . ., II, p. 157.

157 See sources listed in note 77.

158 Storch, *Tableau historique* . . ., II, pp. 157–9; Tooke, op. cit., III, pp. 541–2, and on exports Oddy, op. cit., p. 205.

159 Clarke, op. cit., I, pp. 204–5.

160 See p. 281 below, and for method of calculation note 77 above.

161 Tooke, op. cit., III, p. 365.

162 See p. 260 above.

163 C. de Rechberg and G.B. Denning, *Les peuples de Russie* (Paris, 1812–14); Clarke, op. cit., I, , pp. 236–44; G. Forster, *A Journey from Bengal to England* (London, 1798), II, p. 258; E. Craven, *A Journey through the Crimea to Constantinople in a series of letters* (London, 1789), p. 176.

164 Calculated as described in note 98.

165 *Gosudarstvennaya torgovlya v raznikh eye vidakh, 1802–1807 gg* (St Petersburg, 1802–7).

166 Newman, 'Fruit trade'.

167 Based on trade statistics referred to in note 165.

168 Wilmot, op. cit., p. 81; Coxe, op. cit., p. 433.

169 Based on trade statistics and production statistics in Isaev, op. cit., p. 135, and Pazhitnov, op. cit., pp. 27–9.

170 Similarly based on trade statistics and production ones in Isaev, op. cit., pp. 146, 156.

171 See trade statistics and on the nature of this trade in the 1780s, and the British contribution in particular, Newman 'The English contribution to the economic revolution . . .'. On domestic production estimates pp. 245–6 above.

172 Calculations on the direction and volume of trade undertaken in the manner

indicated in note 98 and data on changes in the institutional framework derived from Mironov, *Vnutrenni rinok . . .*, pp. 62–3, 74–5.

173 ibid. pp. 74–5.
174 The estimates on the impact of the grain trade on regional grain availability are made on the basis of the data referred to in notes 61, 122, and 126.
175 Calculated by adding or subtracting traded grain from regional production.
176 Mironov, *Khlebnie tseni . . .* pp. 52ff.
177 Koval'chenko, *Russkoe krepostnoe krest'yanstvo . . .*, p. 386, and Kabuzan, op. cit., pp. 159–65.
178 Whitworth, op. cit., p. 178.
179 Clarke, op. cit., I, pp. 236–4.
180 The comparative materials used here are, for France P. O'Brien and C. Keyder, *Economic Growth in Britain and France 1780–1914. Two Paths to the Twentieth Century* (London, 1978), and for Britain N.F.R. Crafts, *British Economic Growth during the Industrial Revolution* (Oxford, 1985). In each instance their 'series' have been recast in terms of constant 1762 prices and converted, in terms of 1764 silver roubles, by contemporary exchange rates on which see H. Storch, *Supplementband zum funften, sechsten und siebenten Theil des historisch statistischen Gemaldes des Russischen Reichs* (Leipzig, 1803), table 3.
181 Quoted in P. Putnam (ed.), *Seven Britains in Imperial Russia, 1698–1812* (1952), p. 197, and J. Carr, *A Northern Summer* (London, 1805), p. 291.
182 Coxe, op. cit., I, p. 437; Wilmot, op. cit., pp. 146–7.
183 Holderness, op. cit., pp. 76, 125.

Appendix 1   Russia. Precious-metal and copper production

1 See for instance A. Attman, *Dutch Enterprise in the World Bullion Trade, 1550–1800* (Göteborg: Acta Regiae Societatis Scientiarum et Litterarum Gothoburgensis, Humaniorum 23, 1983), p. 90n.
2 Provides the basis for estimates in B.N. Mironov, *Khlebnie tseni v Rossii za dva stoletiya, XVIII–XIX vv.* (Leningrad, 1985), p. 116.
3 Employed with data from Storch's work to provide production figures for the 1770s and 1800s in W.H. Parker, *An Historical Geography of Russia* (London, 1968), pp. 174–5. Unfortunately, apart from an incorrect printing of the data in the table on p. 174, the author has also misread Coxe, including as an annual production figure for 1772 that author's figure of total production for the years 1749–71.
4 This data is utilized as the basis for table 3.7 in A. Kahan, *The Plow, the Hammer and the Knout. An Economic History of Eighteenth Century Russia* (Chicago, 1985), pp. 84–5. Again unfortunately, although the author clearly realized that these figures were at variance with those presented by Danilevskii and Karpenko, not to mention Pavlenko (whose works and sources will be considered below), his untimely death prevented their correction, a task which was also overlooked by Richard Hellie who posthumously edited the work.
5 See for instance the structure of the labour force at the Salairsk mine in 1795 in B.F.J. Hermann, *Sochineniya o sibirskikh rudnikakh i zavodakh* (St Petersburg, 1797–1801), Chast III, pp. 60–2.
6 H. Storch, *Tableau historique et statistique de l'empire de Russie à la fin du dix-huitième siècle*, (Paris and Basle, 2 vols, 1801), II, pp. 384–5.
7 On the territorial limits of the two mining regions see for Nerchinsk A. Ermann, *Travels in Siberia, including excursions northward, down the Obi to the Polar*

Circle and Southward to the Chinese frontier, trans. William Desborough Cooley (London, 2 vols, 1848), II, pp. 296–300, and Kolyvan H.M. Renovantz, *Mineralogisch-geographische und andere vermischte Nachrichten von den Altaischen Geburgen Russische Kanserlichen Antheil* (Reval, 1788), map 1.

8 In the case of the Kolyvan *nachal'stva* the boundaries had finally been stabilized by 1760 and administrative change thereafter took place in a stable territorial unit which, encompassing some 390,600 square verst, was as large as a number of western European countries (e.g. the Netherlands or England). On the administrative reorganization of 1779/1783 see *Polnoe sobranie zakanov rossiskoi imperii* = PSZ, 1st series, t. XX, no. 14868 and t. XXI, no. 15857 and on that, associated with the division of Siberia into two provinces in 1796, PSZ, 1st series, t. XXIV, no. 17634 and 17862.

9 Storch, op. cit., II, p. 384–5; J.P. Falk, *Beitrage zur topographischen Kentniss des russischen Reichs* (St Petersburg, 3 vols, 1785), I, p. 305.

10 Falk, op. cit., I, p. 334.

11 PSZ, 2nd series, t. III, no. 1960.

12 See N.I. Pavlenko, 'Materiali o razvitii ural'skoi promyshlennosti v 20–40 kh godakh XVIII v.', *Istoricheskii Archiv*, IX (1953), pp. 208–9 which places administration 1704–19 under the Siberian governor or particulars.

13 PSZ, 1st series, t. V, no. 3464.

14 Storch, op. cit., II, pp. 388–9.

15 N.I. Pavlenko, *Razvitie metallurgicheskoi promyshlennost: Rossii v pervoi polovine XVIII veka. Promyshlennaya politika i upravlenie* (Moscow, 1953), p. 112, quoting ukaz of 19 August 1720 TsGADA f. 271, Kn 2.

16 ibid., p. 86.

17 See pp. 176f. above.

18 On successive commissions of enquiry during the 1730s see R. Portal, *L'Oural au XVIIIe siècle. Étude d'histoire économique et sociale* (Paris: Collection historique de l'institut d'études slaves, t. XIV, 1950), pp. 105–6 on the commissions appointed by the Senate and College of Finance in 1730, and the monetary commission of 1731.

19 See p. 395 n. 9 below.

20 R. Portal, op. cit., pp. 107–10 on the Golovkine commission of 1734/5, the inquiry into the Demidov's affairs, and the resultant Shafirov commission of 1735. For the report of the Golovkin commission and the Shafirov commission reproduced in almost their entirety see V. Rozhkov, 'Berg-Kompaniya na Magnitnii gore', *Gornii zhurnal*, 6–8 (1885).

21 PSZ, 1st series, t. IX, no. 7047.

22 Portal, op. cit., pp. 98ff.

23 Portal, op. cit., pp. 113–14 on the second Shafirov commission of 1738.

24 PSZ, 1st series, t. X, no. 8571.

25 Portal, op. cit., p. 114; Pavlenko, op. cit., p. 200.

26 Portal, op. cit., p. 114.

27 Storch, op. cit., II, p. 384.

28 ibid., p. 388.

29 ibid. pp. 384–6.

30 PSZ, 1st series, t. XXVI, no. 198000.

31 N.I. Pavlenko, *Istoriya metallurgii v Rossii XVIII veka. Zavodi i zavadovladel'tsa* (Moscow, 1962) and the same author's, *Razvitie metallurgicheskoi Rossii v pervoi polovine XVIII veka* (Moscow, 1953); V.V. Danilevskii, *Russkoe zoloto* (Moscow, 1959); Z.P. Karpenko, *Gornaya i metalurgicheskaya promyshlennost' zapadnoi Sibiri v 1700–1860* (Novosibirsk, 1963).

32 Together with the more general B.F.J. Hermann, *Statistische Schilderung von Russland in Rucksicht auf Bevoelkerung, Landesbeschaffenheit, Naturprodukte, Landwirthshaft, Bergbau, Manufacturen und Handel* (St Petersburg and Leipzig, 1790) which, ibid.., pp. 318f, seems to include the first presentation of data on production at Kolyvan collected on his return to St Petersburg in 1789 which subsequently was published in full and brought up to date in his paper 'Sur l'exploitation des mines de L'empire de Russie', *Nova Acta Academiae Scientiarum Imperialis Petroplitanae*, 11 (1798) and other works published in 1797/8.

33 W. Coxe, *Travels into Poland, Russia, Sweden and Denmark, Interspersed with Historical Relations and Political Enquiries*, (London, 2 vols, 1784), II, p. 286n.

34 All data published before 1810 relates only to unrefined silver (blicksilver).

35 B.B. Kafengauz, *Istoriya khozyaistva Demidovikh v XVIII–XIX vv. Opit issledovaniya po istorii ural'skoi metallurgii*. T. 1 (Moscow, 1949).

36 B.F.J. Hermann, *Mineralogische Reisen in Siberien vom Jah 1783 bis 1796* (St Petersburg, 1797–8), I, p. 315.

37 Coxe, op. cit., II, p. 284, and Ermann, op. cit., I, pp. 267–71.

38 B.F.J. Hermann, *Statistische Schilderung von Russland* (St Petersburg and Leipzig, 1790), p. 316.

39 Which commanded a 3.5 per cent premium over the unrefined product.

40 1726–9 *c.* 60 per cent of total output, 1730–57 less than 25 per cent and thereafter about 10 per cent.

41 PSZ, 1st series, t. V., no. 3464.

42 Kafengauz, op. cit., I, p. 186.

43 Pavlenko, *Razvitie metallurgicheskoi promishlennosti. . .*, pp. 56, 78, 222.

44 See pp. 302–3 above.

45 See M. Popov, *Tatischev i ego vremia* (St Petersburg, 1861) pp. 150–2.

46 See pp. 302–3 above.

47 See ch. 4 note 57.

48 See ch. 4 notes 73, 80, and 114.

49 See pp. 92–5 above.

50 See pp. 90–2 above.

51 See pp. 92–5 above.

52 See pp. 95–8 above.

53 See p. 97 above.

54 See p. 315 above.

55 See pp. 116–17 above.

56 See pp. 196–7 above.

57 See pp. 319–22 above.

58 See pp. 199–200 above.

59 See p. 322 above.

60 See p. 303 above.

61 See p. 324 above.

62 Based on plant histories referred to in notes to table A1.9.

63 ibid.

64 Derived from information contained in files of claims for war damage quoted by Portal, op. cit., pp. 331–5.

65 Plant histories.

66 Pavlenko, *Istoriya metallurgii v Rossii. . .*, p. 462; Storch, op. cit., II, p. 381n; Coxe, op. cit., II, pp. 287–8.

67 Information on sectoral output derived from sources listed in note 66 above. Total output calculated from data in B.F.J. Hermann, *Die Wichtigkeit des russischen Bergbaues* (St Petersburg, 1810), table 10 gg–hh in manner described on pp. 321–2 above.

Appendix 2 Russia. Monetary stocks, prices, transactions demand, and national income

1 See B.N. Mironov, ' "Revolyutsiya tsen" v Rossii v XVIII veke', *Voprosi Istorii*, 11 (November 1971), pp. 54–5, and *Khlebnie tseni v Rossii za dva stoletiya, XVIII–XIX vv.* (Leningrad, 1985), pp. 114–17.

2 B.N. Mironov, 'O metodike obrabotki istochnikov po istorii tsen', *Arkheograficheskii ezhegodnik za 1968* (Moscow, 1970); 'Faktory dinamiki rossiiskikh khlebnykh tsen v XIX nachale XX vv'. *Matematicheski metody v issledovannakh po sotsial'no-ekonomicheskoi istorii* (Moscow, 1975).

3 B.F.J. Herman, *Statistische Schilderung von Russland,* (St Petersburg and Leipzig, 1790), p. 457.

4 I.I. Kaufman, *Serebryanii rubl' v Rossii ot ego vozniknoveniya do kontsa XIX veka* (St Petersburg, 1910), pp. 151–9, 168–9, 177, and G.M. Romanov's studies: *Moneti tsarstvovaniya imperatora Petra II* (St Petersburg, 1892), t. 1, pp. 152–3; *Moneti tsarstvovaniya imperatritsi Elizabeti I i imperatora Petra III*. (St Petersburg, 1896) t. 1, pp. 272–7; *Moneti tsarstvovaniya imperatritsi Anni Ioannovni i imperatora Ioanna VI* (St Petersburg, 1901), pp. 212–16; *Moneti tsarstvovaniya imperatritsi Ekaterini II*. (St Petersburg, 1894), vip. 1, 335–41; and *Moneti tsarstvovaniya imperatora Pavla I i imperatora Aleksandra I* (St Petersburg, 1891), vip. 1, pp. 22–3; vip II, pp. 75–9.

5 P.N. Miliukov, *Gosudarstvennoe khozyaistvo Rossii v pervoi chetverti XVIII stoletiya i reforma Petra Velikogo* (St Petersburg, 1892), p. 203.

6 Population within 1701 boundaries see V.M. Kabuzan, *Narodonaslenie Rossii v XVIII–pervoi polovine XIX veka* (Moscow, 1963):

1719–1722	(First Revision)	14.1 million
1762	(Third Revision)	19.4 million
1795	(Fifth Revision)	28.8 million

7 See pp. 345–6 below.

8 A fuller description of the 'calling-down' of the currency on these occasions will be found on pp. 340–1 below.

9 Table A2.3. For sources of mint output see note 4 above and on mint prices see Mironov, *Khlebnie tseni. . .*, tale 1, p. 36. Whilst giving a general outline of mint prices for silver and gold, however, it affords no information on copper prices or on those ukazi passed subsequent to the main price changes reflecting the effectiveness of implementation.

Copper prices were steadily enhanced during the reign of Peter I from 12.8 roubles per pud in 1700–1 to 15.4 in 1702–3 and 20 in 1704–18. On the occasion of the general enhancement of 1718 they were again raised to 40 roubles per pud, at which level they remained, in spite of an attempted 'calling down' in 1731, until 1735 (K. Heller, *Die Geld-und Kreditpolitik des Russischen Reiches in der Zeit der Assignaten, 1768–1839/43* (Wiesbaden, 1983), pp. 11–12). Only in 1735 was the government able to reduce the price of newly-minted copper to 10 roubles per pud and it was left to Elizabeth's reign for the existing coinage to be called down, the existing 5-kopeck coins being set, during the years 1744–6, at a nominal value which was steadily reduced from 4 to 2 kopecks before minting finally ceased in 1752 (B.F.J. Hermann, *Die Wichtigkeit des Russischen Bergbaues* (St Petersburg, 1810), table 10gg, and *Polnoe sobranie zakonov rossiiskoi imperii = PSZ*, 1st series, t. XIII, nos 8939, 9185, 9297). Only in 1755 was reform finally completed when, by the ukaz of the 7/9 March, reminting of existing coins began at the reduced rate of 8 roubles per pud (*PSZ*, 1st series, t. XIV, nos 10,370, 10,374), a process which only lasted some two years before rates were again raised to 16 roubles per pud at which level they remained, save for the year 1762, thereafter until 1809 (B.F.J. Hermann, *Die Wichtigkeit. . .*, table 10gg–hh).

In the case of silver, the first problems arise on the occasion of the 1711 reform, data upon which is completely omitted in S.G. Strumilin's study 'Oplata truda v Rossii', *Planovoe Khozyaistvo*, nos 4, 7–8 (1930), reprinted in *Ocherki ekonomicheskoi istorii Rossii i SSSR* (Moscow, 1966), p. 40, and about which B.N. Mironov, *Khlebnie tseni. . .*, p. 36, suggests there was an enhancement in price of 20 per cent from the pre-existing level of 16.45 roubles per funt. An examination of the decree instituting the reform (*PSZ*, 1st series, t.IV., no. 2371) reveals, however, that on this occasion the government set prices at 12 roubles per funt for pre-1698 roubles and Joachimstalers brought to the mint and 13 roubles for new silver or *c.* 20 per cent below the prevailing price, and it was not until 1718, when roubles, struck in the form of 1699–1710 but to a fineness of 70 rather than 84 per cent, were issued that prices were raised to 19.75 roubles per funt or 20 per cent above the level of 1699–1710. Nor was the seeming stability of silver prices to 1731 and, after a reduction of 10 per cent, their continuing stability to 1763, indicated in Strumilin, op. cit., p. 40, and Mironov, *Khlebnie tseni. . .*, p. 36, a realistic picture of mint pricing. In only five years after the reform of 1718, because of the availability of imported silver, an order was sent to the mint (*PSZ*, 1st series, t.VII, no. 4278) allowing it to accept silver at a rate of 19.2 roubles per funt, and when abundance gave way to acute shortages during the years 1725–9 another was despatched by Peter II ordering it to accept silver at current, greatly enhanced, prices. Similarly the attempted reform of 1731, which attempted to reduce silver prices from 19.0 to 17.25 roubles per funt (*PSZ*, 1st series, VIII, no. 5726), was no more successful in the case of silver than it was for copper, and in 1732 the mint was allowed to pay 18.75 roubles per funt instead of the 17.25 set in the previous year, a rate which continued until 1741 (Kaufman, op. cit., p. 153, and *PSZ*, 1st series, t.VIII, no. 5726). Indeed, it was only by an ukaz of 8 June 1741 (*PSZ*, 1st series, t.XI, no. 8395) that silver was finally called down to 18.25 roubles per funt, at which level it remained until 1764 (*PSZ*, 1st series, t.XII, no. 9187, t.XIII, no. 9751).

10 See pp. 165–6 above. There is a high correlation between the levels of specie imports represented in table A2.3 and the level of bi-metallic premiums on the Amsterdam bourse reported in J. Newman, *Russian Foreign Trade, 1680–1780: The British Contribution . . .*, table 7.3/1, pp. 361–2.

11 Judging by the level of pre-1698 minting which averaged some 348,387 'old' roubles per year it is unlikely that the stock in 1699, a product of the output cycle of 1682–98, was greatly in excess of the 6.2 million roubles which Miliukov reports as being reminted in the first decade of the eighteenth century. Accordingly it is unlikely that the enhancement in the price offered for these pieces, amounting to *c.* 5 per cent, generated much of a supply of them to the mint.

12 On the Amsterdam exchange rate see note 10 above and on direct evidence A. Semenov, *Izuchenie istoricheskikh svedenii o rossiiskoi vneshei torgovle i promishlennosti s polovini XVII–go stoletiya po 1858 god.* (St Petersburg, 3 vols, 1859), III, 23f.

13 Kaufman, op. cit., pp. 169, 177.

14 *PSZ*, 1st series, t.VIII, no. 5726; Heller, op. cit., p. 12.

15 See pp. 176–7 above.

16 A. Kahan, *The Plow, the Hammer and the Knout. An Economic History of Eighteenth Century Russia* (Chicago, 1985), p. 313.

17 Table A2.4 which is derived from table A2.3.

18 Table A2.5. On minting see table A2.4 and on mint supply appendix 1.
19 The data on specie imports are residuals whose validity has been cross-checked against direct and indirect evidence: See pp. 164f. above.
20 Table A2.6 derived from A2.5 (supply accumulation) estimate cross-checked against A2.3 (mint output) estimate.
21 See ch. 4 note 16.
22 T:B.F.J. Hermann, *Statistische Schilderung.* . ., pp. 454–7. See p. 348 below. P:B.N. Mironov, *Khlebnie tseni.* . . pp. 68–9. M: Muscovy, pp. 339–40 above. Baltics estimated from pp. 343–4 above.
23 See the first section of this appendix.
24 This involved identifying Professor Mironov's price regions within each zone and the corresponding population region in V.M. Kabuzan, *Narodonaselenie Rossii v XVIII- pervoi polovine XIX v. po materialam revizii* (Moscow, 1963), and adjusting the former to bring it into conformity with the latter on the basis of the listing, ibid., prilozhenie, 3. Subsequently, the regional price series were weighted in a composite zonal index in relation to the region's share of zonal population.
25 See ch. 4 note 17.
26 ibid.
27 On price data seen B.N. Mironov, *Khlebnie tseni.* . ., ch. 4: Pribaltic region, pp. 84–6; western region, pp. 86–8; Ukraine region, pp. 89–90; and south-western region, pp. 90–2, as well as relevant data contained in prilozhenie, pp. 189ff.
28 ibid., ch. 4, pp. 67–84, 92–8, and prilozhenie.
29 See tables A2.3, 6, and 7.
30 B.N. Mironov, *Khlebnie tseni.* . ., tables 5, pp. 46–7, and 9, pp. 54–5, series B. Base 1762 price is a weighted index (based on zonal populations) of grain prices = 15.66 specie kop. per pud.
31 Kabuzan, op. cit., table 18, pp. 164–5.
32 M.M. Shcherbatov, 'Sostoyanie Rossii v rassuzhdenii deneg i khleba v nachale 1788g, pri nachale Turetskoi voini', *Sochineniya* (St Petersburg, 1896), 1, pp. 683–720.
33 Kabuzan, op. cit., table 18, pp. 164–5.
34 On the new issues since 1840 when the monetary stock had stood at 161 million roubles see Kashkarov, op. cit., II, prilozhenie, pp. 13–15.
35 Table A2.15B. Mulhall, 1881, 1884, 1892, and 1896.

Bibliography

In exploring new areas of research the historian must perforce tread cautiously. Failing detailed modern monographic studies based on archival materials it was necessary to fall back for this study on contemporary works, a large number of which exist describing both mining activity and the state of the Russian economy in the eighteenth century. The mysterious and exotic lands of the Tsars attracted a myriad of visitors, many of those in Peter's reign themselves becoming involved in his grandiose schemes to transform society, many a later traveller stopping to wonder at the changes under way even at the time of their visit. In their perception of these changes, moreover, they benefited not only from their observations but also from the generosity of their Russian compatriots who made official documents available to them and from discussions with Russian men of letters who themselves evinced a profound interest in their society. Not least amongst this latter group were that group of foreign mining engineers, domiciled in Russia, who formed part of an international mining brotherhood whose works during the years 1650–1850 created a strong contemporary literature on mining matters which laid the foundations for the brilliant discussions evoked by the bi-metallic crisis of the late nineteenth century. In using these works, inevitably the historian is made only too aware that literary standards, in compilation and printing, have changed for the better, and yet whilst this involves him in much labour correcting the mistakes in ill-set tables or correlating editions to identify mistakes of transcription, the ultimate inestimable value of the works makes it all worthwhile, for these contemporary authors were unconstrained by the preconceptions of later historians and explored and commented upon, with an open mind, many areas now neglected by historians. As it is precisely these areas of investigation which have emerged as important in this study the contemporary studies which discuss them play a central role in its construction.

Accordingly, in the bibliography below, which is divided between works on mining history (used in chapters 1–4) and on Russian economic history (used in chapters 4–5) first place is always given to those contemporary works used. Secondary studies then follow. Comparative materials and unpublished works other than theses, whether working papers, conference communications, or articles still at the time of writing in press, which due to the generosity of their authors have been made available to me, are not included in this brief bibliography but are cited fully in the notes and references.

I. International mining history

The production and distribution of Central and South American specie

Contemporary works and printed sources

F. Bernaldez and R. Rua Figueroa, *Memoria sobre los minas de Almaden y Almadenejos* (Madrid, 1861).

F. Fonseca and C. de Urrutia, *Historia generale de real hacienda* (Mexico City, 6 vols, 1845–53).

J.M. Hoppensack, *Ueber den Bergbau in Spanien uberhaupt und den Quecksilverbergbau zu Almaden inbesondere* (Weimar, 1796).

A. von Humboldt, *Political Essay on the Kingdom of New Spain* (London, 3 vols, 1822).

J. Moreno y Castro, *Arte de nuevo modo de beneficiar los metales de oro y plato* (Mexico, 1758).

M.E. Rivero y Ustariz, *Coleccion de memorias cientificas, agricolas e industriales* (Brussels, 1857).

F. Sonneschmid, *Tratado de amalgamacion de Nuevo Espani* (Mexico and Paris, 1825).

H.G. Ward, *Mexico in 1827* (London, 2 vols, 1828).

Secondary works

A. Attman, *The Russian and Polish Markets in International Trade 1500–1600* (Göteborg: Meddelanden fron Ekonomisk-historiska institutionem vid Göteborgs universitet, 26, 1973).

—— , *The Struggle for Baltic Markets. Powers in Conflict, 1558–1618* (Göteborg: Acta Regiae Societatis Scientiarum Litterarum Göthoburgensis. Humaniora 14, 1979).

—— , *Dutch Enterprise in the World Bullion Trade 1550–1800* (Göteborg: Acta Regiae Societatis Scientiarum Litterarum Göthoburgensis. Humaniora 23, 1983).

—— , *American Bullion in the European World Trade, 1600–1800*, (Göteborg; Actae Regiae Societatis Scientiarum Litterarum Göthoburgensis. Humaniora 26, 1986).

P.J. Bakewell, *Silver Mining and Society in Colonial Mexico, Zacatecas 1545–1700* (Cambridge, 1971).

D.A. Brading, *Merchants and Miners in Bourbon Mexico* (Cambridge, 1971).

D.A. Brading and H.E. Cross, 'Colonial silver mining: Mexico and Peru', *Hispanic American Historical Review*, LII (1972).

J. Burkhart, 'Memoradia de la explotacion de las minas de Pachuca y Real del Monte', *Annales de la Miniera* (1861).

W.L. von Eschwege, *Pluto Brasiliensis* (Berlin, 1833).

V. Magalhaes Godhino, 'Le Portugal, les flottes du sucre et les flottes de l'or 1660–1760', *Annales: économies, sociétés, civilisations*, 5 (1950).

M.H. Kuss, 'Memoire sur les mines et usines d'Almaden', *Annales des Mines*, 7th series, XIII (1878).

M.O. Mendizabel, 'Los minerales de Pachuca y Real del Monte en la epoca colonial', *El trimestre economico*, VIII (1941).

I. Mohoric, *Rudnik zivego srebra v Idrizi* (Idrizi, 1960).

M. Morineau, *Incroyables gazettes et fabuleux metaux: les retours des tresors americains d'apres les gazettes hollandaises (XVI-ᵉ–XVIII-ᵉsiècles)* (Cambridge, 1985).

A. Nöggarath, 'Mittheilungen uber die Quecksilberbergwerke zu Almadén und

Almadéjos in Spanien nebst einen Ueberblick den Vorkommnisse von Quecksilver in Allgemeinen', *Zeitschrift für das Berg, Hütten- und Salinwesen in dem Preussischen Staate*, X (1862).

R.W. Randell, *Real del Monte. A British Mining Venture in Mexico* (Austin, Texas, 1972).

A.J.R. Russell-Wood, 'Technology and society: the impact of gold mining on the institution of slavery in Portuguese America', *Journal of Economic History*, XXXVII (1971).

A. Soetbeer, *Edelmetall-Producktion und Werthverhaltniss zwischen Gold und Silver seit der Entdeckung America's bis zur Gegenwart* (Gotha; Erganzhungsheft No. 57 zu Petermann's Mittheilungen, 1879).

A. Szasdi, 'Preliminary estimate of gold and silver production in America 1501–1610', in H. Kellenbenz (ed.), *Precious Metals in the Age of Expansion* (Nürnberg; Beiträge zur Wirtschaftsgeschichte, Bd 2. 1981).

A. Matilla Tascon, *Historia de las minas de Almaden* (Madrid, 1958).

G. Lohmann Villena, *Las minas de Huancavelica en los siglos XVII* (Seville, 1949).

R.C. West, *The Mining Community of Northern New Spain, The Parral Mining District* (Berkeley and Los Angeles, 1949).

A.P. Whitaker, *The Huancavelica Mercury Mine* (Cambridge, Mass., 1941).

European mining and metallurgical industries

Contemporary works and printed sources

C. Gatterer, *Anteilung den Harz und andere Bergwerke mit nuzen zu bereisen* (Gottingen, 3 vols 1785–90) and (Nüremberg, 2 vols, 1792–3).

J.F. Gmelin, *Beitrage zur Geschichte des teutschen Bergbau* (Halle, 1783).

J. von Sanger, *Beitrage zur Geschichte des Bergbaues in Tirol* (Innsbruck: Sammler für Geschichte und Statistik in Tirol, Bd 1, 1807).

M. von Schwartner, *Statistik des Königreiche Ungarn* (Ofen, 2 vols, 1809–11).

Secondary works

S.E. Åstrom, *From Stockholm to St Petersburg. Commercial Factors in the Political Relations between England and Sweden 1675–1700* (Helsinki: Finnish Historical Society. Studia Historica 2, 1962).

D.B. Barton, *A History of Copper Mining in Cornwall and Devon* (Truro, 1961).

G.E. Benseler, *Geschichte Freibergs und seines Bergbau* (Freiberg, 1843).

B.I. Berg, 'Produktion, Belegschaft und Produktivitat beim Kongsberger Silberbergwerk 1623–1805', in E. Westermann (hrsgb.), *Quantifizierungsprobleme bei der Erforschung der Europaischen Montanwirtschaft des 15 bis 18 Jahrhunderts* (St Katharinen, 1988).

I. Blanchard, 'Commericial crisis and change: trade and the industrial economy of the north-east', *Northern History*, VII (1973).

——, 'Resource depletion in European mining and metallurgical industries 1400–1800', in A. Maczak and W.N. Parker (eds), *Natural Resources in European History* (Washington, RFF Research Papers R.13, 1979).

——, 'Lead mining and smelting in medieval England and Wales', in D. Crossley (ed.), *Medieval Industry* (London: CBA Research Report, no. 40, 1981).

——, 'England and the international bullion crisis of the 1550s', in H. Kellenbenz (ed.), *Precious Metals in the Age of Expansion* (Nürnberg: Beitrage zur Wirtschaftsgeschichte, Bd 2., 1981).

——, 'The British silver–lead industry and its relations with the Continent, 1470–1570', in W. Kroker and W. Westermann (eds), *Montanwirtschaft Mitteleuropas vom 12 bis 17 Jahrhundert. Stand, Wege und Aufgaben der Forschung* (Bochum: Der Anschnitt. Beiheft 2, 1984).

B. Boethius, *Koppar – Bergslagen fram till 1570 – talets genombrott. Uppkomst, medeltid, tidig vasatid* (Uppsala, 1965).

C. Bose, *Generale Haushalts – Principia vom Berg-, Hutten-, Salz-und Forstwesen in Specie vom Harz* (Berlin, 1877).

R. Burt, 'Lead production in England and Wales, 1700–1770', *Economic History Review*, 2nd series, XXII (1967).

—— , *The British Lead Mining Industry* (Redruth, 1984).

R. Gandilhon, 'Une mine d'argent dans la Baronie de Brugny decouverte par le Baron de Beausoleil, 1632', *Histoire des enterprises*, XIII (1963).

K. Glamann, 'The Dutch East India Company's trade in Japanese copper 1645–1736', *Scandinavian Economic History Review*, I, (1953).

—— , *Dutch–Asiatic Trade, 1620–1740* (Copenhagen and the Hague, 1958).

D. Häberle, 'Die bergbaulichen Verhältnisse der Pfalz', *Pfalzer Heimatkunde*, XV (1919).

H. Hamilton, *The English Brass and Copper Industries to 1800* (London, 1926).

G. Hammersley, 'Technique or economy? The rise and decline of the early English copper industry, ca 1550–1650', in H. Kellenbenz (ed.), *Schwerpunkte der Kupferproduktion und der Kupferhandel in Europa, 1500–1650 (*Cologne und Vienna: Kölner Kollogien zur internationalem Sozial-und Wirtschaftsgeschichte, Bd 3,1977).

J.R. Harris, *The Copper King. A Biography of Thomas Williams of Llanidan* (Liverpool, 1964).

P.M. Hembry, *The Bishops of Bath and Wells, 1540–1640* (Oxford, 1967).

E. Henschke, *Landherrschaft und Bergbauwirtschaft. Zur Wirtschafts – und Vertwaltungs – geschichte des Oberharzer Bergbaugebietes im 16 und 17 Jahrhunderts* (Berlin, 1974).

R. Hildebrandt, 'Augsburger und Nürnberger Kupferhandel 1500–1619. Produktion, Marktanteil, und Finanzieurung im vergleich zweier Städte und ihrer wirtschaftslichen Führungsschichten', *Zeitschrift für Wirtschafts – und Sozialwissenschafte*, CXII (1972).

A. Jager, *Beitrage zur Tirolisch – Salzburgischen Bergwerkgeschichte* (Vienna: Archivs fur Osterreichische Geschichte. Bd 53. 1875).

D.T. Kiernan, 'Technological, economic and social change in the Derbyshire lead industry 1540–1600' (unpublished Sheffield Ph.D. thesis, 2 vols., 1985).

Kongsberg, Foredrag i tilknytning til ekskursjon 11 September 1976 (Oslo: Det Norske Viderskap-Akademi, 1976).

H.-J. Kraschewski, 'Der Bergbau des Harz im 16 und zu Beginn des 17 Jahrhundert. Stand und Aufgaben der Forschung', in W. Kroker and E. Westermann (eds) *Montanwirtschaft Mitteleuropas vom 12 bis 17 Jahrhundert* (Bochum:der Anschnitt. Beiheft 2, 1984).

J.F. Kuiper, *Japan en de buitenwereld in de achttiende eeuw* (S'Gravenhage, 1921).

M. Levy, *Der Silber-und Blei-Bergbau zu Przibram* (Vienna, 1875).

W.G. Lewis, 'The Cymsymlog lead mine', *Ceredigion*, II (1952).

A.S. Lindroth, *Gruvbrythyng och Kopparantering vid Stora Kopparberget intill 1800-talets borjan* (Uppsala, 1955).

D. Molenda, *Kopalnie rud olowiu na terenie zlóz śląsko-krakowskich w XVI–XVII w* (Wroclaw, Warsaw, Kraków, and Gdansk, 1972).

B. Neumann, *Die Metalle, Geschichte, Vorkommen und Gewinnung nebst ausfuhrlicher Produktion-und Preis-Statistik* (Halle, 1904).

C.J. Schmitz, *World Non-Ferrous Metal Production and Prices, 1700–1976* (London, 1979).

Herrn Schrader, 'Der Mansfeldsche Kupferschiefer-Bergbau', *Zeitschrift für Berg-Hütten-und Salinenwesen in dem Preussischen Staate*, XVII (1869).

K. Sprauten, 'Die norwegischen Kupferbergwerke als Investionsobjekte 1720–1760. Eine Diskussion uber Mengen, Preise, Gewinne und Zubussen', in Westermann (hrsgb.) *Quantifizierungsprobleme bei der Erforschung der Europaischen Montanwirtschaft des 15 bis 18 Jahrhunderts* (St Katharinen, 1988).

K. Sternberg, *Umrisse einer Geschichte der Bohmischen Bergwerke* (Prague, 2 vols, 1836–7).

Stora Kopparberg Berglag Aktiebolag (Falun, 1949).

L. Suhling, *Der Seigerhuttenprozess, Die Technologie des Kupferseigerns nach dem fruhen metallurgischen Schriften* (Stuttgart, 1976).

J. Vlachovič, *Slovenská Med v 16 a 17 storoči* (Bratislava, 1964).

E. Waschinskii, *Währung, Preisentwicklung und Kaufkraft des Geldes in Schleswig-Holstein, 1226–1864* (Neumünster, 2 vols, 1952–9).

C.G.A. Weissenbach, *Sachsens Bergbau, National-okonomisch betrachtet* (Freiberg, 1983).

E. Westermann, 'Das "Leipziger Monopolprojekt" als Symptom der mitteleuropaischen Wirtschaftskrise um 1527/8', *Vierteljahrschrift für Sozial – Wirtschaftsgeschichte*, LVII (1971).

——, 'Die Bedeutung des Thuringer Saigerhändels für den mitteleuropaischen Handel an der Wende von 15 zum 16 Jahrhundert', *Jahrbuch für Geschichte Mittel-und Ostdeutschlands*, XXI (1972).

——, 'Zur Silber-und Kupferproduktion Mitteleuropas von 15 bis zum fruhen 17 Jahrhundert'', *Der Anschnitt*, 83, 5/6 (1986).

200 vyroce zavendenie nepriamej amalgamacie a zalozenia. I. medzinarodnej vedeckej spolocnosti na svete zbornik (Donovaly-Sklene Teplice-Banska Stiavnica, 1986).

Russian mining and metallurgical industries

Contemporary works and printed sources

(a) Contemporary accounts by mining officials

B.F.J. Hermann, *Versuch einer mineralogischen Beschreibung der uralischen Erzgeburges* (Berlin and Stettin, 1789).

——, *Statistiche Schilderung von Russland* (St Petersburg and Leipzig, 1790).

——, 'Sur l'exploitation des mines de l'empire de Russie', *Nova Acta Academiae Scientiarum Imperialis Petropolitaniae*, II (1798).

——, *Mineralogische Reisen in Sibirien vom Jahr 1783 bis 1796* (St Petersburg, 4 vols, 1797–8).

——, *Sochineniya o sibirskikh rudinakh' i zavodov* (St Petersburg, 3 vols, 1797–1801).

——, *Opisanie zavodov pod vedomostom Ekaterinburgskogo gornogo nachal'stva sostoyavshikh* (Ekaterinburg, 1808).

——, *Die Wichtigkeit des russischen Bergbaues* (St Petersburg, 1810).

——, *Istoricheskoe nachertanie gornogo proizvodstva v rossiiskoi imperii* (Ekaterinburg, 1810).

H.M. Renovantz, *Mineralogische – geographische und andere Vermischte Nacht-richten von den Altaischen Geburgen Kanserlichen Anteil* (Reval, 1788), and the Russian translation, *Mineralogicheskie, geograficheskie i drugie smeshanie izvestiya o Altaiskikh gorakh* (St Petersburg, 1792).

(b) Descriptions by contemporary visitors

J.F. Falk, *Beitrage zur topographischen Kentniss der russischen Reichs* (St Petersburg, 5 vols, 1785).

J.G. Gmelin, *Reise durch Sibirien von dem Jahr 1733 bis 1743* (Göttingen, 4 vols, 1751–2).

P.S. Pallas, *Reise durch verschiedene Provinzen des russischen Reichs in einem ausfuhrlichen Auszuge* (Frankfurt and Leipzig, 3 vols, 1776–8), translated into French, *Voyages de M. PS Pallas en differentes provinces de l'empire de russie et dans l'Asie septentrionale* (Paris, 6 vols, 1778–93).

(c) Other source materials

N.I. Pavlenko, 'Materiali o razvitii ural'skoi promyshlennosti v 20–40kh godakh XVIII V', *Istoricheskii Arkhiv*, IX (1953).

M.A. Pavlov (ed.), *W. de Hennin, Opisanie ural'skikh i sibirskikh zavodov 1735* (Moscow, 1937).

Secondary works

N.B. Baklanov, *Tekhnika metallurgicheskogo proizvodstvo XVIII veka na Urale* (Moscow, 1935).

I. Bogoslovskii, *Istoriko-statisticheskii ocherki proizvoditel'nost Nerchinskogo gornogo okruga s 1703 do 1871* (St Petersburg, 1873).

V.V. Danilevskii, *Russkoe zoloto, Istoriya otkritiya i dochichi do seredini XIX* (Moscow, 1959).

B.B Kafengauz, *Istoriya khozyaistva Demidovikh v XVIII–XIX v. Opit issledovaniya po istorii ural'skoi metallurgii* (Moscow, 1949).

Z.P. Karpenko, *Gornaya i metallurgicheskaya promyshlennost' zapadnoi Sibiri v 1760–1780 godakh* (Novosibirsk, 1966).

P.G. Liubomirov, *Ocherki po istorii rossiiskoi promyshlennosti, XVII, XVIII i nachale XIX veka (Moscow, 1947)*.

N.I. Pavlenko, *Razvitie metallurgichesko promyshlennosti Rossii v pervoi polovine XVIII veka. Promyshlennaya politika i upravlenie* (Moscow, 1953).

——, *Istoriya metallurgii v Rossii XVIII veka. Zavodi i zavodovladel'tsi*, (Moscow, 1962).

R. Portal, *L'Oural au XVIIIe* (Paris, 1950).

B. Rozhov, 'Berg-kompaniya na Magnitnii gore', *Gornii zhurnal*, 6–8 (1885).

——, 'A.N. Demidov na svoikh Kolivano-Voskresenskikh zavodov', *Gornii zhurnal*, III (1891).

S.G. Strumilin, *Istoriya chernoi metallurgii v SSSR. T.1. Feodal'nii period 1500–1860* (Moscow, 1860).

N. Zobin, 'Masterovie altaiskikh gornikh zavodov do osvobozhdeniya', *Sibirskii Sbornik*, II (1892).

——, 'Pripisnie krest'yane na Altae', *Altaiskii Sbornik*, I (1894).

II Russian economic history

General works

Contemporary works and printed sources

P.H. Bruce, *Memoirs of Peter Henry Bruce Esquire* (London, 1782).

C. le Brun, *A new and more correct translation of Mr. Cornelius le Brun's Travels* (London, 1758).

J. Carr, *A Northern Summer* (London, 1805).

E.D. Clarke, *Travels in Various Countries of Europe, Asia and Africa* (London, 1810).

W. Coxe, *Travels into Poland, Russia, Sweden and Denmark* (London, 2 vols, 1784).

E. Craven, *A Journey through the Crimea to Constantinople, in a series of letters* (London, 1789).

A. Ermann, *Travels in Siberia* (London, 2 vols, 1848).

G. Forster, *A Journey from Bengal to England* (London, 1798).

W.C. Friebe, *Physische-okonomisch und statistische Bermerkungen von Lief-und Ehstland oder von den beidem Statthalterschaften Riga und Reval* (Riga, 1794).

J.F. Georgel, *Voyages à St Petersbourg in 1799–1800* (Paris, 1818).

M. Guthrie, *A Tour performed in the years 1795–1796 through the Tauride or Crimea* (London, 1802).

M. Holderness, *New Russia* (London, 1823).

S.P. Krasheninkov, *Opisanie zemli Kamchatka* (Moscow, 1949).

L. Lange, 'Journal of the residence of Mr. de Lange, Agent of his Imperial Majesty of all the Russians, Peter the First, at the Court of Peking during the Years 1721 and 1722', in J. Pinkerton, *A General Collection of the Best and Most Interesting Voyages in all Parts of the World*, VII (London, 1716).

J.J. Oddy, *European Commerce* (London, 1805).

J. Perry, *The State of Russia under the Present Czar* (London, 1716).

R.K. Porter, *Travelling Sketches in Russia and Sweden 1805–1808* (London, 1809).

A.N. Radishev, 'Zapiski puteshestviya po Sibiri' in *Polnoe sobranie sochinenii* (Moscow, 3 vols, 1932–52).

J. Reuilly, *Voyages en Crimée et sur les bords de la Mer Noire* (Paris, 1806).

Russia. Imperatorskaya Kantseliarii, *Polnoe Sobranie zakonov rossiiskoi imperii s 1649 goda*, 1st series, 1649–1825, 46 vols in 43 plus 3 appendices (St Petersburg, 1830).

H. Storch, *Tableau historique et statistique de l'empire de Russie à la fin due dix-huitieme siecle* (Basle and Paris, 2 vols, 1801).

—— , *The Picture of St. Petersburg* (London, 1801).

P.J. Strahlenberg, *Das Nord und ostlich Theil von Europa und Asie* (Stockholm, 1730).

A. Swinton, *Travels into Norway, Denmark and Russia in 1788, 1789, 1790 and 1791* (London, 1792).

W. Tooke, *View of the Russian Empire during the Reign of Catherine II and to the close of the present century* (London, 3 vols, 1799).

F.C. Weber, *Der veranderte Russland* (Frankfurt, 1721).

C. Whitworth, 'An account of Russia as it was in the year 1710', *Fugitive Pieces* (London, 1758).

M. and C. Wilmot, *The Russian Journals of Martha and Catherine Wilmot*, ed. Lady Londonderry and H.M. Hyde (London, 1934).

F.G. Würst, *Bermerkungen uber einige Gegenstande der russischen Staatswirthschaft* (Berlin, 1806).

E. Zyablovskii, *Zemleopisanie rossiiskoi imperii* (St Petersburg, 6 vols, 1810).

Secondary works

A. Kahan, *The Plow, the Hammer and the Knout. An Economic History of Eighteenth-Century Russia* (Chicago and London, 1985).

National incomes estimates

Contemporary works and printed sources

B.F.J. Hermann, *Statistiche Schilderung von Russland* (St Petersburg, 1790).

I. Golitzuin, *Statisticheskaya tablitsi vserossiiskoi imperii ili fizicheskoe, politicheskoe nachertanie Rossii s XIX stolitiya* (Moscow, 1807). (Note: the author of this anonymous work is as it is given in the British Library catalogue, a nineteenth-century transliteration of I.A. Golitsyn.)

A.N. Radishev, 'Pis'mo o kitaiskoi torge', in *Polnoe sobranie sochinenii* (Moscow, 3 vols, 1932–52).

Secondary works

A Vainstein, *Narodnii dokhod Rossii i SSSR* (Moscow, 1969).

The domestic economy

V.A. Aleksandrov, 'Russkoe naselenie Mangazeysko-Turukanskogo kraya v XVII-pervoi polovine XVIII veka', *Institut Etnografii imeni N.N. Mikiukho-Maklaya. Kratkie soobshcheniya*, 35 (1960).

V.K. Andreivich, *Sibir v XIX veka* (St Petersburg, 2 vols, 1889).

T. Armstrong, *Russian Settlement in the North* (Cambridge, 1965).

D.N. Belikov, *Pervie russkie krest'yane-nasel'niki Tomskogo kraya* (Tomsk, 1898).

L.A. Beskrovnii, *Russkaya armiya i flot v XVIII veke, Ocherki* (Moscow, 1958).

J. Blum, *Lord and Peasant in Russia* (New York, 1974).

I.Y. Bulygin, *Polozhenie krest'yan i tovarnoe proizvodstvo v Rossii, vtoraya polovina XVIII veka* (Moscow, 1966).

E.I. Druzhinina, *Yuzhnaya Ukraina 1800–1825gg* (Moscow, 1970).

Y.A. Fedorov, *Pomeshchich'i krest'yane tsentral'no promishlennogo raiona kontsa XVIII- pervoi polovine XIX veke* (Moscow, 1974).

J.R. Gibson, *Feeding the Russian Fur Trade. Provisionment of the Okhotsk Seaboard and Kamkatka Peninsula 1639–1856* (Madison, Wisc., 1969).

S.L. Hoch, *Social Control in Russia. Petrovskoe, a Village in Tambov* (Chicago, 1986).

E.I. Indova, *Krepostnoe khozyaistvo v nachale XIX veka* (Moscow, 1953).

G.S. Isaev, *Rol' tekstil'noi promyshlennosti v genezise i razvitii kapitalizma v Rossii 1760–1860* (Leningrad, 1970).

V.M. Kabuzan, *Narodonaslenie Rossii v XVIII–pervoi polovine XIX veka* (Moscow, 1963).

——, *Izmeniya v razmeschenii naseleniya Rossii v pervoi polovine XVIII veka* (Moscow, 1971).

V.M. Kabuzan and N.M. Shepukova, 'Tabel' pervoi revizii narodonaseleniya Rossii 1718–1727gg', *Istoricheskii Arkhiv*, III (1959).

B.N. Kafengauz, 'Ekonomicheskie svyazi Ukraini s Rossii v kontse XVII–nachale XVIII stoletiya', *Vossoedinenie Ukraini s Rossiei 1654–1954, Sbornik Statei* (Moscow, 1954).

E.I. Kamentseva and N.V. Ustyugov, *Russkaya metrologiya* (Moscow, 1965).

Z.P. Karpenko, 'Promyshlennie goroda i poselki zapadnoi Sibir' v XVIII stoletii', *Materiali po istorii Sibiri. Sibir' perioda feodalizm. I. Sibir' XVII–XVIII vv.* (Novosibirsk, 1962).

M. Klochov, *Nasleniya Rossii pri Petre Velikom po perepisyam togo vremeni* (St Petersburg, 2 vols, 1911).

I.D. Koval'chenko, 'Dinamika urovnaya zemledel'cheskogo proizvodstva Rossii v pervoi polovine XIX v', *Istoriya SSSR*, 1 (1959).
——, *Krest'yane i krepostno khozyaistvo Ryazanskoi i Tambovskoi gubernii v pervoi polovine XIX veke* (Moscow, 1959).
——, *Russkoe krepostnoe krest'yanstva v pervoi polovine XIX veke* (Moscow 1967).
I.D. Koval'chenko and L.V. Milov, *Vserossiiski agrarnii rinok XVIII–nachale XIX veke* (Moscow 1974).
Yu V. Kozhukhov, *Russkie krest'yane vostochnoi Sibiri v pervoi polovine XIX veka 1800–1861* (Leningrad, 1967).
E. Melton, 'Two estates in nineteenth-century Russia', *Past and Present*, 115 (1987).
I.V. Meshalin, *Tekstil'naya promyshlennost' krest'yan moskovskoi gubernii v XVIII i pervoi polovine XIX V.* (Moscow, 1950).
P.N. Miliukin, *Gosundarstvennoe khozyaistvo Rossii v pervoi chetverti XVIII stoletiya i reforma Petra Velikogo* (St Petersburg, 1892).
B.N. Mironov, *Vnutrenni rinok Rossii vo vtoroi polovine XVIII–pervoi polovine XIX V.* (Leningrad, 1981).
K.A. Pazhitnov, *Ocherki istorii tekstil'noi promyshlennosti dorevolyutsionnoi Rossii. Khlopchatobumazhnaya, l'no-pen'kovaya i shelkovaya promyshlennost* (Moscow, 1958).
A.G. Rashin, *Formirovanie promishlennogo proletariata v Rossii* (Moscow, 1940).
——, *Naselenie Rossii za 100 let, 1811–1913* (Moscow, 1956).
N.L. Rubinstein, *Sel'skoe khozyaistvo Rossii vo vtoroi polovine XVIII v. Istoriko-ekonomicheskii ocherk* (Moscow, 1957).
T.I. Rychkov, *Topografia Orenburgskoi gubernii* (Orenburg, 1762).
A.S. Savich, *Ocherki istorii krest'yanskikh volnenii na Urale v XVIII XVIII v.* (Moscow, 1931).
L.N. Semeonova, *Rabochie Petersburga v pervoi polovine XVIII veka* (Leningrad, 1974).
V.I. Semevskii, *Krest'yane v tsartsvovanie imperiatritsi Ekaterini II* (St Petersburg, 1901).
V.N. Sherstoboyev, 'Zemledelie severno predbaikalya v XVII–XVIII vv', in G.D. Grekov (ed.), *Materiali po istorii zemledeliya SSSR,* I (Moscow and Leningrad, 1952).
V.I. Shunkov, 'Geograficheskoie razmeschenie Sibirskogo zemledeliy v XVII veka', *Voprosi geografi,* 20 (1950).
R.E.F. Smith and D. Christian, *Bread and Salt. A Social and Economic History of Food and Drink in Russia* (Cambridge, 1984).
Yu Tikhonov, *Pomeshchich'i krest'yane v Rossii. Feodal'naya renta v XVII i nachale XVIII v* (Moscow, 1974).
M.A. Tsvetkov, *Izmenenie lesistosti evropeiskoi Rossii s kontsa XVII stoletiya po 1914 god* (Moscow, 1957).

Foreign trade

Contemporary works and printed sources
M.D. Chulkov, *Istoricheskoe opisanie rossiskoi kommertsii pri vsekh portakh i granitsakh of drevnikh vremen do nine nastoyashchego i usekh preimushchestvennykh ukazonenii po onoi gosudaria imperatora Ekaterini Velikiya* (St Petersburg, 21 vols in 7, 1781–8).

Russia, Ministerstvo kommertsii, *Gosudarstvennaya torgovlya 1802–1807 goda v raznikh eya vidakh* (St Petersburg, 1802–7).

Secondary works

J. Baszanowski, *Z dziejow handly polskiego w XVI–XVIII wieku. Handel wolami* (Gdansk, 1977).

B.N. Florya, 'Torgovlya Rossii so stranami zapadnoi Evropi v Arkhangel'ske kontse XVI-nachalo XVII V.', *Srednie Veka*, XXXVI (1963).

C.M. Foust, *Muscovite and Mandarin, Russia's Trade with China and its Setting, 1725–1805* (Chapel Hill, 1969).

E. Harder, 'Seehandel zwischen Lübeck und Russland im 17/18 Jahrhundert', *Zeitschrift des Vereins für lübeckische Geschichte und Altertumskunde*, LXI (1961), XLIII (1963).

E. Harder-Gersdorff, 'Handelskonjunkturen und Warenbilanzen im lübeckischen Seeverkehr im 18 Jahrhundert', *Vierteljahrschrift fur Sozial-und Wirtschafts-geschichte*, LVIII (1970).

A. Izyumov, 'Razmeri russkoi torgovli XVII v. cherez Arkhangel'sk v svyazi s neobsledovannimi arkhivnimi istochinkami', *Izvestiya Arkhangel'skogo obshchestv izucheniya Russkogo severea*, VI (1912).

G. Jensch, 'Der Handel Rigas im 17 Jahrhundert', *Mitteilungen aus der livlandischen Geschichte*, XIV (1936).

B.G. Kurts, *Gosudarstvennaya monopoliya Rossii s Kitaem v pervoi polvine XVIII Stoletiya* (Kiev, 1928).

B.N. Mironov, 'Eksport russkogo khleba vo vtoroi polovine XVIII-nachale XIX v.', *Istoricheskii Zapiski*, XCIII (1979).

K.G. Mityaev, 'Oboroti i torgovie svyazi smolenskogo rinka v 70-kh godakh XVII veka', *Istoricheskii zapiski*, XXXII (1942).

J. Newman, *Russian Foreign Trade 1680–1780: The British Contribution* (unpublished Ph.D. thesis, University of Edinburgh, 1985).

——, 'Anglo-Dutch commercial co-operation and the Russia trade in the eighteenth century', *The Interactions of Amsterdam and Antwerp with the Baltic Region 1400–1800* (Leiden: Werken uitgegeven door de vereeniging het Nederlansch Ekonomisch-Historisch Archief gevestig te Amsterdam, 16, 1983).

——, 'The Russian grain trade, 1700–1779', in W. Minchinton (ed.), *The Baltic Grain Trade. Five Essays* (Exeter, 1985).

S.B. Okun, *The Russian–American Company* (Cambridge, Mass., 1951)

Kh. A. Piirimae, 'Tendenciya razvitiya i ob'em torgovli pribaltiiskiikh gorodov v period shvedskogo gospodstva v XVII vek', *Skaninavskii sbornik*, VIII (1964).

N.L. Rubinstein, 'Vneshnaya torgovlya Rossii i russkoe kupechestvo vo vtoroi polovine XVIII v.', *istoricheskie Zapiskie*, LIV (1955).

A. Semeonov, *Izuchenie istoricheskikh svedenii o rossiskoi vneshei torgovle i promyshlennosti s polovini XVII-go stoletiya po 1858* (St Petersburg, 2 vols, 1859).

Monetary systems and prices

M. Kashkarov, *Denezhnoe obrashchenie v Rossii* (St Petersburg, 2 vols, 1898).

I. Kaufman, *Serebryanni rubl' v Rossii of ego vozniknoveniya do kontsa XIX veka* (St Petersburg, 1910).

H. Keller, *Die Geld und Kreditpolitik des russischen Reiches in der Zeit der Assignaten 1768–1839/43* (Wiesbaden, 1983).

A. Mikotajczyk, 'W kwesti znalezisk nowozytnych monet niderlandzich na zemiach koronnych', *Zapiski historyczne*, XIIX (1984).

B.N. Mironov, 'O metodnike obrabotki istochnikov po istorii tsen' *Arkeograficheskii ezhegodnik na 1968* (Moscow, 1970).

——, ' "Revolyutsiya tsen" v Rossii v XVIII veke', *Voprosi Istorii*, 11 (1971).

——, 'Faktory dinamiki rossiiskikh khlebnykh tsen v XIX–nachale XXvv', *Matematicheski metodi v issledovannakh po sotsial'noekonomicheskoi istorii* (Moscow, 1975).

——, *Khlebnie tseni v Rossii za dva stoletiya (XVIII–XIX vv)* (Leningrad, 1985).

W. McKenzie Pinter, *Russian Economic Policy under Nicholas I* (Ithaca, NY, 1967).

V.N. Riabtsevich, 'Monetnie kladi XVII i pervoi chetverti XVIII v na territorii Chernigov-severnoi zemli i vostochnoi Belorossi', *Numizmatika i Sfragistika*, 1 (1963).

G.M. Romanov, *Moneti tsarstvovaniya imperatora Pavla I i imperatora Aleksandrav I* (St Petersburg, 1891).

——, *Moneti tsarstvovaniya imperatora Petra I* (St Petersburg, 1892).

——, *Moneti tsarstvovaniya imperatritsi Ekaterini II* (St Petersburg, 1894).

——, *Moneti tsarstvovaniya imperatritsi Elizabeti I i imperatora Petra III* (St Petersburg, 1896).

——, *Moneti tsarstvovaniya imperatritsi Anni Ioannovni i imperatora Ioanna VI* (St Petersburg, 1901).

Russia, Ministerstvo Finansov, *Ministerstvo finansov, 1802–1902*, (St Petersburg, 2 vols, 1902).

M.M. Scherbatov, 'Sostoyanie Rossii v rassuzhdenii deneg i khleba v nachale 1788g., pri nachale Turetskoi voini', *Sochineniya* (St Petersburg, 1896).

A. Schmit, 'Das russische Geldwesen wahrend der Finanzverwaltung des Grafen Cancrin (1823–1843)', *Russische Revue*, VII (1875).

P.A. Storkh, 'Materiali dlya istorii gosudarstvennikh denzhnikh znakonov v Rossii s 1653 po 1840 god', *Zhurnal ministerstvo narodnogo prosveshcheniya*, CXXXVII (1868).

Index of places

Most places referred to in the text may be found in any good atlas. To facilitate the reader's access to the section on Russia (chapters 3–5), however, cross references to relevant textual maps are given below in bold type.

Index of persons

Index of subjects

Admiralty, Russia, 248. *See* departments of state and industries, Russia.

agricultural and horticultural production, Russian: flax and hemp production, 220, 222–3, 234–5, 240–1, 258, 271; fruit, 220, 226, 229, 240, 268, 271–2, 274–5; garden crops, 111, 118, 130, 132, 147–8, 220, 222–3, 226, 229, 233–5, 240, 241, 256, 264–5, 271; grain production, 65, 67, 78, 111, 148, 220, 222–3, 226, 229, 233–5, 238, 256–60, 262, 264–5, 279–80

agricultural systems, Russian: Baltic provinces, 222, 234; central black earth region, 218, 220, 226–8, 238, 256, 258–9, 280; Moscow region, 220, 238; northern region, 258–9; Siberian, 76–8, 90, 111, 146–8; Ukraine, 222–4, 235, 256, 258; Volga region, 228–9, 238, 256, 258–9, 280

agro-industries, Russian (*see also* handicrafts, kustari, artisanal pursuits; manufactures; industries, Russian.): breweries, 106; charcoal making, 76–81, 125, 138–41, 145, 152; corn mills, 106; fish products (e.g. caviar, oil, isinglass), 242; tallow making, 243, 270; vodka production, 228, 235; wood cutting, 76–8, 138–9, 141–3, 145, 152; wood working, 241–2, 270–2, 274

animal feed, Russia, 235, 239, 243–4, 257, 273

animal husbandry, Russian: cattle, 118, 130, 132, 147–8, 218, 220, 222–3, 226, 228–9, 242–3, 265, 269–70; goats, 220, 268; horses, 148, 218, 228, 242, 269;

sheep, 148, 220, 265, 268, 269

apiary, Russia, 220, 222, 242, 275

apothecaries, 106, 129. *See* medical services, Russia

archives, Russian, 306–8: Central State Archives of Ancient Acts = Tsentral'nogo gosudarstvennogo arkhiva drevnikh aktov, Moscow, 307–8; Central State Historical Archive = Tsentral'nogo gosudarstvennogo istoricheskogo arkhiva v Leningrade, 307–8; State Archive of the Altai Krai = Gosudarstvennogo arkhiva Altaiskogo kraya, 307

armaments production, Russian (*see* industries, Russian.): cannon, 249; small arms, 248–9

assignats, Russian currency, 173, 193, 197–8, 201–12, 342

banks, Russia: artillery and engineering corps, 187, 342; assignat, 199; copper, 187, 342

blasting powder, 23, 152

breweries, Russia, 106. *See* agro-industries, Russian

brickworks, Russia, 106; *See* construction industries, Russian

Cabinet Office, Russia, mining section, 306–9. *See* departments of state, Russia

canals, Russia, 224–6

caravans, Russo-China trade, 67–9. *See* foreign trade, Russian

cattle trades, Russia, 148, 223, 226–7, 233–5, 243–4, 252, 269–70, 274, 281. *See* commercial activity, domestic, and foreign trade, Russian

426